T0205580

Sustainable Textiles: Production, Processing, Manufacturing & Chemistry

Series Editor

Subramanian Senthilkannan Muthu, Head of Sustainability, SgT and API, Kowloon, Hong Kong

This series aims to address all issues related to sustainability through the lifecycles of textiles from manufacturing to consumer behavior through sustainable disposal. Potential topics include but are not limited to: Environmental Footprints of Textile manufacturing; Environmental Life Cycle Assessment of Textile production; Environmental impact models of Textiles and Clothing Supply Chain; Clothing Supply Chain Sustainability; Carbon, energy and water footprints of textile products and in the clothing manufacturing chain; Functional life and reusability of textile products; Biodegradable textile products and the assessment of biodegradability; Waste management in textile industry; Pollution abatement in textile sector; Recycled textile materials and the evaluation of recycling; Consumer behavior in Sustainable Textiles; Eco-design in Clothing & Apparels; Sustainable polymers & fibers in Textiles; Sustainable waste water treatments in Textile manufacturing; Sustainable Textile Chemicals in Textile manufacturing. Innovative fibres, processes, methods and technologies for Sustainable textiles; Development of sustainable, eco-friendly textile products and processes; Environmental standards for textile industry; Modelling of environmental impacts of textile products; Green Chemistry, clean technology and their applications to textiles and clothing sector; Eco-production of Apparels, Energy and Water Efficient textiles. Sustainable Smart textiles & polymers, Sustainable Nano fibers and Textiles; Sustainable Innovations in Textile Chemistry & Manufacturing; Circular Economy, Advances in Sustainable Textiles Manufacturing; Sustainable Luxury & Craftsmanship; Zero Waste Textiles.

More information about this series at https://link.springer.com/bookseries/16490

Ali Khadir · Subramanian Senthilkannan Muthu
Editors

Biological Approaches in Dye-Containing Wastewater

Volume 2

 Springer

Editors
Ali Khadir
Western University
London Ontario, ON, Canada

Subramanian Senthilkannan Muthu
SgT Group and API
Hong Kong, Kowloon, Hong Kong

ISSN 2662-7108   ISSN 2662-7116 (electronic)
Sustainable Textiles: Production, Processing, Manufacturing & Chemistry
ISBN 978-981-19-0528-5   ISBN 978-981-19-0526-1 (eBook)
https://doi.org/10.1007/978-981-19-0526-1

This Springer imprint is published by the registered company Springer Nature Singapore Pte Ltd.
The registered company address is: 152 Beach Road, #21-01/04 Gateway East, Singapore 189721,
Singapore

Contents

About the Editors

Ali Khadir is an environmental engineer and a member of the Young Researcher and Elite Club, Islamic Azad University of Shahre Rey Branch, Tehran, Iran. He has published several articles and book chapters in reputed international publishers, including Elsevier, Springer, Taylor & Francis, and Wiley. His articles have been published in journals with IF of greater than 4, including Journal of Environmental Chemical Engineering and International Journal of Biological Macromolecules. He also has been the reviewer of journals and international conferences. His research interests center on emerging pollutants, dyes, and pharmaceuticals in aquatic media, advanced water, and wastewater remediation techniques and technology.

Dr. Subramanian Senthilkannan Muthu currently works for SgT Group as Head of Sustainability, and is based out of Hong Kong. He earned his Ph.D. from The Hong Kong Polytechnic University, and is a renowned expert in the areas of Environmental Sustainability in Textiles & Clothing Supply Chain, Product Life Cycle Assessment (LCA), and Product Carbon Footprint Assessment (PCF) in various industrial sectors. He has five years of industrial experience in textile manufacturing, research and development, and textile testing and over a decade of experience in life cycle assessment (LCA), carbon and ecological footprints assessment of various consumer products. He has published more than 100 research publications, written numerous book chapters, and authored/edited over 100 books in the areas of Carbon Footprint, Recycling, Environmental Assessment, and Environmental Sustainability.

Dye Removal Using Activated Sludge

Pranay Raut, Dharm Pal, and V. K. Singh

Abstract The textile industry demands substantial volume of water for several activities like dyeing, printing, and finishing. At the end of the processes these industries release huge amount of effluents containing dyes and their components which are not only aromatic but also carcinogenic in nature. Hence the effluents need to be treated for the elimination of dyes and other pollutants before discharge from the industry. High capital cost and less efficiency is the limiting factor for the various physico-chemical processes used for the eradication of dye from wastewater. Activated sludge process is extensively used for the secondary treatment of wastewater as it helps in reducing both the chemical and biological oxygen demand with the help of aeration and dense microbial culture. But the generations of large volume of sludge containing residual biodegradation resistant compounds insist for modification of this process. In this review overview of activated sludge process for dye removal along with its limitation is carried out. This review also highlights recent advancement in the use of other methods like adsorption, aerobic granular technology, sequential chemical, and activated sludge process along with the use of bioflocculant from activated sludge system for dye removal.

Keywords Activated sludge · Bioflocculant · Dye · Granular technology · Adsorption

1 Introduction

Industrial effluent results in enormous amount of suspended solids (SS), biochemical oxygen demand (BOD) and chemical oxygen demand (COD) along with foul odor. These are some of the crucial peculiarities of pollution by wastewater. Industrial wastewater contains various disastrous compounds like heavy metals, chemicals, and dyes. These compounds are released into various water bodies like small or

---

P. Raut · D. Pal (✉) · V. K. Singh
Department of Chemical Engineering, National Institute of Technology, Raipur, Raipur 492010, CG, India
e-mail: dpsingh@nitrr.ac.in

© The Author(s), under exclusive license to Springer Nature Singapore Pte Ltd. 2022
A. Khadir and S. S. Muthu (eds.), *Biological Approaches in Dye-Containing Wastewater*,
Sustainable Textiles: Production, Processing, Manufacturing & Chemistry,
https://doi.org/10.1007/978-981-19-0526-1_1

large river, pond, and lake and thus eventually into sea. Such large level discharge poses major imminence to aquatic flora and fauna. It is the one inimitable reason for contamination of water. One of the prime benefactor to pollution by wastewater in India is synthetic dyes used in textile, cosmetic, leather, and printing industry that are pumped into water basins treated or untreated [1].

Textile industry includes various processes like dyeing of clothes, printing design, and finally finishing the product that demands large volume of water [2]. Different categories of dyes like cataionic, anaionic, azo dyes, and their components are used in above-mentioned industries. They are not only aromatic but also carcinogenic in nature which makes them very much strenuous to obliterate from effluent [3]. Among synthetic chemical dyes used in drug, textile, printing, and food processing industry across the globe, azo dyes comprised of about 70% of the total dyes used. These azo dyes are detrimental, carcinogenic, and challenging to diminish owing to their complex chemical configuration [4, 5].

This dye when let off into water bodies curtails lucidity of water and rate of oxygen transfer. It negatively influences aquatic life of water bodies as well as extent of effluent degradation. Majority of the organic dyes are intractable to reduce in concentration due to high temperature, light, chemical agents, and microbes which makes them almost difficult to break down even at lowest amount [6].

Dyes are liable for many intricacies like irritation of skin, hypersensitivity, eczema, and metastasis in humans. Due to detrimental consequences on Mother Nature and health of mankind, lot of environmental laws and regulations across the world have levied intermissive ceiling on the amount of dyes released in the liquid waste before it gets released into the nearby water living systems [7, 8]. Numerous biological, chemical, and physical treatment techniques have come into existence and used to separate dye from its wastewater [9].

Several accustomed chemical and physical technologies for remediation of dyes comprising their separation by physical and chemical precipitation, sorption, and using membrane turned out to be extensively examined. However, exorbitant investment on capital and costs of running and less efficiency at energy front even at low concentration dye (upto 10 ppm) confines the wide application of these process technologies for effective removal of dye [10].

Conventional biological dye treatment technologies using microbes with and without physicochemical methods have also been explored by many researchers [11]. However advanced biological dye treatment methods have gained attention due to its advantages like economical, ecofriendly, reduced sludge generation, and better efficiency in dye removal [12]. Biological methods used for treatment of effluent from textile manufacturing are principally hinged on biosorption and capability of its biodegradation [13].

Researchers across the scientific fraternity also explored the use of biomaterials which are originated from microbes and plant-based material for wastewater treatment. A great range of biomaterials especially from microbes like algae, fungi, yeast, and bacteria [14, 15] (Iddou and Ouali 2008) are being utilized as an adsorbent of biological origin for the elimination of dyes. However, among above-mentioned microbial treatment of dye, bacteria degrade the azo dye faster than other in the absence of oxygen, micro aerophillic state which normally brings about the generation of aromatic amine which further transform into non-toxic form by oxidation reaction [16–18].

Activated sludge process is mostly employed for the secondary treatment of effluent. It results in the depletion in biological and chemical oxygen demand with the help of supply of air as well as availability of dense microbial culture. The structure and formulation of the activated sludge may vary considerably incumbent on the system and region also rely upon the components of the effluent along with the operating parameters with which process is carried out [19, 20].

Biological degradation of dyes using aerobic condition of activated sludge process is cognized as well structured and self-sustaining technology. Process of activated sludge comprised of plenty of microorganisms for instance protozoa and bacteria. It carries negative charge on its surface which makes it suitable for dye treatment [21]. Nevertheless, generation of copious volume of sludge that contains residual compounds which are resistant to biodegradation, has got grave consequences on environment [22, 23].

The sludge generated by activated sludge process composed of many minerals like nitrogen, potassium, dyes, and metal ions [24]. Molecules of dye are formulated such that they are impervious to impulsive deterioration in surrounding [25]. The release of additional sludge waste having large quantity of dyes accumulated on the surface, originating from industrial textile wastewater treatment plant, inside natural or manmade receiving source of medium will lead to critical destruction to the habitat owing to its pernicious side effect [26, 27].

Microorganisms present in activated sludge from industrial waste are capable of creating a highly acclimatized consortium medium which is capable of breakdown of molecules of virulent dye [28–30]. Hence, treatments by acclimatized activated sludge have to be better restorative as well as properly optimized. Enhancement of dyes biodegradation efficacy and reduction in the quantity of remanent adsorbed dye can be studied based on various conditions that were found in the waste sludge.

Dye degradation using activated sludge process alone cannot suffice but it can further be modified or used with other method so as to increase the removal efficiency of dye. Numerous efforts have been made to improve this technique by designing various configurations to have better performance characteristics. In this review we have tried to gather scattered information about various aspects of dye decolorization using activated sludge process. This review discusses about techniques available for utilization of activated sludge for dye treatment. Adsorption utilizes activated sludge directly in dried form or in the form of carbon for dye eradication. This paper also reviews about dye degradation using bacteria isolated from sludge which can further be used for preparation of granules or bioflocculants aerobically or anaerobically. Further efficiency of fungus isolated from sludge for dye removal is also discussed. This review paper also discussed about amalgamation of advanced oxidation process and activated sludge process for dye removal.

2   Dye Degradation by Adsorption

Large amount of dyes are the prime source of pollution in the wastewater released from textile industry. It not only influences the color of effluent but also causes the reduction in the transfer of light as well as amount of dissolved oxygen in wastewater along with its ability to degrade biologically [31].

Adsorption is one of the simplest physical processes that have been used for the treatment dye. In a particular process of biosorption, dead bacterial cells [32], yeast [33], white rot fungi [34], and microalgae [35] turned out to be utilized as adsorbents in order to eradicate many dyes. Adsorption utilizes sludge either directly or by converting it into carbon which can further be used for dye treatment.

2.1   Adsorption of Dye onto Sludge-Based Activated Carbon

Commercial powdered carbon activated by thermal or chemical treatment is among the most ideal sorbent materials used for dye removal because of its large area of surface and large capacity of adsorption but high cost of starting material limits the application of commercial activated carbon for dye eradication process. This limitation and constraint has lead the researchers to explore various low cost sources for the generation of activated carbon using rice and coffee husk [36], date stone [37], and spent tyre [38].

Industrial sludge consists of high organic components and its disposal calls for multiple process which constitute about 20–60% of the operational cost of treatment of wastewater [39]. So reutilization of sludge waste as a source for manufacturing activated carbon can provide better substitute for its ultimate disposal by using standard process like landfill [40]. Some researchers have produced activated carbon by using sewage sludge and biological sludge [40–42].

Production of activated carbon by using activated sludge served dual ecological significance as it contributes to the management of solid waste by conversion of sludge to activated carbon and also the developed activated carbon act as adsorbent to discard dyes available in colored wastewaters. Typical process of making of activated carbon from sludge is represented in Fig. 1.

Sludge from two type's industrial waste like waste from paper mill [43] and waste from beverage industry [44] have been utilized to generate activated carbon. This work summarizes activated carbon properties from various sources as given in Table 1.

Li et al. [43] have synthesized activated carbon using waste sludge from paper mill to investigate its ability to eradicate cationic Methylene Blue and anaionic Reactive Red 24 dye from its aqueous solution. Paper mill waste sludge high in carbon and silicon dioxide was dried and obtained carbon was carbonized at 300 °C at inert environment and thermally activated at 850 °C in the presence of steam.

Fig. 1 Typical process of generation of activated carbon from sludge

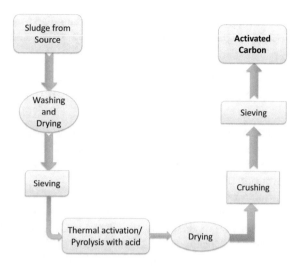

Table 1 Adsorption of dyes onto surface of sludge-based activated carbon

S. No.	Source of sludge	Activation of carbon	Dye removed	Conditions	Obtained adsorption capacity	References
1	Paper Mill waste	Thermal	Methylene Blue Reactive Red 24	pH 1–3 20 °C 30 min	263.16 mg/gm 34.36 mg/gm	[43]
2	Beverage waste	Pyrolysis	Allura Red Crystal Violet	pH 2 55 °C pH 8 55 °C	287.1 mg/gm 640.7 mg/gm	[44]

Activation of this carbon obtained from sludge can also be accomplished by either physical or chemical activation. Chemical agents like phosphoric acid [45], potassium hydroxide [46] as well as zinc chloride [47] have been employed as dehydrator and oxidizing agent which have an impact on decomposition by devolatilization process and also hinders the generation of tar, escalating the yield of carbons [48].

Physical activation to produce activated carbon comprised of two steps. First step includes the carbonization of a precursor of carbonaceous material at 450–800 °C under an inert environment that may result in loss of some carbon. It is succeeded by the activation of the carbon at raised temperature using CO_2, steam, or air that impedes diffusion of activating agents through the porous network. Reaction with carbon is endothermic in nature [49, 50].

To improve efficiency of color removal using adsorption, different chemicals and adsorbents generally incorporated precisely into the activated sludge systems. Color removal upto 90% has been obtained for acid orange 7 with the inclusion of activated carbon inside the large rectangular tank provided with air. However, increase in particle size of activated carbon reduces efficiency of color removal [51].

2.2 Adsorption of Dye onto Dried Activated Sludge

Only 50–90% of wastewater expelled from residential and commercial facilities throughout the world can be recycled by generating activated sludge, which then shifted into sedimentation tanks [52]. Landfill sites are most commonly utilized for disposing off the remaining sludge.

Some researchers utilized dried biomass of activated sludge directly for eradication of dyes by using adsorption process. Activated sludge with negative charge and Chemical oxygen demand reduced to 20 mg/L was used to remove basic dyes. Adsorption capacity upto 157 mg/gm was achieved for Basic blue 43 dyes [53].

Some additives like organic flocculant, powdered activated carbon, bentonite, and commercial clay were added in activated sludge and its efficiency for dye removal was observed. Organic flocculants and powdered activated carbon has shown effective elimination of dye color from wastewater from cotton textile industry [54].

Activated sludge from sewage treatment plant contains particles that comprised of various living organisms like bacteria, fungi, yeast, and protozoa whose cell walls consist of functional groups that can well connect with organic and inorganic substances. Modification in the company of magnetic flux along with heat treatment on activated sludge has shown better separation efficiency of dye like aniline blue, safranin O, and nile blue from waste compared to untreated activated sludge. This is extensively applied for separations of components in complex and difficult-to-handle media [55]. Table 2 shows that the dye removal efficiency can be enhanced by giving treatment to sludge.

Table 2 Effectiveness of sludge for various dye removal

Sr. No.	Source of sludge	Treatment to sludge	Dye removed	Conditions	Obtained adsorption capacity	References
1	Sewage treatment plant	Thermal + magnetic modification	Safranin O Nile blue Bismarck brown Y Aniline blue	pH 7 pH 8 pH 8 90 min pH 4–5	326.8 mg/gm 515.1 mg/gm 246.9 mg/gm 768.2 mg/gm	[55]
2	Wastewater treatment plant	Drying	Red bimacid (E5R)	pH 5 25 °C 10 min	36 mg/gm	[56]
3	Wastewater treatment plant	Drying	Basic red-29	30 °C	224.72 mg/gm	[53]

3 Aerobic Granular Technology

Treatment of dye containing wastewater using bacteria exhibits lot of advantages like anaerobic degradation of bond in azo dye and aerobic degradation of resulting aromatic amine over other biological [57] and many physiochemical treatment methods [58].

Dye containing wastewaters can be effectively reduced by employing potentially enticing method that uses integrated anaerobic and aerobic bacterial process [59]. This process is customarily executed sequentially but it can also be conducted concurrently using aerobic granular technology [60]. This can be shown from Fig. 2.

The first report of using aerobic granules employed for the reduction of dye from textile wastewater was that in a subsequent batch reactor controlled and managed under aerobic–anaerobic cycles by applying a combination of activated and anaerobic sludge as inoculums. Dye removal from synthetic effluent using generated granules gave rise to better scrapping of chemical oxygen demand but only 63% of the color reduction took place [60].

Another probable benefit of the application of aerobic granules is the soaring resistance of aerobic granules to toxic features [61].

However properties of aerobic granules can be changed through optimization of the hydraulic retention time and the span of the anaerobic stage of the sequential batch reactor cycle. This results in increase in efficiency of color removal to 95%. The use of aerobic granules both in the presence or absence of dye was also carried out in order to eliminate azo dyes like reactive blue 59 [62], Acid Red 14 [63], and Acid Red 18 [64].

The availability of dye or other contaminants does not hinder the formation of granules in synthetic media for short period of time whereas formation of granules in dye holding synthetic effluent offer benefits that population of bacteria inside the granules gets acclimatized to dye and other components of wastewater in the course of granule generation period itself [31].

Successful formation of aerobic granules in fabricated media involving dyes [60, 63] and its application for the dye removal from actual textile wastewaters have till so far been explored very little [65, 66]. Aerobic granules have been produced by species of bacteria secluded from sludge of textile origin or soil contaminated with wastewater from textile industry and sterilized activated sludge in actual textile

Fig. 2 Granular technology for dye treatment

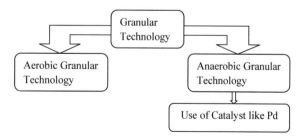

wastewater. Longer time taken for the generation of granules in real wastewater compared to the synthetic media is the only drawback. The granules have generated in sterilized wastewater originated from textile industry, in all probability to eradicate intervention of microbial flora of the wastewater on the process of aerobic granule generation [31].

Anaerobic granular sludge which consists of granular shaped multilayered structure (outer layer of acidogens, syntrophic microcolonies in central layer, and acetic lastic methanogens at the central core) of microbial aggregates has also shown to provide effective solution for dye eradication from wastewater. Some of the peculiarities of this technology are concise structure, sizeable microbial solidity, composite microbial culture, and robust resistance to noxius compounds [67, 68]. Low degradation rate of azo dye causes longer hydraulic retention time when anaerobic granular sludge is employed in reactor (Silva et al. 2012). Use of palladium nanoparticles as catalyst to discard recalcitrant dyes like methyl orange and Evans blue from wastewater has been reported [69]. Biotic synthesis of palladium nanoparticles with the help of microbes such as *Desulfovibrio* [70], *Shewanellaoneidensis* [71], and *Clostridium butyricum* [72] have also been carried out to reduce the use of toxic compounds via chemical route. These catalyst nanoparticles produced by microbes need to be immobilized in order to save their leaching and to have proper separation. Loss of activity is one of the pitfalls in the immobilization of biologically synthesized nano catalyst which further can be improved by self-immobilization into biogranules. Bacterial population available in activated sludge has been explored anaerobically in situ for the generation of nano-palladium particles and their subsequent application for degradation of azo dye like Orange II, Evans Blue and Congo red. It showed that availability of formic acid and formate as hydrogen donor accelerates the dye decolorization [3].

4 Sequential Chemical and Activated Sludge Process

Recently, advanced oxidation process is the foremost favorable technique for treatment of wastewater which can oxidize an extensive range of compounds that are onerous to degrade [73]. Use of non-toxic reagents like Fe^{2+} and hydrogen peroxide for degradation and decolorization of dyes makes oxidization using Fenton's reagent very captivating [74, 75].

A lot of studies regarding the dye removal using Fenton's oxidation [74, 76] and its capability to enhance organic biodegradability of toxic or non-biodegradable wastewaters have been performed [77]. High cost of reagent and generation of sludge waste that consists of iron, demands the proper management and safe disposal are some of the flaws of this process [78].

Most of the research has highlighted the use of a pretreatment with the help of chemical route thereafter a treatment using biological route with the help of a particular strain of microbes or microbial conglomeration like biofilms that are capable of achieving absolute breakdown of dye [79, 80]. Surprisingly, an amalgamation of advanced oxidation processes and activated sludge process has explored very little. A few studies recommended about a treatment using Fenton reagent integrated with a biological treatment using oxygen might be a compelling alternative that it would be possible to remove dyes [81]. Furthermore, the effect of azo dyes upon the population of microbes in activated sludge processes in the wake of contact with dye is also very little known.

Partial degeneration of synthetic dye with azo group like Reactive Violet 5 for concentration of more than 500 mg L^{-1} was achieved using Fenton's oxidation process which helped them to make the dye biodegradable. Complete degradation of azo dye was obtained using biological process such as activated sludge process in which exposure of azo dye Reactive Violet 5 to activated sludge required incubation period of up to 168 h that changed its microbial composition from diverse to very specific including *Trichosporon, Aspergillus,* and *Clostridium* species [28, 30].

5   Dye Treatment Using Fungus

Remediation of dye biologically is considered to be more promising with reference to its long standing gains and for comprising close to negligible deleterious consequences on Mother Nature. Most of the white rot fungi have found to be effective to get rid of textile dyes from their wastewater. Dyes are eliminated from synthetic waste using fungi by using biosorption [82–84] and mortification with the help of enzymes like peroxidase.

Aspergillus Niger strain was secluded from diluted sludge of wastewater from textile industry. It was then multiplied in agar made up from dextrose at 28 °C for 8 days and further adapted to different concentrations of dye by sub culturing onto mineral salt agar plate at neutral pH. It was then tested for acid, dibasic, di azo, and mono azo dye removal. These dyes get adsorbed onto fungal biomass and maximum removal was obtained for acidic dye [85].

6   Dye Treatment Using Biofloculant Isolated from Activated Sludge

Dyes account for only a little fraction of the whole magnitude of waste released in textile industry. Typical microbial process find it difficult to remove these dye molecules completely and it can also be very much ruinous to microbial population that may result in reduced effectiveness of process [86, 87]. Bioflocculants have an ability that can be employed in several industrial units such as wastewater treatment plants from pharmaceutical, textile industries, brewery, as well as domestic wastewater treatment units [88, 89].

Zhang et al. [90] disclosed that bioflocculants generated using species of bacteria xn7 + xn11 were very much successful in removing basic fuchsin dye ($100 \, \text{mg} \, \text{L}^{-1}$) but not so much as efficacious in eliminating reactive black ($50 \, \text{mg} \, \text{L}^{-1}$), demonstrating a bleaching capability of 93 and 35%, individually. Inexpensive and easy-to-operate biological method, appeared to be the center of attention of contemporary research on dye eradication. 16S rRNA gene sequencing analysis can be employed to ascertain bacterial isolates [91].

Bioflocculants originating from isolates were capable of eliminating several dyes arising out of industrial wastewater in different proportion, perhaps due to agglomeration of molecules and cells due to coupling and neutralization of charge [92].

Bridging takes place when flocculant arising from particle's surface gets dissolved into the solution for a gap exceeding the distance on the top of which interparticle repulsion operates [93, 94].

7   Activated Sludge Parameters

Various parameters used for evaluation using the activated sludge method are:

- Chemical oxygen demand (COD) and total dissolved solids (TDS)
  Microorganisms from activated sludge can tolerate up to 5000 mg/L and some up to 1000 mg/L if they encounter saline conditions. If the TDS concentrations are higher than this, there will be huge decrease in COD removal and it directly affects the treatment efficiency. A low TDS concentration does not have remarkable effect on the capability of the activated sludge process and COD eradication.
- Mixed Liquor Suspended Solids (MLSS) and Mixed Liquor Volatile Suspended Solids (MLVSS)
  MLSS and MLVSS concentrations are measured to know the growth of microorganisms during the activated sludge method. If the salinity is high in contaminated dye water for a particular dye concentration, there will be decreasing trend in MLSS and MLVSS concentrations and hence a decrease in population growth of microorganisms. But if the salinity is low, there is an increase in MLSS concentration and there will be an increase in degradation of dye by microbes.

  This suggests that if the salinity is high in dye contaminated water, there will an inevitable impact on MLSS concentrations due to changes in microorganisms.
- pH parameters
  When the activated sludge method is performed in the pH range of 6.50–8.50, there will be no significant changes in the dye removal efficiency. This is because the microorganisms strive to sustain their environment pH in neutral region by consuming energy.

8   Conclusion

Dye from textile industry offers a lot of difficulty in terms of its disposal due to its recalcitrant and aromatic nature which makes it major source of pollution in water. Numerous chemical, physical and biological methods came to be put in place to obliterate dye from wastewater. Most of the methods have sludge generation problem during dye treatment which pose another threat to nature in terms of disposal. Use of activated sludge to prepare its activated carbon or the use of dried activated sludge besides the fact that it minimizes the sludge disposal issue moreover help it to treat dye containing wastewater more efficiently. Activated sludge can also be good source of fungi, bacteria, or yeast which can be isolated and used for biological treatment of dyes. Use of activated sludge along with chemical process like advanced oxidation process offers good efficiency for dye removal. Modification of activated sludge process by using either bioflocculant or its use along with sequential oxidation process can prove to be more effective to achieve higher degree of dye degradation from industrial discharge.

References

1. Mahmoodi NM, Salehi R, Arami M, Bahrami H (2011) Dye removal from colored textile wastewater using chitosan in binary systems. Desalination 267:64–72
2. Nigam P, Banat IM, Singh D, Marchant R (1996) Microbial process for the decolorization of textile effluent containing azo, diazo and reactivedyes. Process Biochem 31(5):435–442
3. Quan X, Zhang X, Xu H (2015) In-situ formation and immobilization of biogenic nanopalladium into anaerobic granular sludge enhances azo dyes degradation. Water Res 78:74–83. https://europepmc.org/article/med/25912251
4. Dos Santos AB, Cervantes FJ, van Lier JB (2007) Review paper on current technologies for decolourisation of textile wastewaters: perspectives for anaerobic biotechnology. Biores Technol 98(12):2369–2385
5. Huang J, Chu S, Chen J, Chen Y, Xie Z (2014) Enhanced reduction of an azo dye using henna plant biomass as a solid phase electron donor, carbon source, and redox mediator. Biores Technol 161:465–468
6. De Luna MDG, Flores ED, Genuino DAD, Futalan CM, Wan M-W (2013) Adsorption of Eriochrome Black T (EBT) dye using activated carbon prepared from waste rice hulls—optimization, isotherm and kinetic studies. J Taiwan Instit Chem Eng 44:646–653. https://doi.org/10.1016/j.jtice.2013.01.010
7. Sadaf S, Bhatti HN, Nausheen S, Amin M (2015) Application of a novel lignocellulosic biomaterial for the removal of Direct Yellow 50 dye from aqueous solution: batch and column study. J Taiwan Inst Chem Eng 47:160–170
8. Santhi T, Smitha T, Sugirtha D, Mahalakshmi K (2009) Uptake of cationic dyes from aqueous solution by bioadsorption onto granular cucumissavita. J Appl Sci Environ Sanit 4:29–35
9. Haddad M, Abid S, Hamdi M, Bouallagui H (2018) Reduction of adsorbed dyes content in the discharged sludge coming from an industrial textile wastewater treatment plant using aerobic activated sludge process. J Environ Manag 223:936–946. https://doi.org/10.1016/j.jenvman.2018.07.009
10. Volesky B (1990) Biosorption and biosorbents. In: Biosorption of heavy metals. CRC Press, Florida, pp 3–6

11. Garg SK, Tripathi M (2017) Microbial strategies for discoloration and detoxification of azo dyes from textile effluents. Res J Microbiol 12:1–19
12. Chen KC, Wu JY, Liou DJ, Hwang SCJ (2003) Decolorization of the textile dyes by newly isolated bacterial strains. J Biotechnol 101:57–68
13. Mullai P, Yogeswari MK, Vishali S, Tejas Namboodiri MM, Gebrewold BD, Rene ER, Pakshirajan K (2017) Aerobic treatment of effluents from textile industry, current developments in biotechnology and bioengineering biological treatment of industrial effluents. In: Biological treatment of industrial effluents. Elsevier, pp 3–34
14. Djafer A, Kouadri Moustefai S, Iddou A, Si Ali B (2014) Study of bimacid dye removal from aqueous solution: a comparative study between adsorption on pozzolana, bentonite, and biosorption on immobilized anaerobic sulfatereducer cells. Desalin Water Treat 52:7723–7732
15. Henini G, Laidania L, Souahi F (2014) Study of the kinetics and thermodynamics of adsorption of Red Bemacid on the cords of Luffa cylindrical. Desalin Water Treat 57:3741–3749
16. Carvalho MC, Pereira C, Goncalves IC, Pinheiro HM, Santos AR, Lopes A, Ferra MI (2008) Assessment of the biodegradability of a monosulfonated azo dye and aromatic amines. Int Biodeter Biodeg 62:96–103
17. Franciscon E, Zille A, Fantinatti-Garboggini F, Silva IS, Cavaco-Paulo A, Durrant LR (2009) Microaerophilic-aerobic sequential decolourization/biodegradation of textile azo dyes by a facultative Klebsiella sp. strain VN-31. Process Biochem 44:446–452
18. Tripathi A, Srivastava SK (2011) Ecofriendly treatment of azo dyes: biodecolorization using bacterial strains. Int J Biosci Biochem Bioinform 1:37–40
19. Henze M, Harremoes P, LaCour J, Arvin E (2000) Wastetwater treatment: biological and chemical processes, 3rd edn. Springer, Berlin Heidelberg, New York
20. Van H, Catunda PFC, Araujo L (1998) Biological sludge stabilization, Part 2-Influence of the composition of waste activated sludge on anaerobic digestion. Water SA 24:231–236
21. Vijayaraghavan K, Yun Y (2007) Utilization of fermentation waste (Coryne bacterium glutamicum) for biosorption of Reactive Black 5 from aqueous solution. J Hazard Mater 141:45–52
22. Guo WQ, Yang SS, Xiang WS, Wang XJ, Ren NQ (2013) Minimization of excess sludge production by in-situ activated sludge treatment processes—a comprehensive review. Biotechnol Adv 31:1386–1396
23. Manai I, Miladi B, El Mselmi AL, Smaali I, Ben Hassen A, Hamdi M, Bouallagui H (2016) Industrial textile effluent decolourization in stirred and static batch cultures of a new fungal strain Chaetomium globosum IMA1 KJ472923. J Environ Manag 170:8–14
24. Roy Choudhury AK (2017) Sustainable chemical technologies for textile production. Sustainable fibres and textiles. Woodhead Publishing, pp 267–322
25. Casas N, Blánquez P, Vincent T, Sarrà M (2013) Mathematical model for dye decoloration and laccase production by Trametes versicolor in fluidized bioreactor. Biochem Eng J 80:45–52
26. Sohaimi KSA, Ngadi N, Mat H, Inuwa IM, Wong S (2017) Synthesis, characterization and application of textile sludge biochars for oil removal. Environ Chem Eng 5:1415–1422
27. Wang Q, Luan Z, Wei N, Li J, Liu C (2009) The color removal of dye wastewater by magnesium chloride/red mud (MRM) from aqueous solution. J Hazard Mater 170:690–698
28. Meerbergen K, Crauwels S, Willems KA, Dewil R, Van Impe J, Appels L, Lievens B (2017) Decolorization of reactive azo dyes using a sequential chemical and activated sludge treatment. J Biosci Bioeng. https://doi.org/10.1016/j.jbiosc.2017.07.005
29. Lade H, Kadam A, Paul D, Govindwar S (2015) Biodegradation and detoxification of textile azo dyes by bacterial consortium under sequential microaerophilic/aerobic processes. Excli J 14:158–174
30. Meerbergen K, Crauwels S, Willems KA, Dewil R, Impe JV, Appels L, Lievens B (2017) Decolorization of reactive azo dyes using a sequential chemical and activated sludge treatment. J Biosci Bioeng 124:668–673
31. Manavi N, Kazemi AS, Bonakdarpour B (2016) The development of aerobic granules from conventional activated sludge under anaerobic-aerobic cycles and their adaptation for treatment of dyeing wastewater. Chem Eng J. https://doi.org/10.1016/j.cej.2016.11.155

32. Nacera Y, Aicha B (2006) Equilibrium and kinetic modelling of methylene blue biosorption by pretreated dead streptomycesrimosus: effect of temperature. Chem Eng J 119:121–125. https://doi.org/10.1016/j.cej.2006.01.018
33. Yu JX, Wang LY, Chi RA, Zhang YF, Xu ZG, Guo J (2013) A simple method to prepare magnetic modified beer yeast and its application for cationic dye adsorption. Environ Sci Pollut Res 20:543–551. https://doi.org/10.1007/s11356-012-0903-3
34. Aydogan MN, Arslan NP (2015) Removal of textile dye reactive black 5 by the cold-adapted, alkali- and halotolerant fungus Aspergillus flavipes MA-25 under non-sterile conditions. Desalin Water Treat 56:2258–2266. https://doi.org/10.1080/19443994.2014.960463
35. Hernandez-Zamora M, Cristiani-Urbina E, Martinez-Jeronimo F, Perales-Vela H, Ponce-Noyola T, Montes-Horcasitas MD, Canizares-Villanueva RO (2015) Bioremoval of the azo dye Congo red by the microalga Chlorella vulgaris. Environ Sci Pollut Res 22:10811–10823. https://doi.org/10.1007/s11356-015-4277-1
36. Paredes-Laverde M, Salamanca M, Diaz-Corrales JD, Flórez E, Silva-Agredo J, Torres-Palma RA (2021) Understanding the removal of an anionic dye in textile wastewaters by adsorption on ZnCl₂ activated carbons from rice and coffee husk wastes: a combined experimental and theoretical study. J Environ Chem Eng 9(4):105685
37. Hijab M, Parthasarathy P, Mackey HR, Al-Ansari T, McKay G(2021) Minimizing adsorbent requirements using multi-stage batch adsorption for malachite green removal using microwave date-stone activated carbons. Chem Eng Process Process Intensif 108318
38. Jones I, Zhu M, Zhang J, Zhang Z, Preciado-Hernandez J, Gao J, Zhang D (2021) The application of spent tyre activated carbons as low-cost environmental pollution adsorbents: a technical review. J Clean Prod 312:127566
39. Muazu ND, Jarrah N, Zubair M, Alagha O (2017) Removal of phenolic compounds from water using sewage sludge–based activated carbon adsorption: a review. Int J Environ Res Public Health 14:1–34
40. DjatiUtomo H, Ong XC, Lim SMS, Ong GCB, Li P (2013) Thermally processed sewage sludge for methylene blue uptake. Int Biodeterior Biodegrad 85:460–465
41. Hadi P, Xu M, Ning C, Sze C, Lin K, McKay G (2015) A critical review on preparation, characterization and utilization of sludge–derived activated carbons for wastewater treatment. Chem Eng J 260:895–906
42. Nunthaprechachan T, Pengpanich S, Hunsom M (2013) Adsorptive desulfurization of dibenzothiophene by sewage sludge-derived activated carbon. Chem Eng J 228:263–271
43. Li W-H, Yue Q-Y, Gao B-Y, Ma Z-H, Li Y-J, Zhao H-X (2011) Preparation and utilization of sludge-based activated carbon for the adsorption of dyes from aqueous solutions. Chem Eng J 171:320–327
44. Streit AFM, Côrtes LN, Druzian SP, Godinho M, Collazzo GC, Perondi D, Dotto GL (2019) Development of high quality activated carbon from biological sludge and its application for dyes removal from aqueous solutions. Sci Total Environ 660:277–287
45. Molina-Sabio M, Rodriguez-Reinoso F, Caturta F, Selles MJ (1995) Porosity in granular carbons activated with phosphoric acid. Carbon 33:1105–1113
46. Fengchen W, Ruling T, Rueyshin J (2005) Preparation of highly microporous carbons from fire wood by KOH activation for adsorption of dyes and phenols from water. Sep Purif Technol 47:10–19
47. Nabarawy TH, Mostafa MR, Youssef AM (1997) Activated carbons tailored to remove different pollutants from gas stream and from solution. Adsorpt Sci Technol 15:61–68
48. Ahmadpour A, Do DD (1996) The preparation of active carbons from coal by chemical and physical activation. Carbon 34:471–479
49. Smith KM, Fowler GD, Pullket S, Graham NJD (2009) Sewage sludge-based adsorbents: a review of their production, properties and use in water treatment applications. Water Res 43:2569–2594
50. Rio S, Faur-Brasquet C, Coq LL, Courcoux P, Cloirec PL (2005) Experimental design methodology for the preparation of carbonaceous sorbents from sewage sludge by chemical activation—application to air and water treatments. Chemosphere 58:423–437

51. Marquez MC, Costa C (1996) Biomass concentration in PACT process. Water Res 30(9):2079–2085
52. Tebbutt THY (1998) Principles of water quality control. Butterworth-Heinemann, Oxford
53. Chu H-C, Lin L-H, Liu H-J, Chen K-M (2013) Utilization of dried activated sludge for the removal of basic dye from aqueous solution. Desalin Water Treat 51(37–39):7074–7080. https://doi.org/10.1080/19443994.2013.772540
54. Pala A, Tokat E (2002) Color removal from cotton textile industry wastewater in an activated sludge system with various additives. Water Res 36:2920–2925
55. Maderova Z, Baldikova E, Pospiskova K, Safarik I, Safarikova M (2016) Removal of dyes by adsorption on magnetically modified activated sludge. Int J Environ Sci Technol. https://doi.org/10.1007/s13762-016-1001-8
56. Djafer A, Djafer L, Maimoun B, Iddou A, Kouadri Mostefai S, Ayral A (2016) Reuse of waste activated sludge for textile dyeing wastewater treatment by biosorption: performance optimization and comparison. Water Environ J. https://doi.org/10.1111/wej.12218
57. Saratale RG, Saratale GD, Chang JS, Govindwar SP (2011) Bacterial decolorization and degradation of azo dyes: a review. J Taiwan Inst Chem Eng 42:138–157
58. Masigol MA, Moheb A, Mehrabani-Zeinabad A (2012) An experimental investigation into batch electrodialysis process for removal of sodium sulfate from magnesium stearate aqueous slurry. Desalination 300:12–18
59. Bonakdarpour B, Vyrides I, Stuckey DC (2011) Comparison of the performance of one stage and two stage sequential anaerobic-aerobic biological processes for the treatment of reactive-azo-dye-containing synthetic wastewaters. Int Biodeterior Biodegrad 65:591–599
60. Muda K, Aris A, Razman M, Ibrahim Z, Yahya A, Van Loosdrecht MCM, Ahmad A, Zaini M (2010) Development of granular sludge for textile wastewater treatment. Water Res 44:4341–4350
61. Liu Y, Tay J (2004) State of the art of biogranulation technology for wastewater treatment. Biotechnol Adv 22:533–563
62. Kolekar YM, Nemade HN, Markad VL, Adav SS, Patole MS, Kodam KM (2012) Decolorization and biodegradation of azo dye, reactive blue 59 by aerobic granules. Bioresour Technol 104:818–822
63. Franca RDG, Vieira A, Mata AMT, Carvalho GS, Pinheiro HM, Lourenço ND (2015) Effect of an azo dye on the performance of an aerobic granular sludge sequencing batch reactor treating a simulated textile wastewater. Water Res 85:327–336
64. Sadri Moghaddam S, Alavi Moghaddam MR (2016) Aerobic granular sludge for dye biodegradation in a sequencing batch reactor with anaerobic/aerobic cycles. CLEAN–Soil Air Water 4:438–443
65. Kee TC, Bay HH, Lim CK, Muda Z K (2015) Ibrahim, Development of bio-granules using selected mixed culture of decolorizing bacteria for the treatment of textile wastewater. Desalin Water Treat 54:132–139
66. Ibrahim Z, Amin MFM, Yahya A, Aris A, Muda K (2010) Characteristics of developed granules containing selected decolourising bacteria for the degradation of textile wastewater. Water Sci Technol 61:1279–1288
67. Hulshoff Pol LW, de Castro Lopes SI, Lettinga G, Lens PNL (2004) Anaerobic sludge granulation. Water Res 38:1376–1389. https://doi.org/10.1016/j.watres.2003.12.002
68. Lim SJ, Kim T-H (2014) Applicability and trends of anaerobic granular sludge treatment processes. Biomass Bioenergy 60:189–202. https://doi.org/10.1016/j.biombioe.2013.11.011
69. Johnson A, Merilis G, Jason Hastings M, Palmer E, Fitts JP, Chidambaram D (2013) Reductive degradation of organic compounds using microbial nanotechnology. J Electrochem Soc 160:G27
70. Baxter-Plant VS, Mikheenko IP, Macaskie LE (2003) Sulphatereducing bacteria, palladium and the reductive dehalogenation of chlorinated aromatic compounds. Biodegradation 14(2):83–90
71. Hennebel T, Simoen H, De Windt W, Verloo M, Boon N, Verstraete W (2009) Biocatalyticdechlorination of trichloroethylene with bio-palladium in a pilot-scale membrane reactor. Biotechnol Bioeng 102(4):995–1002. https://doi.org/10.1002/bit.22138

72. Hennebel T, Van Nevel S, Verschuere, Simon De Corte S, De Gusseme B, Cuvelier C, Fitts JP, van der Lelie D, Boon N, Verstraete W (2011) Palladium nanoparticles produced by fermentatively cultivated bacteria as catalyst for diatrizoate removal with biogenic hydrogen. Appl Microbiol Biotechnol 91:1435–1445. https://doi.org/10.1007/s00253-011-3329-9

73. Oller I, Malato S, Sánchez-Pérez J (2011) Combination of advanced oxidation processes and biological treatments for wastewater decontaminationda review. Sci Total Environ 409:4141–4166

74. Babuponnusami A, Muthukumar K (2014) A review on Fenton and improvements to the Fenton process for wastewater treatment. J Environ Chem Eng 2:557–572

75. Kuo WG (1992) Decolorizing dye wastewater with Fenton's reagent. Water Res 26:881–886

76. Nidheesh PV, Gandhimathi R, Ramesh ST (2013) Degradation of dyes from aqueous solution by Fenton processes: a review. Environ Sci Pollut Res 20:2099–2132

77. Chamarro E, Marco A, Esplugas S (2001) Use of Fenton reagent to improve organic chemical biodegradability. Water Res 35:1047–1051

78. Azbar N, Yonar T, Kestioglu K (2004) Comparison of various advanced oxidation processes and chemical treatment methods for COD and color removal from a polyester and acetate fiber dyeing effluent. Chemosphere 55:35–43

79. Kiran S, Ali S, Asgher M (2013) Degradation and mineralization of azo dye reactive blue 222 by sequential photo-Fenton's oxidation followed by aerobic biological treatment using white rot fungi. Bull Environ Contam Toxicol 90:208–215

80. Punzi M, Anbalagan A, Börner RA, Svensson BM, Jonstrup M, Mattiasson B (2015) Degradation of a textile azo dye using biological treatment followed by photo-Fenton oxidation: evaluation of toxicity and microbial community structure. Chem Eng J 270:290–299

81. Lodha B, Chaudhari S (2007) Optimization of Fenton-biological treatment scheme for the treatment of aqueous dye solutions. J Hazard Mater 148:459–466

82. Conatao M, Corso CR (1996) Studies of adsorptive interaction between Aspergillus niger and the reactive azo dye procion blue MX-G. Eclet Quim 21:97–102

83. Fu YZ, Viraraghavan T (2000) Removal of a dye from aqueous solution by the fungus Aspergillus niger. Wat Qual Res J Can 35:95–111

84. Paymann MA, Mehnaz MA (1998) Decolorization of textile effluent by Aspergillus niger (marine and terrestrial). Fresen Environ Bull 7:1–7

85. Ali N, Hameed A, Ahmed S, Khan AG (2008) Decolorization of structurally different textile dyes by Aspergillus niger SA1.World J Microbiol Biotechnol 24:1067–1072

86. Ogawa T, Shibata M, Yatome C, Idaka E (1988) Growth inhibition of Bacillus subtilis by basic dyes. Bull Environ Contam Toxicol 40:545–552

87. Sanayei Y, Ismail N, Teng TT, Morad N (2010) Studies on flocculating activity of bioflocculant from closed drainage system (CDS) and its application in reactive dye removal. Int J Chem 2:168–173

88. Gong W-X, Wang S-G, Sun X-F, Liu X-W, Yue Q-Y, Gao B-Y (2008) Bioflocculant production by culture of Serratia ficaria and its application in wastewater treatment. Bioresour Technol 99:4668–4674

89. Wang SG, Gong WX, Liu XW, Tian L, Yue QY, Gao BY (2007) Production of a novel bioflocculant by culture of Klebsiella mobilis using dairy wastewater. Biochem Eng J 36:81–86

90. Zhang CL, Cui Y, Wang Y (2012) Bioflocculant produced from bacteria for decolorization, Cr removal and swine wastewater application. Sustain Environ Res 22:129–134

91. Sirianuntapiboon S, Srisornsak P (2007) Removal of disperse dyes from textile wastewater using bio-sludge. Bioresour Technol 98:1057–1066

92. Salehizadeh H, Shojaosadati SA (2001) Extracellular biopolymeric flocculants: recent trends and biotechnological importance. Biotechnol Adv 19:371–385

93. Hantula J, Bamford DH (1991) The efficiency of the protein dependent flocculation of Flavobacterium sp. Appl Microbiol Biotechnol 36:100–104

94. Levy N, Magdasi S, Bar-Or Y (1992) Physico-chemical aspects in flocculation of bentonite suspensions by a cyanobacterial bioflocculant. Water Res 26:249–254

95. Boonnorat J, Techkarnjanaruk S, Honda R, Angthong S, Boonapatcharoen N, Muenmee S, Prachanurak P (2018) Use of aged sludge bioaugmentation in two-stage activated sludge system to enhance the biodegradation of toxic organic compounds in high strength wastewater. Chemosphere 202:208–217. https://doi.org/10.1016/j.chemosphere.2018.03.084
96. Muda K, Aris A, Razman M, Ibrahim Z, Van Loosdrecht MCM, Ahmad A, Zaini M (2011) The effect of hydraulic retention time on granular sludge biomass in treating textile wastewater. Water Res 45:4711–4721

Fundamental Concepts of Dye-Containing Textile Wastewater Treatments: Microbial and Enzymatic Approaches

Grazielly Maria Didier de Vasconcelos,
Vanessa Kristine de Oliveira Schmidt, Stefane Vieira Besegatto,
Éllen Francine Rodrigues, Wagner Artifon, Lidiane Maria de Andrade,
Luciana Prazeres Mazur, Ana Elizabeth Cavalcante Fai, Débora de Oliveira,
Antônio Augusto Ulson de Souza,
Selene Maria de Arruda Guelli Ulson de Souza,
and Cristiano José de Andrade

Abstract The world consumption of dyes has been increasing, mainly due to the textile industry (colorization of fibers). The textile industry generates a massive amount of wastewater. The incorrect disposal of colored effluent into the environment leads to the derangement of aquatic life. Several techniques have been applied to reduce this impact, including adsorption, coagulation, and filtration, among others that, on the one hand, are efficient; on the other hand, require additional management (e.g., a large volume of sludge). In this sense, specific biological pathways for dye degradation and wastewater discoloration have drawn attention to the industry since they can increase wastewater treatment yields. Therefore, this chapter describes the fundamental concepts of dye-containing textile wastewater treatments, particularly

G. M. Didier de Vasconcelos · V. K. de Oliveira Schmidt · S. V. Besegatto · É. F. Rodrigues ·
W. Artifon · L. P. Mazur · D. de Oliveira · A. A. Ulson de Souza ·
S. M. de Arruda Guelli Ulson de Souza · C. J. de Andrade (✉)
Department of Chemical Engineering and Food Engineering, Technological Center, Federal
University of Santa Catarina, Florianópolis, SC 88040-970, Brazil

S. V. Besegatto
e-mail: stefane.besegatto@posgrad.ufsc.br

D. de Oliveira
e-mail: debora.oliveira@ufsc.br

A. A. Ulson de Souza
e-mail: antonio.augusto.souza@ufsc.br

L. M. de Andrade
Department of Chemical Engineering, Polytechnic School, University of São Paulo, R. do Lago,
250 - Butantã, São Paulo, SP 05338-110, Brazil

A. E. C. Fai
Department of Basic and Experimental Nutrition, Institute of Nutrition, Rio de Janeiro State
University (UERJ), Rio de Janeiro, RJ 20550-013, Brazil

microbial and enzymatic approaches, including the most usual textile wastewater treatments and their trends (modern technology).

Keywords Azo dyes · Bioremediation · Bacterial · Fungal · Microalgae · Genetically modified organisms · Combined treatments systems · Biofilms · Resource recovery strategy · Machine learning

1 Introduction

Since the first commercially successful synthesis, dyes have been massively applied for the chemical, food, textile, paper, and other industries [1].

Dyes can be classified taking into account their chromophore group (e.g., anthraquinone, azo, indigo, nitrated, nitro, phthalein, triphenyl) and/or their ionic-fiber interaction (e.g., acid, azoic, direct, disperse, reactive, sulfur, vat) [2]. Azo dyes (N=N–) represent an environmental concern due to their high demand and toxicity, including carcinogenic [3].

The textile industries are the most dye-consumers that inherently produce many effluents with high biochemical oxygen demand, turbidity, and low degradability since 10–15% of dyes do not remain on textile fibers [4, 5]. These dyes are recalcitrant molecules that are harmful to the environment [6]. Thus, biological, physical, and chemical treatments, including activated sludge, oxidation, ozonation, membrane filtration, coagulation, and their combinations, are extensively applied as textile wastewaters [7–10]. However, there are drawbacks for each approach, such as high costs, inefficient, and generation of secondary polluters (e.g., high volume of sludge).

Regarding remediation of dye-contaminated water bodies, the biological processes by bacteria, fungi, algae, oxidative enzymes, and their combinations have been drawing attention, since they are efficient and eco-friendly, operate mildly, etc. [4, 11–14]. In this sense, phycoremediation, and genetically modified organism oxidoreductase producers are promising biological treatment alternatives. The integrated systems as advanced oxidation processes-*Aeromonas hydrophila* or membrane bioreactors (anoxic bioreactors, aerated bioreactor, UV-unit, and granular activated carbon filter).

Therefore, it is essential to discuss properly the methods for treating dye-contaminated wastewater. In this context, this chapter describes the fundamental concepts of dye-containing textile wastewater, including the structure chemical of dyes, their classification and global market, in particular the azo dye. Furthermore, it correlates the environmental effects of dyes, when improperly discharged, the conventional textile wastewater treatments technologies, particularly microbial and enzymatic approaches methods, including the most usual textile wastewater treatments and their trends (modern technology).

2   Chemical Structure of Dyes

Dyes and pigments are chemical compounds widely used to impart color to different substrates (fabric, fiber, leather, plastic, paper, etc.) [15, 16]. Color is described as the qualitative perception of light discriminated by the eyes and brain [17, 18]. These compounds absorb light radiation within the range of the visible region spectrum (380–750 nm), at the same time that they reflect or diffuse that radiation [19].

Synthetic dyes are complex organic substances colored fluorescent [20]. On the other hand, pigments are solid or particle structures composed of organic or inorganic moiety. They can be colored red, colorless (opaque), or fluorescent. It is worth noting that the last one is chemically unstable and shows low water solubility [21, 22]. In a broader sense, the difference between these compounds lies precisely in how they interact with the substrate.

Dyes consist of two main components, chromophore, and auxochrome [21, 23]. The chromophore is an unsaturated atomic group, in which the arrangement of single and double bonds allows the absorption of light, that is, the chromophore group is responsible for adding color to the substrate, examples are: sulfide ($-C=S$), azo ($-N=N-$), carbonyl ($-C=O$), and nitro ($-NO_2$) [21, 23, 24]. On the other hand, the auxochromic group contains many functional groups, such as sulfonic ($-SO_3H$), amine ($-NH_2$), hydroxyl ($-OH$), and carboxylic ($-COOH$), which are substituents bounded to the chromophore. They improve the color to the material (fiber) and affect the solubility in water, by either donating (auxochrome) or receiving (antiauxochrome) electrons [25, 26]. Additionally, the third component of the dye corresponds to the matrix, composed of benzene, anthracene, and perylene rings, among others [19, 21, 23].

2.1   Global Dye Market

Dyes have been used since prehistory, such as color surfaces, objects, and fabrics [27]. Initially, pigment production was limited to natural sources as plants, minerals, and insects [28, 29]. Later, the first dye synthesis breakthrough was made in 1743 by Barth (indigo carmine) [30]. However, in 1856, William Perkin revolutionized the dye industry by accidentally finding a Mauveine dye chemical route and synthesizing (aniline purple, CI 50245) [31, 32]. Perkin's discovery was crucial to accelerating synthetic dyes' production at a global scale [33]. Synthetic dyes were introduced into the market, they were also applied to other industrial areas, such as paper, food, and pharmaceutical, among others [34].

Since the invention of the first synthetic dye, it is estimated that more than 10,000 synthetic organic dyes became commercially available, with an annual production surplus of 700 thousand metric tons [24, 35, 36]. In the early twentieth century Europe was responsible for the global production of dyes. Asian countries, particularly China and India, are the largest dye producers worldwide [34].

2.2   Classification of Textile Dyes

Since there are many synthetic dyes and their formulations, a systematic classification is essential to enhance the entire textile industry.

2.2.1   Chromophore Structure and Color Index (CI)

A wide variety of textile dyes can be classified according to their source, solubility, chemical structure, and application, among others [33]. The chemical structure-based classification should be correlated in particular to chromophore groups as azo dyes, anthraquinone, nitro, xanthenes, and arylmethane [19, 37]. Table 1 shows some widely used synthetic organic dyes.

In this sense, the solubility-based classification is often used, since it is low cost and, highly related to application, since different textile substrates are evaluated [16, 21, 41], for instance, water-soluble dyes are acidic, basic, reactive, and direct [16, 42], whereas insoluble dyes can be classified as vat, dispersed, sulfur, and azoic [35, 36, 42, 43].

Alternatively, textile dyes can be classified according to the Colour Index (C.I.) [44], in which each dye is assigned a Generic Colour Index Name based on the application class, color, and identification code [45]. The C.I. Number is composed of 5 digits as shown in Table 1 [19, 45].

3   Azo-Based Textile Dyes

Azo compounds contain at least one nitrogen-nitrogen double bond (N=N), where the nitrogen is bounded to aromatic ring groups, such as benzene, naphthalene, and heterocyclic rings [46–48]. However, many chemical structural configurations are possible. These dyes can be synthesized from the diazotization of an aromatic amines and coupling reaction with electro-rich nucleophiles [49]. As the number of azo bonds increases in the same molecular structure, they can be named Disazo, Trisazo, and Polyazo [50]. Azo dyes are brightly colored compounds, highly chemically stable, including sunlight exposure, biodegradation, and wash fading [47, 51].

3.1   Global Azo Dye Market

Azo dyes correspond to 70% out of the total synthetic dye market [37, 52], in which the textile industry uses 80%, that is, 56% out of total [16, 34, 53–55].

Table 1 Synthetic organic dyes

Chromophore	Chemical structure	C.I. name C.I. number	Name	References
Azo		C.I. Acid Orange 7 C.I. 15510	Acid Orange II	[38]
Anthraquinone		C.I. Reactive Blue 4 C.I. 61205	Procion Blue MX-R	[39]
Indigo		C.I. Acid Blue 74 C.I. 73015	Indigo carmine	[40]

3.2 Effects of Azo Dyes on Environment

The dyeing processes are inherently related to dye (concentration, type, etc.), the textile material (shape, type, etc.), and the color desired [56, 57]. Most dyes are soluble organic molecules that adhere to fibers, yarns, or fabrics to impart color, thus, dyes have to be highly stable to many factors such as surface active agents (soaps and detergents) and light exposure, among others [57, 58]. The dyeing procedure can be carried out in exhaust dyeing (batch), continuous or semi-continuous processes, in general, involving a sequential mechanism, that includes migration, adsorption, and diffusion [57, 59, 60]. Initially, the dye is transported from the dyebath onto the fiber surface, followed by the dye molecule's adsorption at the surface of the fiber, and finally, the dyes are diffused from the surface to the interior of the fiber by fixation [57].

The textile industry consumes substantial amounts of water, especially by dyeing and printing processes, which, in turn, are responsible for approximately 25% out of total water consumption [61]. Furthermore, water is also required for washing dyed textile material, generating high volumes of colored effluents [61–64]. It is worth noting that dyes and many other chemicals such as enzymes, mineral and organic acids, alkalis, surfactants, salts, and oxidizing agents are required in dyeing processes [63]. Hence, the textile wastewaters are highly complex and recalcitrant that are unfeasible to treat by conventional methods [42, 61].

The textile industries can be responsible for massive environmental impacts since their effluents are composed of enzymes, mineral and organic acids, alkalis, surfactants, salts, oxidizing agents, and dyes, all of them at high concentrations [63, 64]. The wastewater generates in the dyeing mill comprises the most significant amount of the total wastewater of the textile industry. This wastewater comes from the dye preparation, spent dye bath, and washing processes [63]. It is estimated that over 7×10^5 tons of synthetic dyes are annually produced worldwide, and during the coloration process of the textile industry, up to 3.10^5 tons of this amount are lost to wastewater [59, 65]. According to Samsami et al. [66], the textile industry is responsible for 54% of the total discharge of dyes into the environment [66].

The dye wastewaters are strong colored, possess high biological oxygen demand (BOD), high dissolved solids (DS), low suspended solids (SS), high salt content, alkalinity, and low concentration of heavy metals [63].

Many dyes are easily visible (naked eye) even at low concentration. Thus, when incorrect disposal into the water bodies, they will become visually unpleasant and affect aquatic life [61, 65, 67].

Textile dyes are recalcitrant compounds. Thus, they can reach the aqueous ecosystem as pollutants. In addition, dyes can harm water quality in terms of total organic carbon (TOC), chemical oxygen demand (COD), and BOD [51, 68]. Furthermore, some dyes, as azo-type textile dyes, can be bioaccumulated in the food chain (biomagnification) [68, 69]. Many synthetic textile dyes and their metabolic intermediate products can have direct and indirect toxic effects (mutagenic and carcinogenic agents) [67, 70]. Thus, azo dyes can be correlated to animal, and human diseases such as allergies, tumors, cancers, dysfunction of the liver, reproductive system, kidney, brain, and central nervous system, and suppression of the immune system [65, 70–72]. In this sense, it is worth noting that the toxicity of azo dyes is mainly related to aromatic amines that can be biodegradation products [65].

Despite some azo dyes (for example, Acid Orange 7 and Reactive Red 195) have been reported as non-toxic [24, 47, 73], the most of them are environmental hazardous substances [36]. It is worth noting that these azo dyes can reach the drinking water supply systems [24]. As already mentioned, the textile industry is the major azo dye polluter [74, 75]. In this sense, it is estimated that between 15 and 50% of azo dyes eluates (do not get adsorb) from textile fibers [69, 76].

The incorrect disposal of azo dye-containing azo is visually unpleasant [69] and, also decrease the penetration of sunlight and consequently affects the aquatic biota, leading to lower oxygen levels [77–79].

The synthetic origin and molecular arrangement of azo dyes lead to high chemical stability, including sunlight exposure and biodegradation [80–82]. Therefore, in recent years, the contamination of these substances and their degradation products have been intensively investigated [16, 21, 48, 83–85].

The accumulation of azo dyes in plants [79], animals [86], water [87, 88], and soil [89] can be absorbed (oral or inhalation) by human beings and, consequently, impact their health as blood disorders, colic, allergies, and skin irritation [69, 90]. It is also known that prolonged exposure to these substances can induce carcinogenic and mutagenic effects in animal cells [86, 91] including human [24, 90–93]. Furthermore, some azo dyes reveal mutagenic potential at the chromosomal level [93, 94] inducing DNA and RNA damage. The Table 2 briefly illustrates some toxic effects related to synthetic dye exposure.

Therefore, the commercial relevance of azo dyes is undeniable. However, their toxicity and environmental impact should be carefully evaluated.

Table 2 Azo dyes and their adverse environmental and health impacts

C.I. name of dye	Effects	References
Acid Orange 7	Molecular, cellular, and organism level toxicity; chromosomal abnormalities and reduced mitotic index in *Allium cepa* bioassay	[73]
Direct Blue 15	Mutation in microalgae, cladocerans, and zebrafish embryos	[95]
Disperse Orange 1	Mutation in human lymphocyte and human hepatoma cells;	[94]
Disperse Blue 291	Mutation in mouse bone marrow cells; DNA fragmentation, genotoxic and mutagenic in a human hepatoma cell line	[96, 97]
Disperse Red 1	Chromosome aberrations and primary DNA damage in liver	[98]

4   Textile Wastewater Conventional Treatments

During dyeing and printing processes, the fixation of reactive dyes on fabrics occurs by forming a covalent bond between the dye molecule containing the electrolytic reactive group and nucleophilic fiber residues [99]. However, the linkage is insufficient,—emphasizing the azo-type textile dyes which, around 15–50%, do not bind to the fabric—and are released into wastewater [100]. About 1000–3000 m^3 of sewage is generated per day to produce 12–20 tons/day of a fabric product [100, 101]. Table 3 illustrates the consumption of water, chemicals, and energy for the production of 1 kg colored fabric. In this sense, the textile industry is recognized among all industrial sectors as the most polluters, due to the volume of water discharged and the environmental load generated (e.g., high consumption of energy and chemicals, CO_2 emissions, and effluent production with a high rate of impurities) [102]. The effluent composition of various stages of a textile manufacturing industry is shown in Table 4. The high concentrations of BOD, COD, color, pH, and metals hamper the treatment of these wastewaters.

Therefore, it is necessary to combine different treatment processes to enable textile wastewater disposal. Different physical–chemical and biological treatments such as separation and concentration, decomposition, degradation, and exchange processes are used in textile manufacturing industries. In summary, three main treatment processes guide textile wastewater remediation: primary, secondary, and tertiary (Fig. 1). According to [104], suspended solid waste, excessive amounts of oil, and granular materials present in the effluent are removed by primary treatment. In the

Table 3 Global inventory for the dyeing process (functional unit: 1 kg of colored fabric) [103]

System unit	Input	Conventional process
Pre-treatment	Water	20 L
	Solvent	120 kg
	Energy	0.13 kWh
Dyeing	Water	20 L
	Solvent	20.98 g
	Dyestuff	10 g
	Auxiliary	25 g
	Energy	3.82 kWh
Washing	Water	10 L
	Washing agents	5 g
	Energy	0.16 kWh
Drying	Energy	0.04 kWh
	Gas	0.27 kWh

Table 4 Major textile polluters at stages of manufacturing [5]

Process	Constituents	Wastewater characteristics/typical concentrations
Sizing	Yarn waste and unused starch-based sizes	High BOD and medium COD
Desizing	Enzymes, starch, waxes, and ammonia	BOD (34–50% of total), high COD, and temperature 70–80 °C
Scouring	Disinfectants and insecticides residues, NaOH, surfactants, and soaps	Oily fats, BOD (30% of total), high pH, temperature 70–80 °C, and dark color
Bleaching	H_2O_2, Adsorbable Organic Halogen, NaOCl, and organics	High pH and TDS
Mercerization	NaOH	Low BOD (less than 1% of total), TDS, and oil and grease
Dyeing	Color, metals, salts, acidity/alkalinity, and formaldehyde	High toxicity, BOD (6% of total), high dissolved solids, and high pH
Printing	Urea, solvents, color, and metals	High toxicity, high COD, high BOD, high dissolved solids, high pH, and intense color
Finishing	Chlorinated compounds, resins, spent solvents, softeners, waxes, and acetate	Low alkalinity, low BOD, and high toxicity

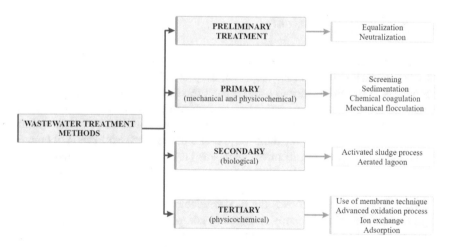

Fig. 1 Classification of wastewater treatments

secondary treatment process under aerobic or anaerobic conditions, there is a reduction in BOD, phenol, oil contents, and color control in the effluents. Electrodialysis, reverse osmosis, and ion exchange—briefly described below—are examples of tertiary treatments.

4.1 Primary Treatment

Before primary treatment, the textile manufacturing industry effluent processing, pre-treatment, or preliminary treatment must be applied to equalize the effluent, to homogenize it in terms of pollution load, pH, and temperature [105]. Then, primary treatments based on screening, sedimentation, and flotation processes are used to remove sedimentable organic and inorganic solids.

After primary treatment, fine suspended particles and colloidal particles are not removed efficiently, therefore, the processes of mechanical flocculation, and chemical coagulation are effectively employed. These processes have the function of destabilizing colloidal particles by adding coagulating agents (e.g., lime [$Ca(OH)_2$], alum, ferrous sulfate ($FeSO_4$), and ferric chloride ($FeCl_3$), among others), which have a high charge/mass ratio with colloidal particles significantly increasing their size. This factor contributes to sedimentation and the effluent can then be processed through a flocculator and settling tank [105]. Treating textile wastewater with coagulating agents helps to reduce color (insoluble dyes, 70–90%), suspended solids (50–75%), BOD (25–50%), COD (50–60%), and oils and grease (65%). Although, the addition of a coagulating agent to the wastewater stream increases the cost of this process and the sludge formed requires further treatment [106].

4.2 Secondary Treatment

In secondary treatment, biodegradation has become a promising method for treating textile effluents compared to physicochemical processes. Besides the lower cost of the process, it is unnecessary to use chemical products and the sludge produced— a result of cell proliferation—has a low content of chemical compounds [107]. In this process, dissolved organic compounds, remaining colloidal particles, and the color present in wastewater are removed and/or reused in the stabilization of organic matter, due to the presence of microorganisms [108]. The activated sludge system stands out as one of the most commonly used treatment methods and is divided into two categories: aerobic and anaerobic treatment processes.

The aerobic process removes the biodegradable components from the effluent, (e.g., carbohydrates and readily degradable compounds). In other words, in this process occurs the: (I) coagulation and flocculation of colloidal matter, (II) oxidation of organic matter dissolved in carbon dioxide, and (III) degradation of nitrogenous organic matter into ammonia, which is then converted into nitrite and eventually into nitrate. However, after this process, more complex xenobiotic compounds, such as dyes and surfactants, remain in the effluent [99, 105]

In anaerobic treatment, factors such as pH, temperature, waste load, absence of oxygen and toxic material directly influence the efficiency of the process, which is mainly used for digestion of activated sludge [105].

Most dyes are generally recalcitrant to aerobic degradation but can undergo reductive discoloration under anaerobic conditions [107]. In a treatment plant, the two treatment systems are usually used together. The anaerobic system is followed by an aerobic system, producing better results, for example, the color reduction is significantly more significant than using the isolated aerobic process (88 vs. 28%), as well as the reduction of Total Organic Carbon (79–90%) [64, 105, 108]. It is essential to mention that textile effluent contains significant amounts of non-biodegradable chemicals (e.g., amino benzene sulfonic or naphthyl amine sulfonic acids and heavy metals) [99, 105, 109]. Since these conventional treatment systems are not very effective in removing pollutants, an efficient tertiary treatment process is needed.

4.3 Tertiary Treatment

Environmental quality standards for the release of textile effluents into surface water bodies demand tertiary treatment. In addition to conventional treatments, textile effluents have tertiary processes to remove specific contaminants (e.g., various types of dyes such Congo Red, Methyl Orange, Methylene blue, C.I. Direct red 80, C.I. Reactive Red 17, and C.I. Direct Yellow 50, among others), as well as complete removal of solids and organic matter, reduce color or degrade recalcitrant compounds, reduce nutrients (e.g., ammonia and phosphorus) and assist in effluent disinfection [99, 105].

The most common tertiary treatment solutions are removing residual color organic compounds by adsorption and removal of dissolved solids by membrane filtration. Wastewater is also treated with ozone or another oxidizing agent to destroy many contaminants and minimize textile effluent disposal problems [105].

5 Aerobic and Anaerobic Microbial Degradation of Dyes: Bacterial, Fungal, and Microalgae

The biological treatments have several advantages, such as implementation and operation low costs and high yields (relatively), among others. It can be carried out under aerobic and/or anaerobic conditions, reaching the complete mineralization [110–112].

Aerobic and anaerobic bacteria, fungi, yeasts, and algae can be used for dye degradation. These microorganisms play essential roles in environmental maintenance since they are responsible for metabolizing organic, inorganic, natural, or xenobiotic compounds [113]. The degradation of dyes can be carried out by pure cultures or microbial consortium [114]. Mandal et al. [115] reported that microbial consortium present higher degradation yield, when compared to pure cultures, due to synergistic metabolic actions, and also higher stability (temperature, and pH).

5.1 Aerobic Microbial Degradation of Dyes

The required oxygen for aerobic microbial degradation can be supplied by atmospheric air (dissolution), and/or pure oxygen, and/or biochemical sources. Conventional aerobic treatment systems are compacted bed reactors, biological filters, aerobic stabilization ponds, and activated sludge [116, 117].

The activated sludge system is the most widely used aerobic treatment for effluents in the textile industry since it shows shorter hydraulic holding time, and higher operational flexibility when compared to anaerobic treatments. However, it generates high biomass volumes (sludge) containing (adsorption) residual compounds that are resistant to biodegradation [118].

The microorganisms commonly used for the decolorization of dyes in aerobic conditions are fungi and bacteria. Bacteria can grow faster than fungi, however, the microorganisms that are more efficient at degrading synthetic dyes are white-rot fungi. White-rot fungi include *Polyporus sanguineus*, *Daedalea flavida*, *Dichomitus squalens*, and *Irpex flavus* [116, 119, 120].

Kodam et al. [121] realized a study showing the use of a bacterium isolated from the textile effluent to degrade azo sulfonated dyes. In this study, the researchers isolated and purified a pure culture called KMK 48 from the sludge collected in a textile dyeing industry located in India. The pure culture KMK 48 was efficiently degraded Reactive Red 2, Reactive Red 141, Reactive Orange 4, Reactive Orange7, and Reactive Violet 5.

Balamurugan et al. [119] evaluated the color degradation of dye-containing textile effluent and the reduction of COD through anaerobic digestion using *Halomonas variabilis* and *Halomonas glacier*. Experiments were carried out at 30 °C in a CO_2 incubator. The maximum degradation was reached after 144 h. Nevertheless, considering the COD, the aerobic treatment did not reach a high yield [116].

Subramanian et al. [122] studied the degradation of triphenylmethane dye malachite green by litter decomposing fungi. The authors isolated and identified fungi (*A. flavus*, *A. niger*, *A. terreus*, *F. oxysporum*, *Penicillium* sp.) in mangrove regions. Among the isolated microorganisms, *Aspergillus flavus* showed greater decolorization capacity (83%) after 9 days.

The aerobic fungal and bacterial degradation of textile effluents has several advantages, such as low cost and high yields (mineralization). Nevertheless, white-rot fungi, present some whiteout fungi disadvantages, such as prolonged growth phase and retention times (up to 14 days) [116, 119].

5.2 Anaerobic Microbial Degradation of Dyes

The main advantage of anaerobic biological processes is the wide variety of microorganisms in the anaerobic consortium. In addition, the anaerobic processes generate biogas (energy source) [116, 123].

Dye degradation under anaerobic conditions is a combination of biological, and chemical mechanisms. Anaerobic biological treatment involves an oxidation–reduction reaction with hydrogen, allowing the azo dye and other water-soluble dyes to be discolored. The decolorization of azo dyes under anaerobic conditions occurs with a redox potential <50 mV. The dye acts as an electron receptor, the double bond (–N=N–) is broken and promotes the formation of an intermediate hydrazo, which has the bond (–N–NH–), which undergoes a reductive cleavage and produces amines. The chemical contribution to reducing and reducing azo dyes is attributed to biogenic reducers such as sulfide, cysteine, ascorbate, and Fe^{2+} [110, 113, 124].

$$R_1 - N = N - R_2 + 2e^- \rightarrow R_1 - HN - NH - R_2 \quad \text{(Intermediate hydrazo)}$$

$$R_1 - HN - NH - R_2 + 2e^- \rightarrow R_1 - NH_2 + R_2 - NH_2 \quad \text{(Aromatic amines)}$$

Conventional anaerobic treatment systems are anaerobic lagoons, septic tanks, anaerobic filters, and high-rate reactors, capable of receiving greater amounts of organic load per volumetric units, such as UASB reactors (upflow anaerobic sludge blanket). Among the reactors, the anaerobic filter, and UASB show high color removal efficiency [125].

Anaerobic degradation of azo dye Acid Orange 7 was performed using a continuous upflow packed bed reactor containing biological activated carbon. High conversion rates (up to 99%) of the azo dye were obtained using a continuous upflow packed bed reactor containing biological activated carbon. The use of a continuous upflow packed bed reactor containing biological activated carbon has proved to be an effective and promising treatment system for the degradation of anaerobic azo dyes [126].

A new method to increase anthraquinone dye reactive blue 19 degradation using the anaerobic system has been proposed by Cai et al. [127] using resuscitation-promoting factors (RPF), which has been proven to resuscitate and stimulate bacterial growth. Resuscitation-promoting factors were efficient in the decolorization of anthraquinone dye reactive blue 19. When compared to the conventional methods, there was an increase in dye decolorization by an additional 20% with the addition of the RPF process. The addition of RPF also resulted in a greater decolorization efficiency of anthraquinone dye reactive blue 19 using microorganisms of the *Peptostreptococcaceae* family.

Nguyen et al. [128] evaluated the potential of submerged anaerobic deflector reactor with sponges (SS-ABR) to improve the processing performance of azo dye-containing wastewaters. The dye-degrading microorganisms present in the SS-ABR were *Clostridium* sp. and sulfate-reducing bacteria *Desulfomonile* sp. and *Desulfovibrio* sp. The efficiency of color and COD removal were 65 ± 3 and $83 \pm 2\%$. In addition, SS-ABR showed great stability.

5.3 Sequential Aerobic–Anaerobic Treatment

The combination of anaerobic–aerobic treatments using different microorganisms has shown promising results for the degradation of dyes present in textile effluents. Wastewater treatment indicates that aeration improves COD and acts as an important complement to anaerobic color removal [129, 130].

You and Teng [131] used a sequential batch anaerobic reactor combined with an anaerobic membrane bioreactor for the treatment of azo dye (Reactive Black 5). In addition, sixty anaerobic bacteria degrading Reactive Black 5 dye were isolated from the sludge taken from the sequential batch anaerobic reactor. The COD removal was 92.3 and 5.2%, whereas the true color removal was 74.6%, and 9.1% using sequential batch anaerobic reactor combined with anaerobic membrane bioreactor, respectively. Five subspecies of *Lactococcus lactis* were sequenced and showed the ability to degrade 99% of Reactive Black 5 dye after 5.5 h. Twenty subspecies of *Lactobacillus casei* showed the ability to degrade more than 99% of Reactive Black 5 dye after 36 h.

Naimabadi et al. [132] studied the decolorization and biological degradation of reactive Red 2 using an anaerobic–aerobic sequential process. A deflector anaerobic reactor on a laboratory scale and a fixed activated sludge reactor were used. The reactors were operated with different organic loads and hydraulic retention times. The experiments were carried out in continuous mode and the effluent from the deflector anaerobic reactor was used as feed for the fixed activated sludge reactor. The removal efficiency of the COD was 54.5% using 1 day-hydraulic retention in the anaerobic reactor. Regarding color, the removal was 89.5%. The results obtained demonstrated that the anaerobic/aerobic sequential system was efficient in the degradation of the reactive azo dye Red2.

5.4 Phycoremediation

An interesting approach for the decolorization and removal of dyes in textile effluents is the phycoremediation (algae) [133].

Although some azo dyes are highly toxic to fish, they do not significantly inhibit algal growth. In this sense, *Chlorella vulgaris*, *Oscillatoria*, and *Chlorella pyrenoidosa* degrade azo dyes [134]. It is worth noting that algal biomass can also be used in the production of methane and/or biodiesel.

The main advantage of phycoremediation is the algae metabolic versatility: (1) Photoautotrophic: the energy (ATP) is produced from light sources and CO_2 (photosynthesis); (2) Heterotrophic: the energy is produced from the oxidation of organic molecules; (3) Myxotrophic: photosynthesis and oxidation of organic molecules occur concurrently; and (4) Photoheterotrophic: the energy is produced from light, and oxidation of organic molecules [133–136].

A *Cosmarium* species (green alga) was studied as a viable biomaterial for the biological treatment of triphenylmethane dye, Malachite Green (MG). The effects of temperature, pH, dye concentration, and algae concentration on dye decolorization were studied. The highest discoloration rates were at temperatures above 45 °C. The ideal pH was 9. *Cosmarium* showed high efficiency of decolorization, reuse, and stability [137].

El-sheekh et al. [138] evaluated the decolorization and removal of ethyl red, orange II, G-Red (FN-3G), basic cationic, and basic fuchsin using *Chlorella vulgaris*, *Lyngbya lagerlerimi*, *Nostoc lincki*, *Oscillatoria rubescens*, *Elkatothrix viridis*, and *Volvox aureus*. The removal efficiency ranged from ~4 to 95%. The basic cationic fuchsin and basic fuchsin dyes showed the greatest capacity for decolorization and removal by all tested algae. *Elkatothrix viridis* was able to remove cationic basic fuchsin (91.6%) and basic fuchsin (90.7%) dyes after 3 *days* of incubation. Basic Fuchsin dye removal rate was up to 95% using *Oscillatoria rubescens*. However, *Volvox aureus* removed only 5.02 and 3.25% of the orange II and G-Red dyes.

6 Enzymatic Degradation of Dyes

Enzymes are well-known as effective biocatalysts that promote specific bioconversion of substances under mild conditions [139]. Thus, taking into account the recalcitrant properties of dyes from the textile industry, the enzymatic process is a promising alternative in reducing pollutants from industrial wastewaters.

The enzymatic degradation of dyes can be carried out by using crude and purified. Purified enzymes present high activity per dosage, however, the presence of redox mediators at low concentration may significantly impact the feasibility of crude and purified approaches [140]. The purification costs are also a relevant drawback when applying enzymes to the degradation of dyes. In this sense, enzyme immobilization is an interesting alternative since it eases recovery, protects the biocatalyst by enhancing its resistance, and maintains the enzyme's ability as a catalyst [141].

A wide range of enzymes has been used in general wastewater treatment. However, due to the recalcitrant characteristic of dyes, the oxidoreductase class is the most cited in the literature for textile industry field. This class includes enzymes that require an oxidizing agent (e.g., hydrogen peroxide, chlorine and potassium permanganate) to promote the catalysis of the reaction. Azoreductases, laccases, and peroxidases are examples of oxidoreductases commonly applied to dye degradation. While the first one is used specifically on the degradation of azo dyes, the last two present activity on several classes of dyes. In this way, this section brings a brief description related to these enzymes, their role on dye decomposition, and some relevant studies.

6.1 Laccases

Laccases are multicopper oxidase proteins [55]. Laccases are oxidoreductases that play important role in biotechnological applications in the bioremediation field due to their nonspecific oxidation capability, no cofactors demand, and ability to use molecular oxygen as electron acceptor [142]. Laccases have attracted relevant attention in the degradation of dyes from textiles wastewater [143]. When in presence of redox mediators, laccases present higher activity, [144]. Fungi or plants biosynthesize these enzymes, nevertheless, some bacteria are also laccase producers [145].

Since laccases catalysis uses molecular oxygen as an electron acceptor, laccases do not show high specificity (when compared to other enzymes). Thus, laccases oxidize phenolic an-phenolic compound also a wide range of substances, including azo dye-containing textile wastewaters [146, 147]. Laccases degrade azo dyes by a nonspecific free radical mechanism that forms phenolic products and inhibit potentially risky aromatic amine formation [148]. The reaction oxidation of aromatic pollutants, such as anilines and phenols, by laccases, yields the formation of phenoxy radicals from its original compounds; these reactions additionally develop a phenolic polymer on polymerization or, by further laccase action, produce quinine [149]. Laccases are related to the removal of hydrogen atoms from hydroxyl groups of *ortho* and *para* mono and polyphenolic substances and aromatic amines resulting in depolymerization, demethylation, or quinine synthesis [150].

The benefits of enzyme immobilization are well-known by researchers also in the dye degradation field. Studies involving enzyme encapsulation on several supports for wastewater decolorization report enhances enzyme activity, thermal and pH stability, reusability, and resistance to metals and organic solvents [151–153]. Several researchers have reported the activity of laccases from a variety of sources on the degradation of recalcitrant industrial dyes used in the textile manufacturing, as reported in Table 5. Zhuo et al. [154] suggest possible pathways for the degradation of Malachite Green (MG) and Remazol Brilliant Blue R (RBBR) based on the intermediates generated. Regarding laccase-MG, there are two possible pathways: (I) laccase mediates successive demethylation which results in the decrease of m/z value; and (II) MG is hydroxylated, then occurs the ring removal and the demethylation of the dye. Based on the intermediates of RBBR degradation, the proposed laccase-mediated pathway is the break of the molecule into two sub-products followed by deamination, hydroxylation, and oxidation finally opening the ring.

Sun et al. [155] reported the immobilization of laccase in chitosan grafted polyacrylamide hydrogel and its application on the degradation of Acid Orange 7 (AO7, an azo dye) and Malachite Green (MG). For the first, the cationic hydrogel used as enzyme support promoted extra adsorption of the anionic AO7 which increased the substrate concentration for the encapsulated laccase and enhanced activity. For the second, the cationic MG did not interact with the surface of the hydrogel and the degradation occurred exclusively due to enzyme activity. Jaiswal et al. [152] developed an immobilized laccase in chitosan beads and tested its activity on indigo carmine. The results showed complete decolorization of a 50 μm mL^{-1} solution

Table 5 Lacasse-based decolorization of dyes

Form/source	Dye	Elapsed time (h)	Decolorization (%)	References
Encapsulated laccase/*Trametes versicolor*	Malachite Green	6.5	>90	[155]
	Acid Orange 7		>80	
Immobilized Laccase/Papaya	Indigo Carmine	8	100	[152]
Laccase/recombinant *Yarrowia lipolytica*	Bromocresol Purple	1	43	[156]
	Safranin		54	
	Malachite Green		55	
	Kristal Violet		49	
	Bromothymol Blue		56	
	Nigrosine		53	
	Phenol Red		37	
Laccase/*Arthrospira maxima*	Reactive Blue 4	96	89	[157]
	Remazol Brilliant blue R		49	
Immobilized Laccase/*Trichoderma harzianum*	Malachite Green	100	16	[158]
	Methylene Blue	90	18	
	Congo Red	60	20	
Immobilized Laccase/*Trametes pubescens*	Reactive Brilliant Blue X-BR	96	60	[159]
	Remazol Brilliant Blue R		61	
	Acid Black 172		77	
	Congo Red		69	
	Methylene Blue		37	
	Neutral Red		48	
	Indigo Blue		56	
	Naphthol Green B		65	
	Crystal Violet		40	
Immobilized Laccase/*Trametes pubescens*	Reactive Brilliant Blue X-BR	48	52	[160]
	Remazol Brilliant Blue R		48	
	Congo Red		54	
	Acid Black 172		68	
	Methylene Blue		25	
	Neutral Red		44	

(continued)

Table 5 (continued)

Form/source	Dye	Elapsed time (h)	Decolorization (%)	References
	Indigo Blue		45	
	Naphthol Green B		37	
	Direct fast Blue FBL 74190		56	
	Crystal Violet		20	
Immobilized Laccase/genetically modified *Aspergillus*	Direct Red 23	1	88	[161]
	Acid Blue 92		48	
Laccase/*Streptomyces ipomoeae*	Acid Black 48	24	22	[162]
	Acid Orane 63		14	
	Reactive Black 5		94	
	Orange II		89	
	Tartrazine		21	
	Azure B		9	
	Indigo Carmine		98	
	Cresol Red		12	
Laccase/*Pleurotus ostreatus*	Malachite Green	24	91	[154]
	Methyl Orange		73	
	Bromophenol Blue		79	
	Remazol Brilliant blue R		85	

within 8 h and the supported enzyme presented 40% of its initial activity after 3 cycles. The decrease in activity may be attributed to the blocking of some pores on the surface of chitosan by substrate oxidation products [163].

6.2 Peroxidases

Peroxidases are widely applied to decolorize textile effluents since they act on phenolic compounds [164]. Peroxidases are hemoproteins that use oxygen peroxide as a mediator to catalyze oxidative reactions of recalcitrant dye compounds yielding insoluble polymeric products [165]. Passardi et al. [166] classified peroxidases into three classes according to their origins. Class I are intracellular peroxidases, which includes yeast cytochrome peroxidase, ascorbate peroxidase, and bacterial catalase peroxidases. Class II comprehends the secretory fungal peroxidases such as lignin and manganese peroxidases that are mainly related to the degradation of lignin. Class

III are secretory plant peroxidases that present numerous functions such as degradation of H_2O_2 and additional toxic substances from chloroplasts and cytosol and auto-defense against wounding.

Sun et al. [153] immobilized horseradish peroxidase in ZnO nanowires/macroporous SiO_2 composite and tested its activity on the decolorization of Acid Blue 113 and Acid Black 10 dyes. The solution concentration range of this study was between 30 and 50 mg L^{-1} and the color removal was 95% for Acid Blue and 90% for Acid Black. Chiong et al. [165] reported the application of soybean peroxidase and *Luffa acutangula* peroxidase for the degradation of azo dye methyl orange from liquid effluents. The first presented degradation of 81% within 1 h and the second a decolorization of 75% after 40 min. The results for crude peroxidases reveal their potential in enzymatic dye treatment.

Bilal et al. [11] proposed a pathway for the degradation of Reactive Black 19 by horseradish peroxidase based on UPLC-MS analysis. The successive enzymatic action on the dye leads to the degradation of the chromophore group and the subsequent disappearance of the color. Bilal et al. [167] also reported a pathway for the degradation of azo dye Methyl Orange based on its UPLC-MS analysis after horseradish peroxidase activity. The decline in spectral shift followed by the absence of new peaks during enzymatic activity indicates the total degeneration of the dye by the cleavage of azo bonds. Ali et al. [168] exposed a potential pathway for the degradation of Reactive Blue 4 by a ginger peroxidase. GC–MS was used in order to elucidate the oxidative reactions by the intermediates. The authors concluded that the dye was first desulphonated, dechlorinated, and deaminated to generate less *m/z* value products, the following reactions lead to the cleavage of the chromophore group and consequent solution decolorization.

The evaluation of toxicity of compounds from enzymatic degradation of dyes is important to verify if the products of the dye decomposition still present relevant risks for the environment. Some authors studied the toxicity of untreated and peroxidase treated textile industrial wastewater and report interesting outcomes. Ali et al. [168] report a reduction in genotoxicity after decolorization of Reactive Blue 4 by a single-strand break in DNA analysis. Maria et al. [169] employed horseradish peroxidase on the degradation of Remazol Turquoise Blue G and Lanaset Blue 2R concluded that the treated wastewater presented less toxicity effect on *Daphnia magna* but no change was detected toward *Artemia salina*.

Baumer et al. [80] tested the viability of horseradish peroxidase on the degradation of Reactive Black 5 (azo), Reactive Blue 19 (anthraquinone), Reactive Red 239 (azo), and Reactive Blue 21 (phthalocyanine) and evaluated the toxicity of the treated wastewater by *Daphnia magna, Euglena gracilisalgae*, and *Vibrio fischeri*. The tests were conducted in batch mode, 125 mL of a dye concentration of 50 mg L^{-1} at 30 °C, and the decolorization attained at 30 min were 87, 96, 17, and 90%, accordingly. The toxicity outcomes showed a significantly higher negative effect on the organisms in the test after enzymatic treatment. The authors attribute these results to the formation of decolorization metabolites more toxic than the original dye molecule and suggest a next successive treatment step for wastewater cleaning. Table 6 presents several

Table 6 Peroxidase-based decolorization of dyes

Form/source	Dye	Elapsed time (min)	Decolorization (%)	References
Immobilized Peroxidase/Horseradish	Reactive Blue 21	30	90	[80]
	Reactive Blue 19		96	
	Reactive Black		87	
	Reactive Red		17	
Immobilized Peroxidase/Horseradish	Acid Blue	35	95	[153]
	Acid Black		90	
Immobilized Peroxidase/Horseradish	Methyl Orange	300	100	[151]
Peroxidase/soybean	Methyl Orange	40	81	[165]
Peroxidase/*Luffa acutangular*			75	
Immobilized Peroxidase/Ginger	Reactive Blue 4	180	99	[168]
Lignin Peroxidase/*P. ostreatus* and *G. lucidum*	Remazol Brilliant Blue R	1440	100	[170]
Peroxidase/*Irpex Lacteus* F17	Reactive Blue 4	30	79	[171]
	Reactive Blue 5		74	
	Reactive Blue 19		78	
	Direct Sky Blue 5B		81	
	Reactive Black 5		60	
	Acid Red 18		43	
	Reactive Violet 5		92	
	Methyl Orange		33	
	Direct Yellow 8		3	
	Orange G		18	
	Orange Yellow II		29	
	Orange Yellow IV		63	
	Congo Red		0	
	Neutral Red		22	
	Malachite Green		85	
	Basic Fuchsin		45	

researchers who have reported the activity of peroxidases from a variety of sources on the degradation of recalcitrant industrial dyes used in textile manufacturing.

6.3   Azoreductase

Azoreductases are enzymes capable of catalyzing the cleavage reduction of azo groups (–N=N–). These enzymes are produced by bacteria that degrade the azo dyes. As an oxidoreductase, this enzyme can be classified according to its cofactors demand in flavin-dependent and flavin-independent, and even further categorized as a function of the electron donor NADH or NADPH [172].

The demand for coenzyme factors is the main drawback for the azoreductase application in wastewater treatment. Thus, it is necessary to incorporate NAD(P)H recycling enzymes, such as glucose dehydrogenases and formate dehydrogenases [173]. In this sense, the importance of glucose 1-dehydrogenase presence in an integrated system with azoreductase and the improvement of this association on decomposing methyl red dye was reported by Yang et al. [174]. Azoreductases also act on degrading dyes from the azo group, such as Remazol Blue [175], methyl orange [172], and even aminoanthraquinone dyes such as Dispersive Blue [176].

Regarding azoreductase immobilization, silicas functionalized with amino and epoxy groups were applied to immobilize a novel azoreductase from *Rhodococcus opacus*. The enzymatic activity and storability increased significantly when comparing the results to the free enzyme [177]. The association of enzymes may promote further degradation due to the increase in degradation routes possibility. Guo et al. [178] employed a bacterium consortium to verify its ability on producing enzymes to decolorize saline water containing metanil yellow G, an azo dye used in paints. It reports that laccase, peroxidases, and azoreductase are responsible for the dye removal outcomes and the reduction of toxicity associated with the proposed degradation pathway.

The wide range of reactions enzymes can catalyze expands their application and makes them suitable in several fields. In general, the studies here presented show the feasibility of oxidoreductases on reducing dyes with recalcitrant characteristics to compounds that can easily be assimilated during conventional biological treatment. Different approaches related to enzyme purification and immobilization have been tested worldwide to develop cost-effective systems. Nevertheless, further research is required to turn enzymatic systems into a viable treatment of real wastewaters.

7   New Trends on Microbial and Enzyme Degradation of Dye-Containing Textile Wastewater Treatment

Regarding wastewater treatments, there are new approaches to traditional methods such as dye adsorption and biofilms. In addition, there are emergent techniques such as genetically modified organisms, combined treatment systems, membrane bioreactors, microbial fuel cells, support-based nanomaterials, and machine learning.

7.1 Dye Adsorption

Adsorption is a well-known process for textile industry wastewater treatment, with activated carbon the most used adsorbent. Nevertheless, its high cost is encouraging the research of low-cost alternative adsorbents [22, 179180, 181]. Activated sludge adsorbent is an interesting alternative, and can be synthesized by centrifugation, carbonization of sludge, pyrolysis, steam activation, H_2SO_4, and NaOH treatment [22].

Water treatment residuals in the dried form can also be applied as adsorbents on filtration column tests, as explored by Gadekar and Ahammed [180] for the decolorization of real textile dye wastewater. The adsorption process promoted a maximum color removal of 36%, the column operation obtained a decolorization rate in the range of 60–70%. Ravenni et al. [181] developed a comparative study about the dye adsorption properties of waste chars from gasification of wood chips and pyrolysis of wastewater sludge with a commercial activated carbon. Sludge char performed maximum adsorption capacities of 13.4 and 8.4 mg g^{-1} for the anionic and cationic dye, respectively. Steam activation improved these values to 19.6 and 12.3 mg g^{-1}. However, the sludge-based adsorbent has not achieved the efficiency of the commercial activated carbon and wood char. The removal of methylene blue was studied using cow dung biochar (CDB), domestic sludge biochar (SB) subject to slow pyrolysis at 500 °C, and rice husk biochar (RHB) as adsorbents. The dye removal efficiencies by CDB, RHB, and SB in a batch experiment were 97.0–99.0; 71.0–99.0, and 73.0–98.9%, respectively [179]. The application of biomass adsorbent at industrial scale presents some limitations such as the accessibility of adsorbents, adsorption sites, adsorbent stability, desorption rates at specific pH, and low adsorption, which should be improved to make this technology competitive [182, 183].

7.2 Genetically Modified Organisms (GMOs)

The variations of textile effluent composition led to the development of some alternatives to improve its remediation. Some molecular biology methodologies, like cloning, directed evolution, gene recombination techniques, heterologous expression, metagenomics, random mutagenesis, rational design, and site-directed mutagenesis, are being explored to enhance effluents treatment processes. The advances in genetic engineering and molecular genetics enable virtual expression and clone of any gene in a suitable microbial host [90].

Laccase purified from *Pleurotus* sp. MAK-II was tested to diazo dye, congo red, anthraquinone, and remazol brilliant blue R decolorization. The enzyme presents high stability in presence of violuric acid redox mediator [184]. Liu et al. [185] inserted the LacTT gene from *Thermus thermophilus* SG0.5JP17–16 into *Pichia pastoris* promoting a more effective aptitude to degrade congo red, reactive black, reactive black WNN, and remazol brilliant blue R. A thermo-alkali-stable laccase gene purified from *Klebsiella pneumoniae* was cloned into *E. coli* to be applied for the remediation of azo phloxine, bromophenol blue, congo red, cotton blue, malachite green, mordant black 9, reactive brilliant blue X-BR, reactive brilliant blue K-GR, reactive brilliant blue KN-R, and reactive dark blue M-2GE from a textile industry wastewater [186]. Undoubtedly, the main benefit of GMOs is the potential to accelerate decolorization efficiency. However, some disadvantages comprise cross-pollination and damaged environment due to horizontal gene transfer, reduced biodiversity, and uncertain long-term health effects [187].

7.3 Combined Treatments Systems

7.3.1 Bio-advanced Oxidation Process

Advanced oxidation processes (AOPs) combined with biological methods are known for their high efficiency for recalcitrant wastewater treatment and their promising application in industries. Biodegradation and photodegradation are the most promising methods for remediating the pollutants in water, besides enhance water pollutants. The application of biological methods can be applied either as a pre-treatment or as a post-treatment [188]. Biological degradation as a first step acts on the biodegradable fraction of wastewater by a low-priced and more eco-friendly method, demanding less energy and chemical inputs for further degradation of the remaining contaminants by chemical approaches [189], used only for the remediation of remained compounds resistant to biological oxidation [190]. Thanavel et al. [191], applied a combined system of biological and AOPs treatment to study the decolorization of Remazol Red, Reactive Black 5, and Reactive Red 180 by *Aeromonas hydrophila* SK16. The individual treatment was also analyzed and a 72% decolorization rate was reported. However, the combination of AOPs treatment with biological treatment was proved to be more effective than single wastewater treatment, by achieving 100% of decolorization.

Regarding non-biodegradable-colored compounds, chemical processes can assist or induce their biodegradability. Nevertheless, it is also possible that the by-products of dye degradation from chemical oxidation negatively interfere with the metabolic pathways of microorganisms [192]. In this context, the main objective of chemical pre-treatment is the partial oxidation [193]. Shanmugam et al. [194] explored the application of Fenton (H_2O_2 and Fe^{2+}) oxidation and posterior biological treatment for the biodegradation of effluent with Acid Blue 113 azo dye. The AOP reduced the dye concentration by 40% which promoted a maximum dye degradation of 85%, i.e., 45% by biodegradation. In this sense, Gott, (2010) recommended chemical oxidation as pre-treatment according to the BOD5/COD ratio when it is below 0.2.

7.3.2 Membrane Bioreactors

Membrane bioreactors (MBRs) are systems that conjugate membrane filtration with a biological approach [195]. MBR is a simple, cost-effective, and reliable method with strong potential to remove high nutrient load, organic chemicals, and coloring pollutants [196].

Membranes efficiently separate the macro-molecules and microorganisms, but the major drawback is the membrane fouling in the bioreactor. Enhanced membrane bioreactor formed by two anoxic bioreactors, aerated bioreactor, UV-unit, and a granular activated carbon filter obtained 95% of color removal from a synthetic textile industry wastewater [196]. Sepehri and Sarrafzadeh [197], tested and proved the efficiency of a nitrifying-enriched activated sludge to decrease membrane fouling in MBR. An enhancement of 2.5 times in the permeation performance was achieved with the nitrifiers community compared to the conventional activated sludge process, which, consequently, improved MBR yields production, compact nature; high degradation rate inorganic pollutants, nutrient removal, quality of treated effluent; lower sludge generation; reliability; and small footprint, when compared to conventional activated sludge process, are some of the advantages of MBR. Nevertheless, some disadvantages alert the requirement of more study and process improvements, such as aeration limitations, stress on sludge in external MBR, membrane fouling, and higher operation cost [196].

7.3.3 Microbial Fuel Cells

Microbial fuel cells (MFCs) are a bio-electrochemical treatment system composed of various microorganisms used as catalysts to the oxidation of inorganic and organic compounds along with the generation of electrons [198, 199]. Sulfate-reducing consortium in MFC anodic chamber was used for simultaneous dye degradation, electricity generation, and sulfate reduction, in which the dye removal yield achieved $\leq 85\%$. The power generation of 258 ± 10 mW m^{-2} was achieved at stable operating conditions [200].

Yang et al. [201] used bio-electrochemical systems to decrease the concentration of Reactive Black 5 (RB5) from 0.503 to 0.124 mM. The effect of circuit connection was investigated for the azo dye degradation and bioelectricity generation. The closed-circuit system exhibited higher decolorization efficiency (96% of Acid Red 18, 67% of Acid Orange 7, and 60% of Congo Red) compared to the open circuit system. The voltage outputs were ranked in the decreasing order of AR18 > AO7 > CR. This method represents an advance, fast treatment efficiency, novel, and sustainable approach for the efficient treatment of industrial wastewaters along with the power generation with reduced CO_2 emission. However, the high treatment cost and sludge generation are important disadvantages associated [202].

7.3.4   Support-Based Nanomaterials

Nanomaterials have been explored as a promising alternative according to several pilot-scale, laboratory and in situ water treatment studies. These materials gave been drawing attention, since they have a wide range of application such as remediation, drinking, and wastewater treatment. Nanomaterials have remarkable properties due to their small size and consequent unique physicochemical effects, including increased adsorption yields and photocatalysis reactivity [203, 204]. Suganya and Revathi [205] immobilized *P. putida* and *B. licheniformis* on sodium alginate and polyacrylamide gel beads to explore the decolorization potential. According to the results, the sodium alginate support provided better conditions for the process when compared to polyacrylamide gel beads and free cells processes.

7.4   Biofilms

Biofilms are negatively charged layers of microbial conglomerations of one or more species that are trapped to abiotic or biotic surfaces by incorporation in the matrix of extracellular polymeric substances (EPSs) [206]. Compared to free bacterial usage, bacterial biofilms have some advantages, such as the ability to exchange nutrient and genetic materials, bigger tolerance to toxic compounds and different metabolic states, protection from the effects of environmental changes [206]. The dye adsorption mechanism by biofilms outcomes from two stages: the first is the dye molecules transport through the solution to the surface of the biofilm, followed by the adsorption of dye molecules to the active sites of biofilm [207]. Moving bed biofilm reactor (MBBR) is one of the most explored bioreactors and unites the benefits of activated sludge and biofilm [189]. Due to the promotion of biomass growth on a moving support, some advantages can be observed such as the absence of obstruction problems, decrease in hydraulic retention time, easier separation among liquid and solid phases, flexibility in operation, increase in biomass weight, less space only and environmental effects, low sludge production, superior biomass residence time, and superior capability to degrade complex compounds [208]. MBBR may work in anaerobic or aerobic conditions and have been effectively used in dye decolorization and degradation studies with great yields of decolorization [208–210].

7.5   Dye Industry Waste and Resource Recovery Strategy

With the proposal to reduce the amount of waste discharged from industries, its use in the resource recovery process, as input material for recovering value-added products is being widely explored. Mishra et al. [211] have performed the decolorization of mixed dyes reactive red 21 (RR21) and reactive orange 16 (RO16). The authors obtained 83% of color removal, by the use of a microbial fuel cell from dye wastewater with bioenergy generation of 940.61 ± 5 mW m^{-2} power density with 790 ± 5 mV voltage output generation. A single-chamber microbial fuel cell with microalgal biocathodes (with 42% of the cathode surface covered by microalgal naturals) was designed for the simultaneous biodegradation of real dye textile wastewater and the generation of bioelectricity. It was observed that the atmospheric and diffused CO_2 promoted good rates of algal growth and immobilization indicating its operation at air-exposed conditions. The maximum volumetric power density achieved was 123.2 ± 27.5 mW m^{-3} highlighting it as a promising alternative to a Pt cathode [212]. Phytoremediation technology by microalgal strains viz. *Anabaena ambigua*, *Chlorella pyrenoidosa*, and *Scenedesmus abundans* using textile wastewater as a nutrient source were studied to generate electricity. The authors correlated the differences of various UV–Vis in the spectra obtained with the breakdown and formation of compounds, and postulated the biodegradation for the wastewater remediation. Electrical conductivity and redox potential reduced to 2.0 mS and 157.86 ± 1.89 mV, respectively [213].

7.6 Machine Learning

Due to their sensibility to environmental changes, algal systems in complex bioprocesses application are not precisely controlled. Therefore, control systems and real-time monitoring, through machine learning algorithms (MLAs) and dynamic models are presented as a solution to preserve optimum conditions. Their application brought some interesting results for the optimizing study of algal system performance, which provides a better understanding of the biological processes when compared to the conventional kinetic or phenomenological models [214].

MLAs application leads to effective real-time monitoring, defect detection, optimization, and anticipation of uncertainties of complex environmental systems. Besides, disruptions or failures from leaking pipelines, malfunctioning of bioreactors, unexpected variations of flow rate, organic loadings, and temperature are also situations that can be efficiently anticipated by integrating these algorithms with online sensors. Notwithstanding the MLAs application benefits, it is important to consider that to implement these advanced models, sensors, data acquisition, transmission, and storage, have to promote knowledge at decision making to optimize their system performance. Increments could be performed by the implementation of multi-parameter sensors and the internet of things (IoT) for full-scale WWTPs which give plant managers support to identify equipment faults, improve energy usage, and decrease greenhouse gas emissions [215].

8 Conclusion

Regarding dye-containing textile wastewaters, they are often pumped into freshwater bodies or sea, without using a treatment system, in particular developing countries. In this sense, biological processes are efficient and eco-friendly, operate mildly, etc. Anaerobic/aerobic sequential systems are highly efficient on azo dye Red2-degradation. In addition, phycoremediation, and GMO oxidoreductases producers, in particular azoreductases, laccases, and peroxidases are promising alternatives. Despite difficulty in operating control, the integrated systems such as AOPs-*Aeromonas hydrophila* or MBRs (anoxic bioreactors, aerated bioreactor, UV-unit, and granular activated carbon filter) have remarkable potential to remove such high nutrient load, organic chemicals, and dyes. Finally, microbial fuel cells can be associated to wastewater treatments.

References

1. Saratale RG, Saratale GD, Chang JS, Govindwar SP (2011) Bacterial decolorization and degradation of azo dyes: a review. J Taiwan Inst Chem Eng 42:138–157. https://doi.org/10.1016/j.jtice.2010.06.006
2. Benkhaya S, El Harfi S, El Harfi A (2017) Classifications, properties and applications of textile dyes: a review
3. Bafana A, Devi SS, Chakrabarti T (2011) Azo dyes: past, present and the future. Environ Rev 19:350–370. https://doi.org/10.1139/a11-018
4. El Bouraie M, El Din WS (2016) Biodegradation of Reactive Black 5 by Aeromonas hydrophila strain isolated from dye-contaminated textile wastewater. Sustain Environ Res 26:209–216. https://doi.org/10.1016/j.serj.2016.04.014
5. Verma AK, Dash RR, Bhunia P (2012) A review on chemical coagulation/flocculation technologies for removal of colour from textile wastewaters. J Environ Manag 93:154–168. https://doi.org/10.1016/j.jenvman.2011.09.012
6. Baban A, Yediler A, Lienert D, Kemerdere N, Kettrup A (2003) Ozonation of high strength segregated effluents from a woollen textile dyeing and finishing plant. Dye Pigment 58:93–98. https://doi.org/10.1016/S0143-7208(03)00047-0
7. Ahmad R, Guo J, Kim J (2019) Structural characteristics of hazardous organic dyes and relationship between membrane fouling and organic removal efficiency in fluidized ceramic membrane reactor. J Clean Prod 232:608–616. https://doi.org/10.1016/j.jclepro.2019.05.244
8. Camargo N De, Beluci L, Affonso G, Mateus P, Sayury C, Cândido N, Guttierres R, Fagundes-klen MR, Bergamasco R, Marquetotti A, Vieira S (2019) Hybrid treatment of coagulation/flocculation process followed by ultra filtration in TiO₂-modified membranes to improve the removal of reactive black 5 dye. Sci Total Environ 664:222–229. https://doi.org/10.1016/j.scitotenv.2019.01.199
9. Herrera-González AM, Caldera-villalobos M, Peláez-Cid AA (2019) Adsorption of textile dyes using an activated carbon and crosslinked polyvinyl phosphonic acid composite. J Environ Manag 234:237–244. https://doi.org/10.1016/j.jenvman.2019.01.012
10. Li X, Tang S, Yuan D, Tang J, Zhang C, Li N, Rao Y (2019b) Ecotoxicology and Environmental Safety Improved degradation of anthraquinone dye by electrochemical activation of PDS. 177:77–85.https://doi.org/10.1016/j.ecoenv.2019.04.015
11. Bilal M, Rasheed T, Zhao Y, Iqbal HMN (2019) Agarose-chitosan hydrogel-immobilized horseradish peroxidase with sustainable bio-catalytic and dye degradation properties. Int J Biol Macromol 124:742–749. https://doi.org/10.1016/j.ijbiomac.2018.11.220

12. Gou M, Qu Y, Zhou J, Ma F, Tan L (2009) Azo dye decolorization by a new fungal isolate, Penicillium sp. QQ and fungal-bacterial cocultures. J Hazard Mater 170:314–319. https://doi.org/10.1016/j.jhazmat.2009.04.094

13. Ooi T, Shibata T, Sato R (2007) An azoreductase, aerobic NADH-dependent flavoprotein discovered from Bacillus sp.: functional expression and enzymatic characterization. 377–386. https://doi.org/10.1007/s00253-006-0836-1

14. Uppala R, Sundar K, Muthukumaran A (2019) Decolorization of azo dyes using dried biomass of Bacillus cereus RC1 and Kocuria kristinae RC3. J Pure Appl Microbiol 13:1969–1976. https://doi.org/10.22207/JPAM.13.4.08

15. Chequer FMD, De Oliveira GAR, Ferraz ERA, Cardoso JC, Zanoni MVB, de Oliveira DP (2013a) Textile dyes: dyeing process and environmental impact. Eco-friendly Text Dye Finish 6:151–176

16. Varjani S, Rakholiya P, Shindhal T, Shah AV, Hao H (2021) Trends in dye industry effluent treatment and recovery of value added products. J Water Process Eng 39:101734. https://doi.org/10.1016/j.jwpe.2020.101734

17. Loe DL (2017) 15—Light, colour and human response. In: Best JBT-CD (ed) Woodhead publishing series in textiles, 2nd edn. Woodhead Publishing, pp 349–369

18. Xuan Z, Li J, Liu Q, Yi F, Wang S, Lu W, Xuan Z, Li J, Liu Q, Yi F, Wang S, Lu W (2021) Artificial Structural Colors and Applications. Innov 2:100081. https://doi.org/10.1016/j.xinn.2021.100081

19. Benkhaya S, Souad M, El Harfi HA (2020) A review on classifications, recent synthesis and applications of textile dyes. Inorg Chem Commun 115:107891. https://doi.org/10.1016/j.inoche.2020.107891

20. Gürses A, Açıkyıldız M, Güneş K, Gürses MS (2016) Dyes and pigments: their structure and properties BT. In: Gürses A, Açıkyıldız M, Güneş K, Gürses MS (eds) Springer International Publishing, Cham, pp 13–29

21. Berradi M, Hsissou R, Khudhair M, Assouag M, Cherkaoui O, El Bachiri A, El Harfi A (2019) Textile finishing dyes and their impact on aquatic environs. Heliyon 5. https://doi.org/10.1016/j.heliyon.2019.e02711

22. Pavithra KG, P. SK, Jaikumar V, P. SR (2019) Removal of colorants from wastewater: A review on sources and treatment strategies. J Ind Eng Chem 75:1–19. https://doi.org/10.1016/j.jiec.2019.02.011

23. Temesgen F, Gabbiye N, Sahu O (2018) Biosorption of reactive red dye (RRD) on activated surface of banana and orange peels: economical alternative for textile e ffl uent. Surf Interfaces 12:151–159. https://doi.org/10.1016/j.surfin.2018.04.007

24. Rawat D, Mishra V, Sharma RS (2016) Detoxification of azo dyes in the context of environmental processes. Chemosphere 155:591–605. https://doi.org/10.1016/j.chemosphere.2016.04.068

25. Holkar CR, Jadhav AJ, Pinjari DV, Mahamuni NM, Pandit AB (2016) A critical review on textile wastewater treatments: possible approaches. J Environ Manag 182:351–366. https://doi.org/10.1016/j.jenvman.2016.07.090

26. Rehman A, Usman M, Hussain T, Saeed M, Ur A, Siddiq M, Rasheed A, Un M (2020) The application of cationic-nonionic mixed micellar media for enhanced solubilization of Direct Brown 2 dye. J Mol Liq 301:112408. https://doi.org/10.1016/j.molliq.2019.112408

27. Cardon D (2007) Natural dyes. Sources Tradit Technol Sci 268

28. Ferreira ESB, Hulme AN, Quye A, Ferreira E (2004) The natural constituents of historical textile dyes. Chem Soc Rev 329–336

29. Zerin I, Farzana N, Sayem ASM, Anang DM, Haider J (2020) Potentials of natural dyes for textile applications. In: Hashmi S, Choudhury IA (eds) Encyclopedia of renewable and sustainable materials. Elsevier, Oxford, pp 873–883

30. de Keijzer M, van Bommel MR, Keijzer RH, Knaller R, Oberhumer E (2012) Indigo carmine: understanding a problematic blue dye. Stud Conserv 57:S87–S95. https://doi.org/10.1179/2047058412Y.0000000058

31. Abel A (2012) 24—The history of dyes and pigments: from natural dyes to high performance pigments. In: Best JBT-CD (ed) Woodhead publishing series in textiles, 2nd edn. Woodhead Publishing, pp 557–587
32. Hagan E, Poulin J (2021) Statistics of the early synthetic dye industry. Herit Sci 9:1–14. https://doi.org/10.1186/s40494-021-00493-5
33. Tamburini D, Shimada CM, McCarthy B (2021) The molecular characterization of early synthetic dyes in E. Knecht et al's textile sample book "A Manual of Dyeing" (1893) by high performance liquid chromatography–Diode array detector–Mass spectrometry (HPLC-DAD-MS). Dye Pigment 190:109286. https://doi.org/10.1016/j.dyepig.2021.109286
34. Tkaczyk A, Mitrowska K, Posyniak A (2020) Synthetic organic dyes as contaminants of the aquatic environment and their implications for ecosystems: a review. Sci Total Environ 717:137222. https://doi.org/10.1016/j.scitotenv.2020.137222
35. Katheresan V, Kansedo J, Lau SY (2018) Efficiency of various recent wastewater dye removal methods: a review. J Environ Chem Eng 6:4676–4697. https://doi.org/10.1016/j.jece.2018.06.060
36. Sharma J, Sharma S, Soni V (2021) Classification and impact of synthetic textile dyes on Aquatic Flora: a review. Reg Stud Mar Sci 45:101802. https://doi.org/10.1016/j.rsma.2021.101802
37. Ali H (2010) Biodegradation of synthetic dyes—a review. Water Air Soil Pollut 251–273. https://doi.org/10.1007/s11270-010-0382-4
38. Muthirulan P, Devi CN, Sundaram MM (2014) TiO2 wrapped graphene as a high performance photocatalyst for acid orange 7 dye degradation under solar/UV light irradiations. Ceram Int 40:5945–5957. https://doi.org/10.1016/j.ceramint.2013.11.042
39. Nakamura KC, Guimarães LS, Magdalena AG, Angelo ACD, De Andrade AR, Garcia-Segura S, Pipi ARF (2019) Electrochemically-driven mineralization of Reactive Blue 4 cotton dye: on the role of in situ generated oxidants. J Electroanal Chem 840:415–422. https://doi.org/10.1016/j.jelechem.2019.04.016
40. Aleboyeh A, Aleboyeh H, Moussa Y (2003) "Critical" effect of hydrogen peroxide in photochemical oxidative decolorization of dyes: Acid Orange 8, Acid Blue 74 and Methyl Orange. Dye Pigment 57:67–75. https://doi.org/10.1016/S0143-7208(03)00010-X
41. Javaid R, Yaqub U, Ikhlaq A, Zahid M, Alazmi A (2021) Subcritical and supercritical water oxidation for dye decomposition. J Environ Manage 290:112605. https://doi.org/10.1016/j.jenvman.2021.112605
42. Hassan MM, Carr CM (2018) A critical review on recent advancements of the removal of reactive dyes from dyehouse effluent by ion-exchange adsorbents. Chemosphere 209:201–219. https://doi.org/10.1016/j.chemosphere.2018.06.043
43. Burkinshaw SM, Lagonika K (2006) Sulphur dyes on nylon 6, 6. Part 3. Preliminary studies of the nature of dye e fibre interaction. 69. https://doi.org/10.1016/j.dyepig.2005.03.012
44. Wich E (1977) The colour index. Color Res Appl 2:77–80. https://doi.org/10.1002/col.5080020205
45. Gupta VK (2009) Application of low-cost adsorbents for dye removal—a review. Elsevier Ltd
46. Besegatto SV, Campos CEM, Selene M, De Souza AU, González SY (2021b) Perovskite-based Ca-Ni-Fe oxides for azo pollutants fast abatement through dark catalysis. Appl Catal B Environ 284. https://doi.org/10.1016/j.apcatb.2020.119747
47. Chung K (2016) Azo dyes and human health: a review. J Environ Sci Heal Part C 34:233–261. https://doi.org/10.1080/10590501.2016.1236602
48. Didier de Vasconcelos GM, Mulinari J, de Arruda Guelli Ulson de Souza SM, Ulson de Souza AA, de Oliveira D, de Andrade CJ (2021) Biodegradation of azo dye-containing wastewater by activated sludge: a critical review. World J Microbiol Biotechnol 37:101. https://doi.org/10.1007/s11274-021-03067-6
49. Shankarling GS, Deshmukh PP, Joglekar AR (2017) Process intensification in azo dyes. J Environ Chem Eng 5:3302–3308. https://doi.org/10.1016/j.jece.2017.05.057

50. Benkhaya S, M'rabet S, El Harfi A (2020) Classifications, properties, recent synthesis and applications of azo dyes. Heliyon 6:1–26. https://doi.org/10.1016/j.heliyon.2020.e03271
51. Vikrant K, Giri BS, Raza N, Roy K, Kim K-H, Rai BN, Singh RS (2018) Recent advancements in bioremediation of dye: current status and challenges. Bioresour Technol 253:355–367. https://doi.org/10.1016/j.biortech.2018.01.029
52. Balapure K, Bhatt N, Madamwar D (2015) Mineralization of reactive azo dyes present in simulated textile waste water using down flow microaerophilic fixed film bioreactor. Bioresour Technol 175:1–7. https://doi.org/10.1016/j.biortech.2014.10.040
53. Oros G, Forgacs E, Cserha T (2004) Removal of synthetic dyes from wastewaters: a review. Environ Int 30:953–971. https://doi.org/10.1016/j.envint.2004.02.001
54. Sarkar S, Banerjee A, Halder U, Biswas R, Bandopadhyay R (2017) Degradation of synthetic azo dyes of textile industry: a sustainable approach using microbial enzymes. Water Conserv Sci Eng 2:121–131. https://doi.org/10.1007/s41101-017-0031-5
55. Singh RL, Singh PK, Singh RP (2015) Enzymatic decolorization and degradation of azo dyes—a review. Int Biodeterior Biodegrad 104:21–31. https://doi.org/10.1016/j.ibiod.2015.04.027
56. Barker AF (2009) Handbook of textiles. Abhishek Publications, Chandigarh
57. Chakraborty JN (2014) Fundamentals and practices in colouration of textiles, 2nd edn. Woodhead Publishing India Pvt. Ltd., New Delhi
58. Mahapatra N (2016) Textile dyes. Woodhead Publishing India Pvt. Ltd., New Delhi
59. Chequer FMD, Oliveira GAR de, Ferraz ERA, Cardoso JC, Zanoni MVB, de Oliveira DP (2013b) Textile dyes: dyeing process and environmental impact. In: Gunay M (ed) Eco-friendly textile dyeing and finishing. IntechOpen
60. Clark M (2011) Handbook of textile and industrial dyeing: principles, processes and types of dyes
61. Arslan S, Eyvaz M, Gürbulak E, Yüksel E (2016) A review of state-of-the-art technologies in dye—containing wastewater treatment—the textile industry case. In: Kumbasar PA, Körlü AE (eds) Textile wastewater treatment. ExLi4EvA
62. Bisschops I, Spanjers H (2003) Literature review on textile wastewater characterisation. Environ Technol (United Kingdom) 24:1399–1411. https://doi.org/10.1080/09593330309385684
63. Patel H, Vashi RT (2015) Characterization and Treatment of Textile Wastewater. Elsevier Inc.
64. Yaseen DA, Scholz M (2019) Textile dye wastewater characteristics and constituents of synthetic effluents: a critical review. Springer, Berlin Heidelberg
65. Neifar M, Sghaier I, Guembri M, Chouchane H, Mosbah A, Ouzari HI, Jaouani A, Cherif A (2019) Recent advances in textile wastewater treatment using microbial consortia. J Text Eng Fash Technol 5:134–146. https://doi.org/10.15406/jteft.2019.05.00194
66. Samsami S, Mohamadi M, Sarrafzadeh MH, Rene ER, Firoozbahr M (2020) Recent advances in the treatment of dye-containing wastewater from textile industries: overview and perspectives. Process Saf Environ Prot 143:138–163. https://doi.org/10.1016/j.psep.2020.05.034
67. Yagub MT, Sen TK, Afroze S, Ang HM (2014) Dye and its removal from aqueous solution by adsorption: a review. Adv Colloid Interface Sci 209:172–184. https://doi.org/10.1016/j.cis.2014.04.002
68. Dafale N, Wate S, Meshram S, Neti NR (2010) Bioremediation of wastewater containing azo dyes through sequential anaerobic–aerobic bioreactor system and its biodiversity. Environ Rev 18:21–36. https://doi.org/10.1139/A10-001
69. Lellis B, Fávaro-Polonio CZ, Pamphile JA, Polonio JC (2019) Effects of textile dyes on health and the environment and bioremediation potential of living organisms. Biotechnol Res Innov 3:275–290. https://doi.org/10.1016/j.biori.2019.09.001
70. Kadirvelu K, Kavipriya M, Karthika C, Radhika M, Vennilamani N, Pattabhi S (2003) Utilization of various agricultural wastes for activated carbon preparation and application for the removal of dyes and metal ions from aqueous solutions. Bioresour Technol 87:129–132. https://doi.org/10.1016/S0960-8524(02)00201-8

71. Aruna BN, Sharma AK, Kumar S (2021) A review on modified sugarcane bagasse biosorbent for removal of dyes. Chemosphere 268:129309. https://doi.org/10.1016/j.chemosphere.2020. 129309
72. Kadirvelu K, Palanivel M, Kalpana R, Rajeswari S (2000) Activated carbon prepared from agricultural by-product for the treatment of dyeing wastewater. Bioresour Technol 75:25–27. https://doi.org/10.1016/S0960-8524(00)00013-4
73. Rawat D, Sharma RS, Karmakar S, Arora LS (2018) Ecotoxic potential of a presumably non-toxic azo dye. Ecotoxicol Environ Saf 148:528–537. https://doi.org/10.1016/j.ecoenv.2017. 10.049
74. Besegatto SV, da Silva A, Campos CEM, de Souza SMAGU, de Souza AAU, González SYG (2021) Perovskite-based Ca-Ni-Fe oxides for azo pollutants fast abatement through dark catalysis. Appl Catal B Environ 284:119747. https://doi.org/10.1016/j.apcatb.2020.119747
75. Kalyani DC, Telke AA, Dhanve RS, Jadhav JP (2009) Ecofriendly biodegradation and detox-ification of Reactive Red 2 textile dye by newly isolated Pseudomonas sp. SUK1. J Hazard Mater 163:735–742. https://doi.org/10.1016/j.jhazmat.2008.07.020
76. Kakarndee S, Nanan S (2018) SDS capped and PVA capped ZnO nanostructures with high photocatalytic performance toward photodegradation of reactive red (RR141) azo dye. J Environ Chem Eng 6:74–94. https://doi.org/10.1016/j.jece.2017.11.066
77. Albhnasawi A, Yüksel E, Gürbulak E, Duyum F (2020) Fate of aromatic amines through decolorization of real textile wastewater under anoxic-aerobic membrane bioreactor. J Environ Chem Eng 8:104226. https://doi.org/10.1016/j.jece.2020.104226
78. Selvaraj V, Karthika TS, Mansiya C, Alagar M (2021) An over review on recently developed techniques, mechanisms and intermediate involved in the advanced azo dye degradation for industrial applications. J Mol Struct 1224:129195. https://doi.org/10.1016/j.molstruc.2020. 129195
79. Yaseen DA, Scholz M (2016) Shallow pond systems planted with Lemna minor treating azo dyes. Ecol Eng 94:295–305. https://doi.org/10.1016/j.ecoleng.2016.05.081
80. Duarte Baumer J, Valério A, de Souza SMAGGU, Erzinger GS, Furigo A, de Souza AAU (2018) Toxicity of enzymatically decolored textile dyes solution by horseradish peroxidase. J Hazard Mater 360:82–88. https://doi.org/10.1016/j.jhazmat.2018.07.102
81. Garcia-segura S, Centellas F, Arias C, Garrido JA, Rodríguez RM, Cabot PL, Brillas E (2011) Electrochimica Acta comparative decolorization of monoazo, diazo and triazo dyes by electro-Fenton process. Electrochim Acta 58:303–311. https://doi.org/10.1016/j.electacta. 2011.09.049
82. Pandey A, Singh P, Iyengar L (2007) Bacterial decolorization and degradation of azo dyes. Int Biodeterior Biodegrad 59:73–84. https://doi.org/10.1016/j.ibiod.2006.08.006
83. Brüschweiler BJ, Merlot C (2017) Azo dyes in clothing textiles can be cleaved into a series of mutagenic aromatic amines which are not regulated yet. Regul Toxicol Pharmacol 88:214–226. https://doi.org/10.1016/j.yrtph.2017.06.012
84. Rovira J, Domingo JL (2019) Human health risks due to exposure to inorganic and organic chemicals from textiles: a review. Environ Res 168:62–69. https://doi.org/10.1016/j.envres. 2018.09.027
85. Sun W, Zhang C, Chen J, Zhang B, Zhang H, Zhang Y, Chen L (2017) Accelerating biodegra-dation of a monoazo dye Acid Orange 7 by using its endogenous electron donors. J Hazard Mater 324:739–743. https://doi.org/10.1016/j.jhazmat.2016.11.052
86. Mansour HB, Ayed-ajmi Y, Mosrati R, Corroler D, Ghedira K, Barillier D, Chekir-ghedira L (2010) Acid violet 7 and its biodegradation products induce chromosome aberrations, lipid peroxidation, and cholinesterase inhibition in mouse bone marrow. 1371–1378. https://doi. org/10.1007/s11356-010-0323-1
87. Rajaguru P, Vidya L, Baskarasethupathi B, Kumar PA, Palanivel M, Kalaiselvi K (2002) Genotoxicity evaluation of polluted ground water in human peripheral blood lymphocytes using the comet assay. 517:29–37
88. Vacchi FI, de Vendemiatti JAS, da Silva BF, Zanoni MVB, de Umbuzeiro GA (2017) Quantifying the contribution of dyes to the mutagenicity of waters under the influence

of textile activities. Sci Total Environ 601–602:230–236. https://doi.org/10.1016/j.scitotenv.2017.05.103

89. Solís M, Solís A, Inés H, Manjarrez N, Flores M (2012) Microbial decolouration of azo dyes: a review. Process Biochem 47:1723–1748. https://doi.org/10.1016/j.procbio.2012.08.014

90. Sen SK, Raut S, Bandyopadhyay P, Raut S (2016) Fungal decolouration and degradation of azo dyes: a review. Fungal Biol Rev 30:112–133. https://doi.org/10.1016/j.fbr.2016.06.003

91. Parrott JL, Bartlett AJ, Balakrishnan VK (2016) Chronic toxicity of azo and anthracenedione dyes to embryo-larval fathead minnow. Environ Pollut 210:40–47. https://doi.org/10.1016/j.envpol.2015.11.037

92. Balakrishnan VK, Shirin S, Aman AM, De SSR, Mathieu-denoncourt J, Langlois VS (2016) Genotoxic and carcinogenic products arising from reductive transformations of the azo dye, Disperse Yellow 7. Chemosphere 146:206–215. https://doi.org/10.1016/j.chemosphere.2015.11.119

93. Rafii F, Hall JD, Cerniglia CE (1997) Mutagenicity of azo dyes used in foods, drugs and cosmetics before and after reduction by clostridium species from the human intestinal tract. Food Chem Toxicol 35:897–901. https://doi.org/10.1016/S0278-6915(97)00060-4

94. Chequer FMD, Angeli JPF, Ferraz ERA, Tsuboy MS, Marcarini JC, Mantovani MS, de Oliveira DP (2009) The azo dyes Disperse Red 1 and Disperse Orange 1 increase the micronuclei frequencies in human lymphocytes and in HepG2 cells. Mutat Res Genet Toxicol Environ Mutagen 676:83–86. https://doi.org/10.1016/j.mrgentox.2009.04.004

95. Martínez-jerónimo MHF (2019) Exposure to the azo dye Direct blue 15 produces toxic effects on microalgae, cladocerans, and zebra fish embryos. Ecotoxicology 890–902. https://doi.org/10.1007/s10646-019-02087-1

96. Fernandes FH, Umbuzeiro GDA, Maria D, Salvadori F (2019) Genotoxicity of textile dye C. I. Disperse Blue 291 in mouse bone marrow. Mutat Res Genet Toxicol Environ Mutagen 837:48–51. https://doi.org/10.1016/j.mrgentox.2018.10.003

97. Tsuboy MS, Angeli JPF, Mantovani MS, Knasmu S (2007) Genotoxic, mutagenic and cytotoxic effects of the commercial dye CI Disperse Blue 291 in the human hepatic cell line HepG2. 21:1650–1655. https://doi.org/10.1016/j.tiv.2007.06.020

98. Fernandes H, Ventura L, Menezes C, Fernandes-cal JR, Roberto M, Fontes DM, Munari CC, Kummrow F, De AG, Maria D, Paulo S, Paulo S, Paulo S (2018) Research article InVivo genotoxicity of a commercial C. I. disperse Red1Dye. Environ Mol Mutagen 828:822–828. https://doi.org/10.1002/em.22226

99. Queiroz MTA, Queiroz CA, Alvim LB, Sabará MG, Leão MMD, Amorim CC de (2019) Reestruturação na forma do tratamento de efluentes têxteis: uma proposta embasada em fundamentos teóricos. Gestão & Produção 26. https://doi.org/10.1590/0104-530x1149-19

100. Rehman K, Shahzad T, Sahar A, Hussain S, Mahmood F, Siddique MH, Siddique MA, Rashid MI (2018) Effect of Reactive Black 5 azo dye on soil processes related to C and N cycling. PeerJ 2018:1–14. https://doi.org/10.7717/peerj.4802

101. Ilyas M, Ahmad W, Khan H, Yousaf S, Yasir M, Khan A (2019) Environmental and health impacts of industrial wastewater effluents in Pakistan: a review. Rev Environ Health 34:171–186. https://doi.org/10.1515/reveh-2018-0078

102. Alkaya E, Demirer GN (2014) Sustainable textile production: a case study from a woven fabric manufacturing mill in Turkey. J Clean Prod 65:595–603. https://doi.org/10.1016/j.jclepro.2013.07.008

103. Parisi ML, Fatarella E, Spinelli D, Pogni R, Basosi R (2015) Environmental impact assessment of an eco-efficient production for coloured textiles. J Clean Prod 108:514–524. https://doi.org/10.1016/j.jclepro.2015.06.032

104. Islam M, Mostafa M (2019) Textile dyeing effluents and environment concerns—a review. J Environ Sci Nat Resour 11:131–144. https://doi.org/10.3329/jesnr.v11i1-2.43380

105. Kumar A, Choudhary P, Verma P (2012) A ccomparative study on the treatment methods of textile dye effluents. J Chem Pharm Res 4:763–771

106. Kumar P, Prasad B, Mishra IM, Chand S (2008) Decolorization and COD reduction of dyeing wastewater from a cotton textile mill using thermolysis and coagulation. J Hazard Mater 153:635–645. https://doi.org/10.1016/j.jhazmat.2007.09.007

107. Rai HS, Bhattacharyya MS, Singh J, Bansal TK, Vats P, Banerjee UC (2005) Removal of dyes from the effluent of textile and dyestuff manufacturing industry: a review of emerging techniques with reference to biological treatment. Crit Rev Environ Sci Technol 35:219–238. https://doi.org/10.1080/10643380590917932
108. Joshi M, Bansal R, Purwar R (2004) Colour removal from textile effluents. Indian J Fibre Text Res 29:239–259
109. Santoro PH, Cavaguchi SA, Alexandre TM, Zorzetti J, Neves PMOJ (2014) In vitro sensitivity of antagonistic trichoderma atroviride to herbicides. Brazilian Arch Biol Technol 57:238–243. https://doi.org/10.1590/S1516-89132014000200012
110. Bhatia D, Sharma NR, Singh J, Kanwar RS (2017) Biological methods for textile dye removal from wastewater: a review. Crit Rev Environ Sci Technol 47:1836–1876. https://doi.org/10.1080/10643389.2017.1393263
111. Crini G, Lichtfouse E (2019) Advantages and disadvantages of techniques used for wastewater treatment. Environ Chem Lett 17:145–155. https://doi.org/10.1007/s10311-018-0785-9
112. Varjani S, Rakholiya P, Ng HY, You S, Teixeira JA (2020) Microbial degradation of dyes: an overview. Bioresour Technol 314:123728. https://doi.org/10.1016/j.biortech.2020.123728
113. Gao Y, Yang B, Wang Q (2018) Biodegradation and decolorization of dye wastewater: a review. IOP Conf Ser Earth Environ Sci 178:012013. https://doi.org/10.1088/1755-1315/178/1/012013
114. Ebrahimi R, Maleki A, Zandsalimi Y, Ghanbari R, Shahmoradi B, Rezaee R, Safari M, Joo SW, Daraei H, Harikaranahalli Puttaiah S, Giahi O (2019) Photocatalytic degradation of organic dyes using WO3-doped ZnO nanoparticles fixed on a glass surface in aqueous solution. J Ind Eng Chem 73:297–305. https://doi.org/10.1016/j.jiec.2019.01.041
115. Mandal T, Dasgupta D, Datta S (2010) A biotechnological thrive on COD and chromium removal from leather industrial wastewater by the isolated microorganisms. Desalin Water Treat 13:382–392. https://doi.org/10.5004/dwt.2010.996
116. Behera M, Nayak J, Banerjee S, Chakrabortty S, Tripathy SK (2021) A review on the treatment of textile industry waste effluents towards the development of efficient mitigation strategy: an integrated system design approach. J Environ Chem Eng 9:105277. https://doi.org/10.1016/j.jece.2021.105277
117. Samuchiwal S, Gola D, Malik A (2021) Decolourization of textile effluent using native microbial consortium enriched from textile industry effluent. J Hazard Mater 402:123835. https://doi.org/10.1016/j.jhazmat.2020.123835
118. Manai I, Miladi B, El Mselmi A, Smaali I, Ben Hassen A, Hamdi M, Bouallagui H (2016) Industrial textile effluent decolourization in stirred and static batch cultures of a new fungal strain Chaetomium globosum IMA1 KJ472923. J Environ Manag 170:8–14. https://doi.org/10.1016/j.jenvman.2015.12.038
119. Balamurugan B, Thirumarimurugan M, Kannadasan T (2011) Anaerobic degradation of textile dye bath effluent using Halomonas sp. Bioresour Technol 102:6365–6369. https://doi.org/10.1016/j.biortech.2011.03.017
120. Naresh B, Jaydip J, Prabhat B, Rajkumar P (2013) Recent biological technologies for textile effluent treatment. Int Res J Biol Sci 2:77–82
121. Kodam KM, Soojhawon I, Lokhande PD, Gawai KR (2005) Microbial decolorization of reactive azo dyes under aerobic conditions. World J Microbiol Biotechnol 21:367–370. https://doi.org/10.1007/s11274-004-5957-z
122. Subramanian J, Ramesh T, Kalaiselvam M (2014) Degradation of triphenylmethane dye: malachite green by Aspergillus flavus. World J Pharm Pharm Sci 3:44–50
123. Şen S, Demirer G (2003) Anaerobic treatment of real textile wastewater with a fluidized bed reactor. Water Res 37:1868–1878. https://doi.org/10.1016/S0043-1354(02)00577-8
124. Popli S, Patel UD (2015) Destruction of azo dyes by anaerobic–aerobic sequential biological treatment: a review. Int J Environ Sci Technol 12:405–420. https://doi.org/10.1007/s13762-014-0499-x
125. Xu H, Yang B, Liu Y, Li F, Shen C, Ma C, Tian Q, Song X, Sand W (2018) Recent advances in anaerobic biological processes for textile printing and dyeing wastewater treatment: a mini-review. World J Microbiol Biotechnol 34:165. https://doi.org/10.1007/s11274-018-2548-y

126. Mezohegyi G, Kolodkin A, Castro UI, Bengoa C, Stuber F, Font J, Fabregat A, Fortuny A (2007) Effective anaerobic decolorization of azo dye Acid Orange 7 in continuous upflow packed-bed reactor using biological activated carbon system. Ind Eng Chem Res 46:6788–6792. https://doi.org/10.1021/ie061692o

127. Cai J, Pan A, Li Y, Xiao Y, Zhou Y, Chen C, Sun F, Su X (2021) A novel strategy for enhancing anaerobic biodegradation of an anthraquinone dye reactive blue 19 with resuscitation-promoting factors. Chemosphere 263:127922. https://doi.org/10.1016/j.chemosphere.2020.127922

128. Nguyen TH, Watari T, Hatamoto M, Setiadi T, Yamaguchi T (2021) Enhanced decolorization of dyeing wastewater in a sponges-submerged anaerobic reactor. Chemosphere 279:130475. https://doi.org/10.1016/j.chemosphere.2021.130475

129. Ceretta MB, Nercessian D, Wolski EA (2021) Current trends on role of biological treatment in integrated treatment technologies of textile wastewater. Front Microbiol 12. https://doi.org/10.3389/fmicb.2021.651025

130. Shindhal T, Rakholiya P, Varjani S, Pandey A, Ngo HH, Guo W, Ng HY, Taherzadeh MJ (2021) A critical review on advances in the practices and perspectives for the treatment of dye industry wastewater. Bioengineered 12:70–87. https://doi.org/10.1080/21655979.2020.1863034

131. You S-J, Teng J-Y (2009) Anaerobic decolorization bacteria for the treatment of azo dye in a sequential anaerobic and aerobic membrane bioreactor. J Taiwan Inst Chem Eng 40:500–504. https://doi.org/10.1016/j.jtice.2009.01.007

132. Naimabadi A, Movahedian Attar H, Shahsavani A (2009) Decolorization and biological degradation of azo dye reactive red2 by anaerobic/aerobic sequential process. Iran J Environ Heal Sci Eng 6:67–72

133. Bhardwaj D, Bharadvaja N (2021) Phycoremediation of effluents containing dyes and its prospects for value-added products: a review of opportunities. J Water Process Eng 41:102080. https://doi.org/10.1016/j.jwpe.2021.102080

134. Acuner E, Dilek F (2004) Treatment of tectilon yellow 2G by Chlorella vulgaris. Process Biochem 39:623–631. https://doi.org/10.1016/S0032-9592(03)00138-9

135. Krishnamoorthy S, Manickam P (2021) Phycoremediation of industrial wastewater: challenges and prospects. In: Bioremediation for Environmental Sustainability. Elsevier, pp 99–123

136. Sarkar P, Dey A (2021) Phycoremediation—an emerging technique for dye abatement: an overview. Process Saf Environ Prot 147:214–225. https://doi.org/10.1016/j.psep.2020.09.031

137. Daneshvar N, Ayazloo M, Khataee AR, Pourhassan M (2007) Biological decolorization of dye solution containing Malachite Green by microalgae Cosmarium sp. 98:1176–1182.https://doi.org/10.1016/j.biortech.2006.05.025

138. El-Sheekh MM, Gharieb MM, Abou-El-Souod GW (2009) Biodegradation of dyes by some green algae and cyanobacteria. Int Biodeterior Biodegrad 63:699–704. https://doi.org/10.1016/j.ibiod.2009.04.010

139. Nunes Costa F, Alex Mayer D, Valério A, de Souza LJ, de Oliveira D, Ulson de Souza AA (2020) Non-isothermal kinetic modelling of potassium indigo-trisulfonate dye discolouration by Horseradish peroxidase. Biocatal Biotransformation 38:385–391. https://doi.org/10.1080/10242422.2020.1754806

140. Nguyen LN, van de Merwe JP, Hai FI, Leusch FDL, Kang J, Price WE, Roddick F, Magram SF, Nghiem LD (2016) Laccase-syringaldehyde-mediated degradation of trace organic contaminants in an enzymatic membrane reactor: removal efficiency and effluent toxicity. Bioresour Technol 200:477–484. https://doi.org/10.1016/j.biortech.2015.10.054

141. Lima JS, Araújo PHH, Sayer C, Souza AAU, Viegas AC, de Oliveira D (2017) Cellulase immobilization on magnetic nanoparticles encapsulated in polymer nanospheres. Bioprocess Biosyst Eng 40:511–518. https://doi.org/10.1007/s00449-016-1716-4

142. Kalyani D, Dhiman SS, Kim H, Jeya M, Kim I-W, Lee J-K (2012) Characterization of a novel laccase from the isolated Coltricia perennis and its application to detoxification of biomass. Process Biochem 47:671–678. https://doi.org/10.1016/j.procbio.2012.01.013

143. Husain Q (2006) Potential applications of the oxidoreductive enzymes in the decolorization and detoxification of textile and other synthetic dyes from polluted water: a review. Crit Rev Biotechnol 26:201–221. https://doi.org/10.1080/07388550600969936

144. Ravikumar G, Kalaiselvi M, Gomathi D, Vidhya B, Devaki K, Uma C (2013) Effect of laccase from hypsizygus ulmarius in decolorization of different dyes. J Appl Pharm Sci 3. https://doi.org/10.7324/JAPS.2013.30128

145. Claus H (2003) Laccases and their occurrence in prokaryotes. Springer

146. D'Souza DT, Tiwari R, Sah AK, Raghukumar C (2006) Enhanced production of laccase by a marine fungus during treatment of colored effluents and synthetic dyes. Enzyme Microb Technol 38:504–511. https://doi.org/10.1016/j.enzmictec.2005.07.005

147. Sharma P, Goel R, Capalash N (2007) Bacterial laccases. World J Microbiol Biotechnol 23:823–832. https://doi.org/10.1007/s11274-006-9305-3

148. Wong Y, Yu J (1999) Laccase-catalyzed decolorization of synthetic dyes. Water Res 33:3512–3520. https://doi.org/10.1016/S0043-1354(99)00066-4

149. Bollag J-M, Myers CJ, Minard RD (1992) Biological and chemical interactions of pesticides with soil organic matter. Sci Total Environ 123–124:205–217. https://doi.org/10.1016/0048-9697(92)90146-J

150. Gonçalves MLFC, Steiner W (1996) Use of Laccase for bleaching of pulps and treatment of effluents. 197–206

151. Bilal M, Rasheed T, Iqbal HMN, Hu H, Wang W, Zhang X (2018) Horseradish peroxidase immobilization by copolymerization into cross-linked polyacrylamide gel and its dye degradation and detoxification potential. Int J Biol Macromol 113:983–990. https://doi.org/10.1016/j.ijbiomac.2018.02.062

152. Jaiswal N, Pandey VP, Dwivedi UN (2016) Immobilization of papaya laccase in chitosan led to improved multipronged stability and dye discoloration. Int J Biol Macromol 86:288–295. https://doi.org/10.1016/j.ijbiomac.2016.01.079

153. Sun H, Jin X, Long N, Zhang R (2017) Improved biodegradation of synthetic azo dye by horseradish peroxidase cross-linked on nano-composite support. Int J Biol Macromol 95:1049–1055. https://doi.org/10.1016/j.ijbiomac.2016.10.093

154. Zhuo R, Zhang J, Yu H, Ma F, Zhang X (2019) The roles of Pleurotus ostreatus HAUCC 162 laccase isoenzymes in decolorization of synthetic dyes and the transformation pathways Rui. Chemosphere 234:733–745. https://doi.org/10.1016/j.chemosphere.2019.06.113

155. Sun H, Yang H, Huang W, Zhang S (2015) Immobilization of laccase in a sponge-like hydrogel for enhanced durability in enzymatic degradation of dye pollutants. 450:353–360. https://doi.org/10.1016/j.jcis.2015.03.037

156. Darvishi F, Moradi M, Jolivalt C, Madzak C (2018) Laccase production from sucrose by recombinant Yarrowia lipolytica and its application to decolorization of environmental pollutant dyes. Ecotoxicol Environ Saf 165:278–283. https://doi.org/10.1016/j.ecoenv.2018.09.026

157. Afreen S, Anwer R, Singh RK, Fatma T (2018) Extracellular laccase production and its optimization from Arthrospira maxima catalyzed decolorization of synthetic dyes. Saudi J Biol Sci 25:1446–1453. https://doi.org/10.1016/j.sjbs.2016.01.015

158. Bagewadi ZK, Mulla SI, Ninnekar HZ (2017) Purification and immobilization of laccase from Trichoderma harzianum strain HZN10 and its application in dye decolorization. J Genet Eng Biotechnol 15:139–150. https://doi.org/10.1016/j.jgeb.2017.01.007

159. Ma H-F, Meng G, Cui B-K, Si J, Dai Y-C (2018) Chitosan crosslinked with genipin as supporting matrix for biodegradation of synthetic dyes: laccase immobilization and characterization. Chem Eng Res Des 132:664–676. https://doi.org/10.1016/j.cherd.2018.02.008

160. Zheng F, Cui B-K, Wu X-J, Meng G, Liu H-X, Si J (2016) Immobilization of laccase onto chitosan beads to enhance its capability to degrade synthetic dyes. Int Biodeterior Biodegrad 110:69–78. https://doi.org/10.1016/j.ibiod.2016.03.004

161. Kashe S, Mehdi S, Mohammad N (2019) Covalently immobilized laccase onto graphene oxide nanosheets: preparation, characterization, and biodegradation of azo dyes in colored wastewater. 276:153–162. https://doi.org/10.1016/j.molliq.2018.11.156

162. Blánquez A, Rodríguez J, Brissos V, Mendes S, Martins LO, Ball AS, Arias ME, Hernández M (2019) Decolorization and detoxification of textile dyes using a versatile Streptomyces laccase-natural mediator system. Saudi J Biol Sci 26:913–920. https://doi.org/10.1016/j.sjbs. 2018.05.020
163. Daâssi D, Rodríguez-couto S, Nasri M, Mechichi T (2014) Biodegradation of textile dyes by immobilized laccase from Coriolopsis gallica into Ca-alginate beads. Int Biodeterior Biodegrad 90:71–78. https://doi.org/10.1016/j.ibiod.2014.02.006
164. Ulson de Souza SMAG, Forgiarini E, Ulson de Souza AA (2007) Toxicity of textile dyes and their degradation by the enzyme horseradish peroxidase (HRP). J Hazard Mater 147:1073–1078. https://doi.org/10.1016/j.jhazmat.2007.06.003
165. Chiong T, Lau SY, Lek ZH, Koh BY, Danquah MK (2016) Enzymatic treatment of methyl orange dye in synthetic wastewater by plant-based peroxidase enzymes. J Environ Chem Eng 4:2500–2509. https://doi.org/10.1016/j.jece.2016.04.030
166. Passardi F, Bakalovic N, Teixeira FK, Margis-Pinheiro M, Penel C, Dunand C (2007) Prokaryotic origins of the non-animal peroxidase superfamily and organelle-mediated transmission to eukaryotes. Genomics 89:567–579. https://doi.org/10.1016/j.ygeno.2007.01.006
167. Bilal M, Rasheed T, Iqbal HMN, Hu H, Wang W, Zhang X (2017) Novel characteristics of horseradish peroxidase immobilized onto the polyvinyl alcohol-alginate beads and its methyl orange degradation potential. Int J Biol Macromol 105:328–335. https://doi.org/10.1016/j.ijb iomac.2017.07.042
168. Ali M, Husain Q, Sultana S, Ahmad M (2018) Immobilization of peroxidase on polypyrrole-cellulose-graphene oxide nanocomposite via non-covalent interactions for the degradation of Reactive Blue 4 dye. Chemosphere 202:198–207. https://doi.org/10.1016/j.chemosphere. 2018.03.073
169. Souza SMAG, Forgiarini E, Souza AAU (2007) Toxicity of textile dyes and their degradation by the enzyme horseradish peroxidase (HRP). J Hazard Mater 147(3):1073–1078. https://doi. org/10.1016/j.jhazmat.2007.06.003
170. Oliveira SF, da Luz JMR, Kasuya MCM, Ladeira LO, Correa Junior A (2018) Enzymatic extract containing lignin peroxidase immobilized on carbon nanotubes: potential biocatalyst in dye decolourization. Saudi J Biol Sci 25:651–659. https://doi.org/10.1016/j.sjbs.2016.02.018
171. Duan Z, Shen R, Liu B, Yao M, Jia R (2018) Comprehensive investigation of a dye-decolorizing peroxidase and a manganese peroxidase from Irpex lacteus F17, a lignin-degrading basidiomycete. AMB Express 8. https://doi.org/10.1186/s13568-018-0648-6
172. Verma K, Saha G, Kundu LM, Dubey VK (2019) Biochemical characterization of a stable azoreductase enzyme from Chromobacterium violaceum: application in industrial effluent dye degradation. Int J Biol Macromol 121:1011–1018. https://doi.org/10.1016/j.ijbiomac. 2018.10.133
173. Dong H, Guo T, Zhang W, Ying H, Wang P, Wang Y, Chen Y (2019) Biochemical characterization of a novel azoreductase from Streptomyces sp.: application in eco-friendly decolorization of azo dye wastewater. Int J Biol Macromol 140:1037–1046. https://doi.org/10.1016/j. ijbiomac.2019.08.196
174. Yang Y, Wei B, Zhao Y, Wang J (2013) Construction of an integrated enzyme system consisting azoreductase and glucose 1-dehydrogenase for dye removal. Bioresour Technol 130:517–521. https://doi.org/10.1016/j.biortech.2012.12.106
175. Karatay SE, Kiliç NK, Dönmez G (2015) Removal of Remazol Blue by azoreductase from newly isolated bacteria. Ecol Eng 84:301–304. https://doi.org/10.1016/j.ecoleng.2015.09.037
176. Elfarash A, Mawad AMM, Yousef NMM, Shoreit AAM (2017) Azoreductase kinetics and gene expression in the synthetic dyes-degrading Pseudomonas. Egypt J Basic Appl Sci 4:315–322. https://doi.org/10.1016/j.ejbas.2017.07.007
177. Qi J, Anke MK, Szymańska K, Tischler D (2017) Immobilization of Rhodococcus opacus 1CP azoreductase to obtain azo dye degrading biocatalysts operative at acidic pH. Int Biodeterior Biodegrad 118:89–94. https://doi.org/10.1016/j.ibiod.2017.01.027
178. Guo G, Liu C, Hao J, Tian F, Ding K, Zhang C, Yang F, Liu T, Xu J, Guan Z (2021) Development and characterization of a halo-thermophilic bacterial consortium for decolorization of azo dye. Chemosphere 272:129916. https://doi.org/10.1016/j.chemosphere.2021.129916

179. Ahmad A, Khan N, Giri BS, Chowdhary P, Chaturvedi P (2020) Removal of methylene blue dye using rice husk, cow dung and sludge biochar: characterization, application, and kinetic studies. Bioresour Technol 306:123202. https://doi.org/10.1016/j.biortech.2020.123202
180. Gadekar MR, Ahammed MM (2020) Use of water treatment residuals for colour removal from real textile dye wastewater. Appl Water Sci 10:1–8. https://doi.org/10.1007/s13201-020-012 45-9
181. Ravenni G, Cafaggi G, Sárossy Z, Nielsen KTR, Ahrenfeldt J, Henriksen UB (2020) Waste chars from wood gasification and wastewater sludge pyrolysis compared to commercial activated carbon for the removal of cationic and anionic dyes from aqueous solution. Bioresour Technol Reports 100421. https://doi.org/10.1016/j.biteb.2020.100421
182. Li W, Mu B, Yang Y (2019a) Feasibility of industrial-scale treatment of dye wastewater via bio-adsorption technology. Bioresour Technol 277:157–170. https://doi.org/10.1016/j.bio rtech.2019.01.002
183. Zhou YY, Lu J, Zhou YY, Liu Y (2019) Recent advances for dyes removal using novel adsorbents: a review. Environ Pollut 252:352–365. https://doi.org/10.1016/j.envpol.2019. 05.072
184. Manavalan A, Manavalan T, Murugesan K, Kutzner A, Thangavelu KP, Heese K (2015) Characterization of a solvent, surfactant and temperature-tolerant laccase from Pleurotus sp. MAK-II and its dye decolorizing property. Biotechnol Lett 37:2403–2409. https://doi.org/10. 1007/s10529-015-1937-7
185. Liu H, Cheng Y, Du B, Tong C, Liang S, Han S, Zheng S, Lin Y (2015) Overexpression of a novel thermostable and chloride-tolerant laccase from Thermus thermophilus SG0.5JP17-16 in Pichia pastoris and its application in synthetic dye decolorization. PLoS One 10:1–14. https://doi.org/10.1371/journal.pone.0119833
186. Liu W, Liu C, Liu L, You Y, Jiang J, Zhou Z, Dong Z (2017) Simultaneous decolorization of sulfonated azo dyes and reduction of hexavalent chromium under high salt condition by a newly isolated salt-tolerant strain Bacillus circulans BWL1061. Ecotoxicol Environ Saf 141:9–16. https://doi.org/10.1016/j.ecoenv.2017.03.005
187. Kishor R, Purchase D, Saratale GD, Saratale RG, Ferreira LFR, Bilal M, Chandra R, Bharagava RN (2021) Ecotoxicological and health concerns of persistent coloring pollutants of textile industry wastewater and treatment approaches for environmental safety. J Environ Chem Eng 9:105012. https://doi.org/10.1016/j.jece.2020.105012
188. Oller I, Malato S, Sánchez-Pérez JA (2011) Combination of advanced oxidation processes and biological treatments for wastewater decontamination—a review. Sci Total Environ 409:4141–4166. https://doi.org/10.1016/j.scitotenv.2010.08.061
189. Paździor K, Bilińska L, Ledakowicz S, Pa K, Bili L, Paździor K, Bilińska L, Ledakowicz S (2019) A review of the existing and emerging technologies in the combination of AOPs and biological processes in industrial textile wastewater treatment. Chem Eng J 376. https://doi. org/10.1016/j.cej.2018.12.057
190. Ledakowicz S, Żyłła R, Paździor K, Wrębiak J, Sójka-Ledakowicz J (2017) Integration of ozonation and biological treatment of industrial wastewater from dyehouse. Ozone Sci Eng 39:357–365. https://doi.org/10.1080/01919512.2017.1321980
191. Thanavel M, Kadam SK, Biradar SP, Govindwar SP, Jeon BH, Sadasivam SK (2019) Combined biological and advanced oxidation process for decolorization of textile dyes. SN Appl Sci 1:1–16. https://doi.org/10.1007/s42452-018-0111-y
192. Dias NC, Alves TLM, Azevedo DA, Bassin JP, Dezotti M (2020) Metabolization of by-products formed by ozonation of the azo dye Reactive Red 239 in moving-bed biofilm reactors in series. Brazilian J Chem Eng 37:495–504. https://doi.org/10.1007/s43153-020-00046-6
193. Venkatesh S, Venkatesh K, Quaff AR (2017) Dye decomposition by combined ozonation and anaerobic treatment: cost effective technology. J Appl Res Technol 15:340–345. https://doi. org/10.1016/j.jart.2017.02.006
194. Shanmugam BK, Easwaran SN, Mohanakrishnan AS, Kalyanaraman C, Mahadevan S (2019) Biodegradation of tannery dye effluent using Fenton's reagent and bacterial consortium: a biocalorimetric investigation. J Environ Manage 242:106–113. https://doi.org/10.1016/j.jen vman.2019.04.075

195. Kurade MB, Waghmode TR, Xiong JQ, Govindwar SP, Jeon BH (2019) Decolorization of textile industry effluent using immobilized consortium cells in upflow fixed bed reactor. J Clean Prod 213:884–891. https://doi.org/10.1016/j.jclepro.2018.12.218

196. Rondon H, El-Cheikh W, Boluarte IAR, Chang C-Y, Bagshaw S, Farago L, Jegatheesan V, Shu L (2015) Application of enhanced membrane bioreactor (eMBR) to treat dye wastewater. Bioresour Technol 183:78–85. https://doi.org/10.1016/j.biortech.2015.01.110

197. Sepehri A, Sarrafzadeh MH (2018) Effect of nitrifiers community on fouling mitigation and nitrification efficiency in a membrane bioreactor. Chem Eng Process Process Intensif 128:10–18. https://doi.org/10.1016/j.cep.2018.04.006

198. Sayed ET, Shehata N, Abdelkareem MA, Atieh MA (2020) Recent progress in environmentally friendly bio-electrochemical devices for simultaneous water desalination and wastewater treatment. Sci Total Environ 748:141046. https://doi.org/10.1016/j.scitotenv.2020.141046

199. Yuan Y, Zhang J, Xing L (2019) Effective electrochemical decolorization of azo dye on titanium suboxide cathode in bioelectrochemical system. Int J Environ Sci Technol 16:8363–8374. https://doi.org/10.1007/s13762-019-02417-0

200. Miran W, Jang J, Nawaz M, Shahzad A, Lee DS (2018) Sulfate-reducing mixed communities with the ability to generate bioelectricity and degrade textile diazo dye in microbial fuel cells. J Hazard Mater 352:70–79. https://doi.org/10.1016/j.jhazmat.2018.03.027

201. Yang HY, Liu J, Wang YX, He CS, Zhang LS, Mu Y, Li WH (2019) Bioelectrochemical decolorization of a reactive diazo dye: kinetics, optimization with a response surface methodology, and proposed degradation pathway. Bioelectrochemistry 128:9–16. https://doi.org/10.1016/j.bioelechem.2019.02.008

202. Oon YL, Ong SA, Ho LN, Wong YS, Dahalan FA, Oon YS, Teoh TP, Lehl HK, Thung WE (2020) Constructed wetland–microbial fuel cell for azo dyes degradation and energy recovery: Influence of molecular structure, kinetics, mechanisms and degradation pathways. Sci Total Environ 720:137370. https://doi.org/10.1016/j.scitotenv.2020.137370

203. Bouabidi ZB, El-Naas MH, Zhang Z (2019) Immobilization of microbial cells for the biotreatment of wastewater: a review. Environ Chem Lett 17:241–257. https://doi.org/10.1007/s10311-018-0795-7

204. Dasgupta N, Ranjan S, Ramalingam C (2017) Applications of nanotechnology in agriculture and water quality management. Environ Chem Lett 15:591–605. https://doi.org/10.1007/s10311-017-0648-9

205. Suganya K, Revathi K (2016) Decolorization of reactive dyes by immobilized bacterial cells from textile effluents. Int J Curr Microbiol Appl Sci 5:528–532 . https://doi.org/10.20546/ijcmas.2016.501.053

206. Mohapatra RK, Behera SS, Patra JK, Thatoi H, Parhi PK (2020) Potential application of bacterial biofilm for bioremediation of toxic heavy metals and dye-contaminated environments. In: Yadav MK, Singh BP (eds) New and future developments in microbial biotechnology and bioengineering: microbial biofilms. Elsevier, pp 267–281

207. Sun P, Hui C, Wang S, Wan L, Zhang X, Zhao Y (2016) Bacillus amyloliquefaciens biofilm as a novel biosorbent for the removal of crystal violet from solution. Colloids Surf B Biointerfaces 139:164–170. https://doi.org/10.1016/j.colsurfb.2015.12.014

208. Castro FD, Bassin JP, Alves TLM, Sant'Anna GL, Dezotti M (2020) Reactive Orange 16 dye degradation in anaerobic and aerobic MBBR coupled with ozonation: addressing pathways and performance. Int J Environ Sci Technol. https://doi.org/10.1007/s13762-020-02983-8

209. Deng D, Lamssali M, Aryal N, Ofori-Boadu A, Jha MK, Samuel RE (2020) Textiles wastewater treatment technology: a review. Water Environ Res 92:1805–1810. https://doi.org/10.1002/wer.1437

210. Ong C, Lee K, Chang Y (2020) Biodegradation of mono azo dye-Reactive Orange 16 by acclimatizing biomass systems under an integrated anoxic-aerobic REACT sequencing batch moving bed biofilm reactor. J Water Process Eng 36:101268. https://doi.org/10.1016/j.jwpe.2020.101268

211. Mishra S, Nayak JK, Maiti A (2020) Bacteria-mediated bio-degradation of reactive azo dyes coupled with bio-energy generation from model wastewater. Clean Technol Environ Policy 22:651–667. https://doi.org/10.1007/s10098-020-01809-y

212. Logroño W, Pérez M, Urquizo G, Kadier A, Echeverría M, Recalde C, Rákhely G (2017) Single chamber microbial fuel cell (SCMFC) with a cathodic microalgal biofilm: a preliminary assessment of the generation of bioelectricity and biodegradation of real dye textile wastewater. Chemosphere 176:378–388. https://doi.org/10.1016/j.chemosphere.2017.02.099

213. Brar A, Kumar M, Vivekanand V, Pareek N (2019) Phycoremediation of textile effluent-contaminated water bodies employing microalgae: nutrient sequestration and biomass production studies. Int J Environ Sci Technol 16:7757–7768. https://doi.org/10.1007/s13762-018-2133-9

214. Iratni A, Chang BN (2019) Advances in control technologies for wastewater treatment processes: status, challenges, and perspectives. IEEE/CAA J Autom Sin 6:337–363. https://doi.org/10.1109/JAS.2019.1911372

215. Sundui B, Ramirez Calderon OA, Abdeldayem OM, Lázaro-Gil J, Rene ER, Sambuu U (2021) Applications of machine learning algorithms for biological wastewater treatment: updates and perspectives. Clean Technol Environ Policy 23:127–143. https://doi.org/10.1007/s10098-020-01993-x

216. Gottschalk C, Libra JA, Saupe A (2010) Ozonation of water and waste water: a practical guide to understanding ozone and its application. Wiley-VCH, Weinheim, Germany

Role of Microbial Biofilms in Dye Degradation During Textile Wastewater Treatment

Chidi B. Okeke, Kenechi O. Chukwu, Johnson K. Ndukwe, Uchenna S. Okechukwu, Chukwudi O. Onwosi, and Frederick J. C. Odibo

Abstract The treatment of dye wastes effluent has been a re-occurring problem since no single treatment method is capable of effectively removing the dye components as well as intermediate metabolites usually generated during the treatment process. Hybrid treatment procedures have also been employed but not without some limitations. This procedure, though effective, but does not completely mineralize the pollutant or intermediates resulting from wastewater treatment. Microbial remediation of textile dye wastewater using pure cultures or consortia of different microbial species also showed promising results, producing high COD, BOD, and percentage decolouration of above 90% on different dyes used. An integrated system that combines physicochemical and biological methods will enhance dye removal processes. These isolates grown as biofilms will enhance their effectiveness because they are made robust by quorum sensing and the consortium of enzymes produced which improves their bioremediation potentials. Optimization of parameters (such as pH, salinity, dye concentration, etc.) involved in dye wastewater treatments to improve their removal efficiency as well as re-usability of the treated water is necessary for the effectiveness of dyes and their intermediates removal from wastewater. This work, therefore, highlights the different treatment methods employed and further listed the roles microbial biofilms played when employed in the integrated treatment system for effective detoxification, degradation, and complete mineralization of pollutants in dye waste effluents.

C. B. Okeke · F. J. C. Odibo
Department of Applied Microbiology and Brewing, Nnamdi Azikiwe University, Awka, Anambra State, Nigeria

K. O. Chukwu · J. K. Ndukwe · U. S. Okechukwu · C. O. Onwosi (✉)
Department of Microbiology, Faculty of Biological Sciences, University of Nigeria, Nsukka, Enugu State, Nigeria
e-mail: chukwudi.onwosi@unn.edu.ng

K. O. Chukwu · J. K. Ndukwe · C. O. Onwosi
Bioconversion and Renewable Energy Research Unit, University of Nigeria, Nsukka, Enugu State, Nigeria

© The Author(s), under exclusive license to Springer Nature Singapore Pte Ltd. 2022
A. Khadir and S. S. Muthu (eds.), *Biological Approaches in Dye-Containing Wastewater*,
Sustainable Textiles: Production, Processing, Manufacturing & Chemistry,
https://doi.org/10.1007/978-981-19-0526-1_3

1   Introduction

Water contamination is considered a global challenge and has exacerbated water scarcity [162, 168]. This is due to the consistent rise in the growing population leading to expeditious anthropogenic activities (industrialization and urbanization) which impact negatively the availability of drinking water as the waste from these activities are dumped into nearby water bodies [157]. Furthermore, the high dependence on synthetic fertilizers in arable soils, construction of paved roads and buildings, improper discharge of industrial wastewater into the environment, etc. are other contributors to water contamination [168].

Yaseen and Scholz [198] hinted that different protocols in the textile industries (e.g., yarn-to-fibre or the reverse, dye application, etc.) generate a large volume of wastewater that leads to environmental pollution when poorly managed. Besides the loss of the environmental aesthetics via the wastewater generated during fabric making, the chemical residues contained therein such as sequestering agents (e.g., azo dyes), stain removers (e.g., residual chlorine), printing gums, and heavy metals are also a threat to the public and aquatic life [66, 83]. Other persistent compounds found in the wastewater include dye, humic substances, phenols, detergents, pesticides, etc. [58]. As noted by Bahafid et al. [11] and Liu et al. [111], the persistence in the environment, cytotoxicity, and potential bioaccumulation in food form the basis for the characterization of these chemical pollutants.

Although many industrial wastewaters contribute to water contamination, wastewaters from textile industries are the major contributors due to the high amount of dye used during the dyeing. Considering the global discharge of traditional synthetic dyes or colorants in the environment, azo dye, in particular, constitutes >50%[123]. Jonstrup et al. [89] and Meng et al. [123] noted that discharging azo dye-laden wastewater into water bodies affects the aquatic life via restricting oxygen influx, impeding light penetration thereby reducing the photosynthesis rate. Several researchers including Van der Zee and Villaverde [189], Tan et al. [179], De Arãgao Umbuzeiro et al. [42], Garcia-Montano et al. [64], and Rauf and Ashraf [152] have demonstrated that different kinds of azo dyes have mutagenic and carcinogenic effects on human and aquatic life. Hence their presence in the environment should be reduced to the barest minimum. One of the major eco-friendly, cost-effective, and reliable approaches to removing azo dyes from the environment is bioremediation. In this process, microorganisms or plants are harnessed to manage the pollutants. In this section, the role of biofilm-producing microorganisms in dye removal from textile wastewater was reviewed.

2   Characteristics of Dye

On exposure to visible light irradiation, dyes are organic compounds that absorb, remit, and diffuse corresponding light energy [108]. Thus, they are best described as ionizing and coloured organic compounds that are relatively soluble in most aqueous solutions. As noted by Rai et al. [150] textile dyes provide lasting coloration that is resistant to the action of water, detergents, and light when applied on fabrics. They are mainly comprised of the chromophore (colorants), colorant enhancers (auxochromes) as well as conjugate aromatic orientation [14]. Rajasimman et al. [151] added that in the course of textile printing or dyeing, these groups are peculiar and distinct in function, having the colour and propensity to be fastened on textile material.

Refining petroleum hydrocarbon (i.e., the distillation of coal tar) is one of the sources of most compounds used in developing textiles dyes. These dye constituents comprising acridine, cresol, toluene, quinolone, phenol, benzene, etc., are further transformed via varying processes to intermediate products before application [150]. The intermediates are formed via displacement of H_2 atoms by nitro ($-NO_2$), hydroxyl ($-OH$), sulphonic acid ($-OSO_3H$), amino ($-NH_2$), groups to form one of these compounds: aniline ($C_6H_5 \cdot NH_2$), nitrobenzene ($C_6H_5 \cdot NO_2$), β naphthalene sulfonic acid ($C_{10}H_7 \cdot SO_3H$), and β-naphthol ($C_{10}H_7 \cdot OH$) [150].

The presence of chromophores and auxochromes are the two essential constituents that aid coloured compounds to function as dyes. This is because colour formation in dyes largely depends on light to absorption in the near-ultraviolet region by the help of the chromosomes while the auxochromes (stable acidic- ($-OH$, $-COOH$, $-SO_3H$) or basic- ($-NH_2$, $-NHR$, $-NR_2$) chemical bonds) aids in the fixation of dyes on fibre support [155]. Notwithstanding, some dyes such as direct dyes do not require mordant to fix them on fibres. However, fibre-reactive dyes produce brighter colours than their direct counterparts. Again, the direct dyes result in poor wash-fastness and better light-fast than the fibre-reactive dyes [150].

The basis for categorizing dyes is chemical orientation, substrate (e.g., fibres, plastics, leather, paper, etc.) affinity, and application techniques [6, 117, 155].

2.1   Classification Based on Their Chemical Structure

a.   Indigo dyes

Indigo dye is commonly derived from indigo. The oxygen, sulphur, and selenium homologues of indigo blue confer this class the unique hypochromic impression with varying colour ranges —orange to turquoise [6, 19].

b.   **Xanthene dyes**

The uniqueness of this dye family is the ability to form intense fluorescence due to its fluorescein components. Although poorly utilized in the textile dyeing industries, they are exploited markers in maritime industries [17].

c.   **Phthalocyanine dyes**

These are a class of dyes synthesized by the reaction of dicyanobenzene with heavy metals such as cobalt, nickel, platinum, and copper. The excellent light fastening attributes of these dye molecules is due to the phthalocyanine backbone. Due to high chemical stability, the copper phthalocyanine has been adjudged the most commonly utilized in this dye class [197].

d.   **Nitrated/nitrosated dyes**

Although limited in the forms, these dyes are among the oldest classes of dyes utilized in the textile industries. The current usefulness of this class of dyes is predicated on its cost effectiveness as well as the simplicity of its molecular orientation. The structure is distinguished by the situating nitro group at ortho position relative to the –OH or –NH$_2$ (electron donor group) [108].

e.   **Diphenylmethane and triphenylmethane dyes**

These classes of dyes, as noted by Wang et al. [193], are synthesized by the auramine and triphenylmethane backbone. Some of the earliest used dyes such as malachite green and fuchsin are members of these groups of dyes. The metallic and acidic forms are derived by the OH-auxochromes neighbouring carboxyl groups and the sulphonation of triphenylmethane, respectively.

f.   **Polymethine dyes**

Polymethine (also referred to as cyanine) dyes are dyeing agents whose chromogenic structure comprises polymethine chain bearing polar heterocyclic groups of varying length [3, 204].

g.   **Azo dyes**

Azo dyes comprises a substantial class of synthetic dyes (about 70%) utilized in the textile and allied industries. Different forms of azo dyes (e.g., mono-, di- and triazo derivatives) that currently exist are developed by the fusion of hydroxyl or amino groups (auxochrome) to one or more (–N=N–) azo groups [76, 158].

h.   **Anthraquinone dyes**

Apart from the azo dyes, anthraquinone is another commonly used dyeing agent in the textile industries and stands as the second most produced dyes regarding volume (Routoula and Patwardhan 2020). The excellent light resistance exhibited by the dye is a result of the chromophore group [62].

2.2 Classification Based on Affinity to Varying Substrates Their Applicability

This categorization, according to Webster et al. [196] and Mijin et al. [124], is based on the solubility of the dye in the aqueous bath, its affinity to numerous fibres, and the sort of the fixation. Water soluble and water insoluble are the two family proponents of this classification.

2.2.1 Water-Soluble Dyes

a. **Acid or anionic dyes**

This water-soluble dye under an acidic medium, are used in the dyeing of fibres (e.g., silk, polyamide, mod acrylic, and wool), which contain an amino group (NH_2). Santos Pisoni et al. [142] highlighted that the solubility of these dyes in water is a result of the varying sulphonates and chromophore groups. The dyes in this category are mostly azo, triarylmethanes and anthraquinone [6, 191].

b. **Basic (cationic) dyes**

These dyes are comprised of salts with organic bases with the ability to directly dye fibres having anionic sites when treated with metallic salts. The dyes found under this category possess anthraquinone, azo, diphenylmethane structures [146, 185].

c. **Metalliferous dyes**

The common dyes in this group bear phthalocyanine and azo orientation. These are developed by integrating the metal atom (cobalt, copper, nickel, chromium, etc.) into the structure of metallic-acid dyes [85]. This incorporation protocol prevents etching operation during dye application. The dyes in this group are subdivided into metalliferous 1/1 and 1/2 complexes depending on the interaction between the dye and metal atom.

d. **Reactive dyes**

This group exhibits high water solubility and could be widely applied in wool and cotton dyeing. The chromophore moieties from anthraquinone, azo, and phthalocyanine are utilized in the formation of this group of dyes [124]. These dyes form strong covalent bonding with textiles or fibre owing to the very active chemical (e.g., vinylsulphone, triazine) constituents [108].

e. **Direct dyes**

Direct dyes also known as substantive dyes are large molecules with the ability to form electrostatically positive or negative charges with fibres. The plane structural orientation of this group confers their high affinity to cellulosic fibres and distinguishes the group from others in the absence of mordants [108]. However, the wettability of the fibre is improved with surfactants (anionic and non-ionic) when the dye is directly applied to salt baths. These dyes have a wide variety of colours, are easy to apply and are low in price. However, they have a low wet strength.

2.2.2   Water-Insoluble Dyes

a.   **Vat dyes**

This category of dyes is water insoluble when used on fibres, however, their solubility is enhanced in an alkaline medium [32, 73, 144]. These dyes have a high affinity for particular fibres (e.g., silk, linen, wool, etc.) very recalcitrant to degradation agents. Vat green, vat blue 4, and Indigo dye (vat blue 1) for dyeing jeans (or denim) are some of the excellent examples of dyes in this family [22, 73, 88, 160].

b.   **Sulfur dyes**

Apart from the distinct chemical structure with high molecular weight due to the sulphur moieties, sulfur dyes resemble the vat dyes when the utilization protocol is considered [31]. These dyes could be utilized as soluble derivative in an alkaline medium in the presence of a reducing agent, although they are insoluble in water. They exhibit light fastness and good wash when applied on textiles [108].

c.   **Dispersible dyes**

As noted by Chakraborty [32] and Muzamil et al. [129], these dyes are commonly utilized as fine powder in a dye bath, although they are insoluble in water. Other interesting attributes include their ease of diffusion and fixation in synthetic fibres during the dyeing process as well as thermal stability. Dispersible dyes are utilized for dyeing fibres such as polyamide and polyester [108].

d.   **Pigments**

The pigments are water insoluble coloured compounds that do not contain any group that can link with the textile fibre by means of a binder. Since they are insoluble in acidic and basic solvents, pigment dyes are commonly used in pigment printing [131, 205].

3   Dyes: Impact on Environment

Due to poor affinity for textiles surfaces during the dyeing or printing process, a large portion of pigments and dyes are released into the wastewater [77, 98]. The charges present in the dye and fibre strongly affect the affinity of the dye and unlike charges result in strong electrostatic interactions between the dye and fibre [124, 151]. Loose interactions between dye and textiles lead to the discharge of a considerable amount of dyes into the effluents. This dye-laden wastewater eventually ends up in the soil and aquatic environment, if not adequately treated. The complex chemical orientation of most dyes makes them very recalcitrant leading to their persistence in the environment. These adversely disorient the biodiversity and ecological dynamics of the soil [10, 54, 105, 176]. In the aquatic bodies, the photogenic proceedings such as light penetration are seriously impaired due to the presence of wastewater [79, 141, 172]. Other potential impacts of dye-laden wastewater include health challenges during human exposure (e.g., cancers, skin irritation, etc.), loss of aesthetics (colour change and malodour), and algal blooms in water bodies [24–27, 43, 49, 79, 153, 161, 173]. Furthermore, the introduction of dyes and textile pigments wastewater to the soil and aquatic bodies regularly distorts the physicochemical parameters such as pH, chemical oxygen demand, biochemical oxygen demand, total organic carbon, and suspended solids [15, 16, 39, 200]. The presence of dyes could also lead to heavy metal toxicity in the environments since some dyes have metals as one of their major components [26, 108, 141, 172, 202]. The discharge of dyes and pigments into the water bodies results in high oxygen consumption and eventual loss of aquatic lives [96, 108, 154, 191]. Eutrophication could also modify the drinking water production quality [24]. Therefore, it is essential to adequately treat textile dye and pigment effluent before discharging them to the environment in order to reduce their toxicity.

4 Understanding Biofilm Formation

Biofilms, according to Donlan and Costerton [50], Hall-Stoodley, and Stoodley [72], and Yadav et al. [199], are highly-structured and multiplex microbial communities that synthesize an extracellular matrix (extracellular polymeric substances (EPS)) conferring them with the capability to adhere to biotic and abiotic platforms. There are five phases in biofilm development in an aqueous media as outlined by Webb [195]: (a) Initial attachment of planktonic microbes with one another and to a solid surface, (b) Complete adherence of successions of microbial cells to the surface enhanced by EPS and hydrogen bonds [187], (c) Multiplication of the primary colonizers leads to monolayer mini colonies development on the either as attached growth or suspended growth on the platform; (d) Maturation of biofilm via linking with debris from neighbouring surrounding and feeding on new entrants (microorganisms), and (e) Active or passive dispersal of quorum sensing-induced planktonic bacteria formation from EPS matrix-enclosed biofilm cells According to Ansari et al. [5] and Krivorot et al. [99], formation of biofilm depends on numerous parameters like availability of nutrients, pH, temperature, divalent cations, turbulence, velocity, EPS production, hydrodynamics, and the support media surface type. The physicochemical attributes (e.g., roughness, pore orientation, material type, etc.) of the carrier platform also controls biofilm formation [68, 128, 182].

4.1 Biofilm Potential in Dye Removal

El Abed et al. [53], Hoh et al. [75], Asri et al. [7] highlighted that the de-pollution efficiency of microorganisms in biofilm reactors is strongly linked to the characteristics of the carrier platform (i.e., surface energy, surface area, and morphology), the quality of the supporting materials and the adherence potential of biofilm on support surface. The traditionally utilized support materials comprise natural stones, clays (charcoal, zeolites, seashell, etc.), ceramics, sintered glass, plastic materials, fire bricks, sand [170, 182]. Other technologies such as are carbon-based materials (e.g., carbon felt, carbon paper, and carbon cloth) and organic material-based supports could aid in biofilm formation [4, 57, 110, 203]. The hydrophobicity, electronegativity, and hydrophilicity properties of the support's surface are also essential for biofilm formation and efficient wastewater treatment [37, 45, 95, 121]. Kesaano and Sims [95] demonstrated that hydrophobic platforms (e.g., stainless steel, Perspex, and titanium) have excellent adhesion on numerous species of microorganisms. The integration of solid support media in suspended growth bioreactors for wastewater treatments enhances biofilm formation and degradation of pollutants [136]. Lewandowski and Boltz [103] and Martin and Nerenberg [118] hinted at the benefits of deploying biofilm-oriented effluent treatment plants. These include adjustable operation, lower space requirements, reduction in hydraulic retention time (HRT), eco-friendly, potential to stabilize multiplex compounds, etc.

4.2 Biofilm-Based Processes in Dye Removal

Most of the conventional physicochemical and biological processes in dye removal from waste could not be used in full-scale separately, because of cost implication, dyeing wastewater, and operational problems [48, 90]. Currently, several experimental trials involving the modification of the traditional treatment plants and associated physicochemical processes have been undertaken to enhance effluent treatment. As noted by Forrez et al. [60], Lewandowski and Boltz [103], Casas et al. [28], Deng et al. [45], biofilm-based dye effluents treatment technology has attracted a lot of interest recently as it promising it is an economical remediation option. The biofilm-based effluent treatment processes have several benefits over conventional techniques. These comprise low space requirements, high active biomass concentration, low sludge production, stable performance, reduced HRT, and degradation of a large spectrum of organic pollutants due to diverse microbial communities [206]. Among the biotreatment techniques currently available for effective stabilization of different textile effluents containing dyes, biofilm-based technologies have been shown to be indispensable [163]. Some of the biofilm-based technologies such as membrane biofilm reactors, moving-bed sequencing batch biofilm reactor, trickling filter, moving-bed biofilm reactors, microbial fuel cells, fixed biofilm reactor, and fixed-bed sequencing batch biofilm reactors are discussed and the major outcomes of hybrid techniques are presented in Table 1.

Table 1 Colour and COD reduction reported for different hybrid systems

Treatment	Type of dye	% COD removal	% colour removal	HRT	Reference(s)
MBBRs + chemical coagulation	Textile dye wastewater Dye wastewater	95.7 95	73.4 97	48 h 44 h	Park et al. [138] Shin et al. [169]
MBBRs + Ozonation	Textile dye wastewater	94.3	96.3	14 h	Gong [67]
Two MBBRs in series + Ozonation MBBR + Ozonation	Azo dye reactive red 239 (RR 239) Reactive orange 16	90 92	100 97	20 min	Dias et al. [47] Castro et al. [30]
MBBRs + chemical precipitation	Newsprint mill wastewater	95		4–5 h	Broch-Due et al. [20]
MBBRs + Membrane filtration	azo dye reactive brilliant red X-3B-containing wastewater	85	95	5 h	Dong et al. [48]
Microbial fuel cell + Microalgal biofilm	Real dye textile wastewater	92–98	93	30 d	Logroño et al. [115]
Bioelectrochemical system + membrane biofilm reactor	Azo dye (Acid orange 7 (AO7))		>90		Pan et al. [137]
Anaerobic bioreactor + Ozonation	Azo dye (remazol red)	85–90	Over 99		Punzi et al. [147]
Poly tetra fluoro ethylene (PTFE) membrane module + TiO$_2$ slurry photocatalytic reactor	Reactive Black 5 (RB5)	50–85	82–100	4 h	Damodar et al. [40]
Up-flow biological aerated filter + ozonation process	acetate and polyester fibre dyeing effluent	99	96		Azbar et al. [10]
Bacterial consortium (*P. aeruginosa, B. flexus* and *S. lentus*) + Fenton reaction	Industrial textile wastewater	94	90		Shanmugam et al. [166]
aerobic granular sludge + heterogeneous electro-Fenton reaction	Methylene blue	75.7	100		Liu et al. [113]

4.2.1   Membrane Biofilm Reactors (MBR)

The potential of membrane biofilm reactors (MBR) in removing and recovery of pollutants has been proposed as promising biotechnology. In this technology, there is a combination of a biofilm-oriented process, a membrane-based separation unit at the secondary settling point, and biodegradation of the pollutants (e.g., dyes and pigments) in the effluent. MBR technique has both organic and inorganic pollutant removal capability and has preoccupied both industrial and urban wastewater treatment processes [46]. Its good disinfection efficiency, high volumetric loading rate, and effluent discharge variability make it unique. According to Kimura et al. [100], Petrovic et al. [140], Luo et al. [116], MBRs have gained prominence in effluent treatment because of their high removal efficiencies of organic micropollutants, human pathogenic bacteria, and suspended solids. Van Nieuwenhuijzen et al. [188] and Meng et al. (2009) added that MBRs are generally deployed when there is need for very high-quality effluent. However, one shortcoming associated with this technique is biofouling and poor membrane permeability. Two configurations of operations: submerged or immersed are employed in this system [102, 106]. The configuration in the submerged allows for membrane replacement and requires lower energy which means a lower operating cost and easy cleaning encouraging the development of this system. Both MBR configurations have successfully been used for the stabilization of different industrial wastewater. Lin et al. [109] opined that an innovative configuration consisting of air-lift sidestream MBRs has been developed recently. This recent configuration exploited all advantages of both submerged and external configuration MBRs according to reports of Chen and Liu [35] and Shariati et al. [167]. Spagni et al. [175] demonstrated that a system combining anoxic-aerobic MBR and anaerobic biofilter was suitable for stabilizing synthetic textile wastewater containing reactive orange 16 (RO16). In the proposed operation condition, the resulting aromatic amines or azo dyes had no interference with the removal efficiency of COD and nitrogen from the wastewater. While the majority of the aromatic amines from the RO16 were removed under aerobic conditions, the sulfonated types formed under anaerobic conditions were recalcitrant. In a different study, Fernando et al. [56] demonstrated that incorporating MBR into a bioelectrochemical system (BES) effectively degraded acid orange-7 (AO-7) (>90% decolourization efficiency) together with its reduction product (i.e., sulphanilic acid) in simulated dye wastewater. Taken together, the BES-MBR system exhibited lower energy consumption—economical and efficient—when compared to conventional aerobic treatment, suggesting its sustainability in dye wastewater management.

4.2.2   Moving-Bed Biofilm Reactor (MBBR)

Moving-Bed Biofilm Reactors (MBBRs) are indispensable in the stabilization of stock wastewater, landfill leachate, and industrial wastewater due to their compact volume, cost effectiveness, high biomass yield, strong organic loading shock resistance, and overall technical stability [36, 119]. As noted by Kermani et al. [94], the biofilm-forming microbial biomass adhered to a high surface area-small fluidizing carrier is kept in constant motion aided by circulating water inside the reactor. The MBBR is one of the hydraulic-oriented systems developed to retain high biomass volume for sustainable wastewater treatment [133]. Numerous reports have demonstrated that MBBR and its hybrids could be successfully deployed in the stabilization of wastewaters containing polycyclic aromatic hydrocarbon, pesticides, phenol, and aniline [9, 36, 44, 143]. Other investigators including Hapeshi et al. [71], Casas et al. [28], and Tang et al. [180] revealed that MBBRs are effective stabilization of wastewater containing a considerable amount of organic micropollutants. However, carriers used in the reactors could lead to biofilm-imposed variations resulting in fluctuations in pollutants removal capacity. High removal efficiency has been recorded during the treatment of real and synthetic textile dye wastewater by several researchers. In simultaneous nitrification and denitrification (SND)-MBBR system bearing a biofilm-forming *Stenotrophomonas maltophila* DQ01 removed about 94.43% of total nitrogen and realized 94.21% concomitant nitrification and denitrification at a cycle time of 7 h [87]. Mahto and Das [119] noted that removal efficiency of 86.8% for total organic carbon from wastewater within HRT of 24 h was achieved with MBBR harbouring a consortium of biofilm-formers (*Rhodobacter* sp. and *Chryseobacterium* sp,). While simulated wastewater was treated with MBBR, a high Congo red removal capacity and removal efficiency of 214.3 mg/Lday and 99.2%, respectively [174]. Pretreatment of textile wastewater with fluidized bed Fenton followed by MBBR (with *Microbacterium marinilacus* as biofilm-former) resulted in BOD and COD removal efficiencies of 81.5% and 86%, respectively [61]. The operation of MBBR under a continuous horizontal flow model has been demonstrated to be effective in handling high-strength textile dyeing effluent [67]. The series comprises anaerobic MBBR, aerobic MBBR (no. 1, ozonation, and aerobic MBBR (no. 2. In this scheme, the biodegradability of the micropollutants in the raw effluent was enhanced by the anaerobic MBBR, the aerobic MBBRs (no.1 and no. 2 ensured that ammonia and COD would be eliminated while ozonation ensured the biodegradability of effluent aerobic MBBR (no. 1.. The removal efficiencies of 96.3, 85.3, 97.8, and 94.3% for colour, ammonia, suspended solids, and COD, respectively, was achieved under optimal working conditions (HRT of 14 h for anaerobic MBBR and no. 1 aerobic MBBR, ozonation period of 14 min and HRT of 10 h for no. 2 aerobic MBBR) during treatment of dyeing effluent.

4.2.3 Moving Bed-Sequencing Batch Biofilm Reactor (MB-SBBR)

Operating MMBR in a sequencing batch mode has been deployed in the treatment of wastewater. The biological treatment systems under this mode (i.e., MB-SBBR) gain from the numerous benefits of both processes [52]. The merits of sequence batch reactor comprise flexibility of operation, compact volume, ease of controlling operational parameters, etc. [52]. The advantages of MBBR include ease of upgrade of the existing activated sludge without resorting to building new facilities, no intermittent backwashing, low head losses, no filter bed channeling, etc. [139]. Hosseini Koupaie et al. [78] hinted that MB-SBBR deployed in the treatment of wastewater containing Acid Red 18 (azo dye) resulted in removal efficiencies of 80 and 98% for COD of anaerobically formed aromatic metabolites and Acid Red decolourization, respectively. The investigators further noted that the deployed stabilization system realized ≥ 65–72% mineralization of dye total metabolites. Ong et al. [135] reported 97% removal of COD and total biodecolourization during treatment of Reactive Orange 16 containing wastewater with MB-SBBR.

4.2.4 Fixed Bed Reactor (FBR)

Also known as a packed bed reactor or fixed-bed biofilm reactor, a fixed-bed reactor (FRB) is a common biofilm-based reactor with tightly packed solid media supports in which biofilms colonize and provides a high interface between the biofilm mass and the liquid. Fixed-bed reactors can achieve high biofilm mass capable of clogging the fixed bed. Trickle-bed biofilm bioreactor (TBR) is among the oldest types of FBR [103]. Media used in TBR are usually plastic, rock, ceramics, and other materials where biofilm develops. The biofilm forms extracellular products that protect the microorganisms from wastewater toxicity and allow pollutants immobilization on the fixed bed [149]. In TBR, wastewater flows downward via a distribution network over a fixed media bearing a biofilm surface. Pollutants present in the wastewater get metabolized as it diffuses through the biofilms [7]. Oxygen diffuses either upward or downward through the water to reach the biofilm. Net production of suspended solids in TBR requires a liquid–solid separation via a clarifier. The biofilm in the reactors may not have enough feed in certain areas and may cause reduced productivity.

4.2.5 Fixed-Bed Sequencing Batch Biofilm Reactor (FB-SBBR)

The fixed-bed sequencing batch biofilm reactor could be successfully applied in the treatment of textile dyeing wastewater. According to Hosseini Koupaie et al. [80], coupled anaerobic/aerobic FB-SBBR furnished with polyethylene media exhibited high elimination efficiencies of 95 and 92% for azo dye (Acid Red 18) and COD, respectively. The investigators further noted that the performance of FB-SBBR largely depends on the packing media used. For instance, the system operating with polyethylene as packing media outperformed that with volcanic pumice stones in regards to COD elimination and the removal of sulfonated aromatic metabolites especially amines produced under anaerobic condition.

4.2.6 Biofilm-Based Hybrid Systems in Dye Removal

These systems involve the use of microorganisms grown as biofilms in conjunction with other chemical and physical wastewater treatment procedures to effectively treat the effluents. According to Kapdan and Alparslan [90], full-scale application of most biological conventional waste treatment processes is not effective when singly used for this process because of the dyeing water characteristics. There arises the need for the modification of the procedures for improved efficiency in toxic dye wastewater treatment procedures. A hybrid system of the anaerobic/aerobic process in series has been recommended for bacterial dye degradation [107, 125]. A study carried out by Logrono et al. [115] integrated biological methods and advanced physicochemical methods towards the reduction of COD up to 98% as well as heavy metals (54–80% chromium and 98% zinc) in wastewater. Also reported was the cathodic decolourization of chromophores in azo-bonded reactive dyes from textile industries by electron transfer from cyclic voltammetry on hanging drop electrodes [183].

Furthermore, inexpensive adsorbents such as ground shell powder applied in biosorption of textile effluents yielded 94.8% decolourization efficiency in an anaerobic sequential batch reactor [151], and biofilms can adsorb to the surface of these materials. Because of the low cost of these adsorbent materials, they can easily be available to developing and underdeveloped nations where chemical coagulants or adsorbents cannot be easily affordable. Kurade et al. [101] on the other hand, reported efficient textile effluents decolourization by yeast consortium in a triple-layered fixed-bed reactor compared to a single organism. Thus, a combination of conventional chemical or physical treatment methods integrated with biological treatments was able to improve dye decolourization efficacy. Biofilm-based hybrid wastewater biological treatments proved to be cost-effective and eco-friendly methods that degrade hazardous contaminants thus serving as an alternative to other conventional methods [41]. The biofilm also forms a barrier that protects the inhabiting microorganisms from toxic wastewaters to improve the immobilization and degradation of the pollutants [8]. The hybrid systems improve the degradation, coagulation, and flocculation of wastewater pollutants for easy sedimentation and separation from the effluents.

4.2.7 Biofilm-Coagulation Hybrid

In this treatment procedure, coagulants and flocculants are used in conjunction with biological treatment with biofilms. The microbial or enzymatic systems provided by the biofilms which adsorb the dyes in the effluent by the attractive forces created by their cell wall's functional groups with the dye components [86] or employ their enzymes to turn them into harmless and simpler products [159]. The coagulants (metal salts and some natural products like *Moringa oleifera* Lam seed extracts), added here under vigorous mixing conditions reduce or neutralize the charges of the dispersed fine particles [51]. Then, gently mixing with flocculants gather the coagulated particles into larger flocs that can be easily separated by sedimentation [2, 23, 38]. According to Chellam and Sari [34], the electrocoagulation process can be in-situ applied to coagulate particles in the wastewater using metal anodes as coagulant agents or catalysts. This process evolves hydrogen gas which helps to float the coagulated particles to generate clear, colourless, and odourless water. The process is advantageous because it requires no chemicals, treats large volumes of water, works under ambient, is cost-effective, eco-friendly, and produces a lesser amount of sludge [70, 74].

Chemical coagulants such as alums, poly aluminium chloride composite coagulant poly aluminium ferric chloride (PAFC), and poly ferric sulphate have been applied for the coagulation of dye in wastewaters and were effective removing agents during the treatment processes [63, 69]. These synthetic chemicals with non-native nature cause some environmental and health hazards such as leaving behind some eco-destructive residues after utilizing them in wastewater treatments. The use of natural coagulants, however, provides a promising, sustainable, eco-friendly, and cost-effective alternative to chemical coagulants [65, 130]. Natural coagulants are biodegradable and consumable since they are of plant origin and some can perform actively at room temperature compared to chemical coagulants whose flocculation performances are better at warmer temperatures [55, 59, 65].

4.2.8   Biofilm-Oxidation Hybrid

Biofilm-based microorganisms can combine with other different treatment methods to form a hybrid of the biofilm and the physical or chemical treatment methods. Recently, wastewater treatment and recovery are being modified constantly for effectiveness improvement [104]. Studies by Hai et al. [69] suggested that a sequential or concurrent combination of advanced oxidation processes (AOPs) with coagulation was effective in the treatment of dye wastewater. Also, the combination of different quantities of natural and chemical coagulants for wastewater treatment as suggested by Saritha et al. (2020), removed about 99.2% turbidity of the wastewater. The AOPs provide radicals which in turn reduces the organic components of the wastewater into harmless products and this process has been generally recognized as an efficient treatment approach for recalcitrant wastewater [1, 29, 104].

The most widely used electrochemical AOP known as the Electro-Fenton process as reported by Rosales et al. [156], removes organic wastes through a combined two-stage process of oxidation and coagulation. This hybridization of electrolysis and Fenton's oxidizes the organic contents of the wastes and by Fenton's reaction reduces the wastes into harmless components as flocs to be further separated by sedimentation and filtration. According to Khataee et al. [97], these processes produce a lesser amount of secondary pollutants, are thus eco-friendly, do not involve the use of chemicals, and involve good process control. This procedure in combination with biofilms, bring about effective organic substances removal from the dye wastewater.

4.2.9   Biofilm-Denitrification Hybrid

The biofilm-denitrification hybrid involves aerobic and anaerobic removal of nitrogen from the dye wastewater with electro-active microbial biofilm immobilized on particle carriers which have a high affinity for outer-membrane cytochromes of bacteria, an important microbial electron transfer mediator [194]. Different layers of microbial communities formed by the biofilm growth on these carriers carry out their different metabolic activities depending on the substrates available to them. Aerobic organisms that grow at the outer surface of the biofilm carry out aerobic denitrification while those that grow under and closer to the carriers carry out anaerobic denitrification of the nitrogen-containing components of the wastewater. This aerobic and anoxic denitrification in biofilms can take place within the same compartment depending on the dissolved oxygen gradient of the compartment [12, 134]. Hydrogen generated from the anode provides electrons for the autotrophic bacterial during denitrification, while the CO_2 produced at the cathode by oxidation of the organic components of the wastewater serves as a pH buffer as well as an additional carbon source as suggested by Wang and Qu [192].

De-ammonification biological treatment process involving two-step bioconversion of ammonia to nitrogen excluding the need for carbon sources and at low air level than in nitrification, also help to remove nitrogen from wastewaters [134, 208]. These steps are usually referred to as the nitritation process which involves aerobic ammonia-N (NH_4-N) oxidation to nitrate-nitrogen (NO_2-N) by aerobic autotrophic ammonia-oxidizing bacteria (AerAOB) and anaerobic ammonia oxidation (anammox) reaction where NH_4-N is anaerobically oxidized to dinitrogen gas by anaerobic ammonia-oxidizing bacteria (AnAOB) using NO_2-N as electron acceptors [81, 132]. The process of deammonification according to Zhang et al. [208], is environmentally sustainable, economically viable, requires a reduced amount of oxygen for nitrogen removal, and reduced sludge production compared to nitrification–denitrification processes. The process is a complete autotrophic process that does not require additional carbon sources.

4.2.10 Biofilm-Ozonation Hybrid

Ozonation is an advanced oxidative process where recalcitrant and conjugated chains responsible for dye colours are degraded by applying ozone to the dye wastewater. According to Turhan and Ozturkcan [184], molecular ozone, $O_2^{\cdot}$, and OH$^{\cdot}$ radicals are the major components involved in the ozonation process. The process can be applied as a pre-treatment procedure or as a final polishing treatment approach for recalcitrant wastewater. Although complete azo dye mineralization generates stable by-products making their removal difficult [207], the ozonation process decomposes recalcitrant compounds in wastewaters to enhance its biodegradability and reduce treated effluent's toxicity [67]. Being a strong oxidant, ozone rapidly reacts with water yielding no solid residue to decompose recalcitrant substances into easily decomposable ones that can be further mineralized by microorganisms. So, incorporation of biofilm into ozone-treated dye wastewater will completely remove the dye or completely mineralize the ozonation by-products into harmless components such as carbon dioxide and water [18].

The ozonation process produces high-quality water after using it for wastewater treatments however, some refractory by-products in the form of trace organic contaminants such as pharmaceuticals, with high toxicity than the initial compounds in the treated water sometimes appear during the reaction process [177, 181]. This usually presents the need for further polishing of the ozonated water where the ozone completely mineralizes degradable organics in the water, Tang and colleagues also opined. According to Bui et al. [21], ozonation is an effective method of pharmaceuticals removal during wastewater treatment at low operational costs. A hybrid of ozonation and microbial biofilms completely removes organics and any secondary refractory by-products that may appear during the treatment procedures. This combination enhances the process efficiency for the degradation, detoxification, and mineralization of industrial effluents before releasing them into the environment [33, 120].

4.2.11   Dye Reductase and Dehydrogenase in Biofilm

Reductases and dehydrogenases are biological enzymes that catalyze oxidation or reduction reactions through the transfer of electrons during the reactions and are thus generally referred to as oxidoreductases. In dye wastewater treatment, oxidoreductive enzymes with microbial origin proved to be an effective means of detoxification and colour removal to meet the required standard before discharging the treated effluent into the environment. These microbial enzymes according to Behbudi et al. [13], act as biocatalysts for biochemical and metabolic reactions in an eco-friendly manner. Thus, their actions do not deteriorate the environment but completely mineralize the polluting agent in the effluents. The enzymes can be applied either as whole organisms or their enzymes used to neutralize the contaminants during bioremediation processes.

Microorganisms are the primary source of bioremediation enzymes, although plants are sometimes employed to remediate some pollutants [112, 122, 201]. Microbial enzymes show high efficiency and performance leaving little or no by-products which can be easily removed by sedimentation or filtration, also easy to produce and their genetic manipulation of microbial genes enhance their enzyme production and application capacities [112]. Some of these enzymes are substrate specific or show broad activity by transforming or precipitating recalcitrant pollutants into innocuous products, more treatment amenable intermediates, or aid the bioconversion of waste substrates into value-added products [91].

Bioremediation is an eco-friendly and cost-effective method of pollutants removal powered by microbial enzymes especially oxidoreductases (dehydrogenases, hydroxylases, and oxidases) and hydrolases (esterases, glutathiones, and haloalkanes), and catalyze oxidation, reduction, and bond cleavage reactions during the process [13, 92]. It's an important procedure for the removal of recalcitrant organic pollutants and xenobiotics from industrial effluents to make it safe for re-introduction into the ecosystem. This method is simple to implement, fast, more reliable, and cheap for the effective removal of some pollutants recalcitrant to conventional biological and chemical methods [91]. The use of enzymes in bioremediation offers a better option because it can be applied over a wide temperature, pH, and salinity ranges, used on bio-refractory compounds, presents no shock loading effects, operate at high and low contaminant concentrations, ease of process control as well as remove delays in biomass acclimatization when inoculated in the contaminated environment. According to Karigar and Rao [92], the application of enzymes in bioremediation will help in developing advanced biotechnology processes that will aid pollutants toxicity reduction and produce novel useful substances.

4.2.12   Biofilm Beneficiation in Dye Removal

Dye removal using biofilm (i.e., microbial biotechnology) is one of the important procedures for the bioremediation of dye-laden textile wastewater. Microbes naturally exist as biofilms to adapt and survive in their environments conferring on them some robust characteristics. During dye removal, microorganisms growing in biofilms show adsorption properties as well as degradation or modification of the pollutants. For instance, Quintelas et al. [148] reported that bacteria immobilized on activated carbon as biofilm support displayed a synergistic effect to degrade or retain organic compounds during the biosorption procedure. This allows for the re-usability of the enzyme or the microorganisms, requires no chemical input, quickly adapts to wastewater input variations, and is cost-effective [145]. The sequential sorption followed by biodegradation of the pollutants in the dye water effluent will enhance dye removal and detoxification with no further production of secondary pollutants during the process [127].

The application of biofilms in dye removal help to reduce their toxicity, improve human and animal health and preserve the ecosystem [190], thus eliminating the dye's toxicity to the flora and fauna of the ecosystem as well as the carcinogenic and mutagenic effects of these dyes. Dye wastewater pollutants recalcitrant to conventional methods of removal are usually mineralized or made innocuous by treating them with microbial biofilms. A combination of biofilm with other methods of dye removal has shown to be more effective and reliable in its decolorization and detoxification to achieve the desired level of treatment required for the effluent [93, 186]. Immobilization of microbial cells in a biofilm community offers a one-stage removal of carbon and nutrients since they appear in an organized species-rich structure where aerobes grow at the outer surface while the facultative and obligate anaerobes grow in the deeper layer of the biofilm. So, concurrent aerobiosis and anaerobiosis take place in the biofilm to remove a wide range of wastewater contents. Microbial cells immobilized by adsorption or by entrapment offers easy separation of the microbes from effluents after treatment for future re-use, better operational stability, and catalytic efficiency more than planktonic cells [171, 178].

4.3   Trending Approaches in Dye Removal via Biofilms

Numerous studies have been carried out on wastewater treatment using single cells or a combination of pure cultures and other physicochemical methods [164]. The result of this procedure proved to be an effective method but recent advancements using microbial consortium or mixed microbial communities of microbes displayed a better approach for dye effluent treatment [84, 165]. Microorganisms from these combinations were able to tolerate salts and metal ions with other stress conditions in textile effluents making them robust to carry out the desired functions Imran and colleagues further reported. Some researchers also reported the effectiveness of integrating biological treatment methods and advanced oxidation processes for the effective treatment of recalcitrant industrial wastewaters [104]. These combinations bring about concurrent oxidation of the wastes present in the effluent and completely mineralize any residual pollutants that may be generated from the process.

Bioremediation studies using purified microbial enzymes supported on matrix need to be used in the real world and on a commercial scale to determine their effectiveness [84]. Microbial enzymes are more effective and easily adaptive to the environment and operate in a wide range of temperature, pH, and other stress conditions. Thus, they perform better than growing microorganisms on the effluent to be treated and can be used to improve existing technologies for better effectiveness. Also, immobilizing of purified enzymes on nanoparticles enhanced their performance in the decolourization of textile wastewaters [178, 194]. Studies on microbial metabolic pathways involved in textile wastewater biodecolorization will help to identify the genes and metabolites involved for better understanding and manipulation of these pathways to develop effective microbial isolates that can decolorize several synthetic dyes in mixed contaminants.

The microbial fuel cell (MFCs) utilizes the concurrent treatment of effluents and energy generation by electrogenic microbes to remove the pollutant from the effluent [82]. Bioenergy recovery from wastewaters has been the major application of MFC and some pollutants such as azo dyes contained in the effluent help in extracellular electric transfer from the microbial cells to the anode to generate electricity [114, 126].

5   Conclusion

Dye wastewater discharge into water bodies without proper treatments poses a serious threat to these environments. The degradation of the dye components found in wastewaters has been reported to be dependent on the physical properties of the dye such as type of dye, initial concentration, salinity, pH, and the presence of toxic secondary pollutants which may be generated from the degradation process. Various biological and physicochemical treatment methods have been applied by researchers and industries but none of these techniques have proved to be an effective method for the complete mineralization of the dye component. Bioremediation processes have also proved to be an easy, simple, economically viable, and eco-friendly means of dye removal with its intermediate metabolites detoxification and mineralization especially enzyme-mediated procedures. The organisms that grow as biofilms on different carriers offer the advantage of cell wash-off and continuous production of the enzymes needed for dye removal. These microbial biofilms have been used in integrated hybrid systems for effective management of dye wastewater when there's a need for the re-use of treated water from the effluent. A hybrid system considers all the factors necessary for effective dye detoxification and mineralization to recover water of high quality and safe to be re-used for other purposes. Thus, a combination of effluent treatment methods such as bioremediation protocols involving a consortium of microorganisms or their enzymes to remove the non-biodegradable portion of the wastes proved to be an effective and eco-friendly method of dye impacts and intermediate metabolites removal procedures. The use of pure culture and co-culture consortium have been used by different researchers and co-culture consortia showed high colour decolourization efficiency due to their adaptation to stressful and toxic environments created by the effluents to co-metabolize and completely mineralize the pollutant. There is a need for hybrid systems that will involve pre-treatment of the effluents with physicochemical methods and integrated with biological methods to effectively manage dye waste effluents. Further optimization of other process parameters involved in dye decolourization is necessary to efficiently manage pollutants during dye wastewater treatment processes.

References

1. Abiri F, Fallah N, Bonakdarpour B (2017) Sequential anaerobic-aerobic biological treatment of colored wastewaters: case study of a textile dyeing factory wastewater. Water Sci Technol 75:1261–1269
2. Ahmad A, Mohd-Setapar SH, Chuong CS, Khatoon A, Wani WA, Kumar R, Rafatullah M (2015) Recent advances in new generation dye removal technologies: novel search for approaches to reprocess wastewater. RSC Adv 5(39):30801–30818
3. Akiyama S, Nakatsuji S, Nakashima K, Yamasaki S (1988) Diphenylmethane and triphenylmethane dye ethynovinylogues with absorption bands in the near infrared. Dyes Pigm 9:459–466
4. Alatraktchi FA, Zhang Y, Angelidaki I (2014) Nano modification of the electrodes in microbial fuel cell: impact of nanoparticle density on electricity production and microbial community. Appl Energy 116:216–222
5. Ansari FA, Jafri H, Ahmad I, Abulreesh HH (2017) Factors affecting biofilm formation in in vitro and in the rhizosphere. In: Ahmad I, Husain, FM (eds) Biofilms in plant and soil health, 1st ed., chapt. 15, pp 275–290
6. Alkan M, Celikcapa S, Demirbas OZ, Dogan M (2005) Removal of reactive blue 221 and acid blue 62 anionic dyes from aqueous solutions by sepiolite. Dyes Pigm 65:251–259
7. Asri M, Elabed A, El Ghachtouli N, Koraichi SI, Bahafid W, Elabed S (2017) Theoretical and experimental adhesion of yeast strains with high chromium removal potential. Environ Eng Sci 34(10):693–702

8. Asri M, Elabed S, Koraichi SI, El Ghachtouli N (2018) Biofilm-based systems for industrial wastewater. In: Hussain CM (ed) Handbook of environmental materials management treatment. Springer International Publishing AG, pp 1–21
9. Ayati B, Ganjidoust H, Fattah M (2007) Degradation of aromatic compounds using moving bed biofilm reactors. Iranian J Environ Health Sci Eng 4:107–112
10. Azbar N, Yonar T, Kestioglu K (2004) Comparison of various advanced oxidation processes and chemical treatment methods for COD and colour removal from a polyester and acetate fiber dyeing effluent. Chemosphere 55:35–43
11. Bahafid W, Joutey NT, Sayel H, Iraqui-Houssani M, El Ghachtouli N (2013) Chromium adsorption by three yeast strains isolated from sediments in Morocco. Geomicrobiol J 30:422–429
12. Barwal A, Chaudhary R (2014) To study the performance of bio-carriers in moving bed biofilm reactor (MBBR) technology and kinetics of biofilm for retrofitting the existing aerobic treatment systems: a review. Rev Environ Sci Biotechnol 13:285–299
13. Behbudi G, Yousefi K, Sadeghipour Y (2021) Microbial enzymes based technologies for bioremediation of pollutions. J Environ Treat Tech 9(2):463–469
14. Berradi M, Hsissou R, Khudhair M, Assouag M, Cherkaoui O, El Bachiri A, El HarfiA (2019) Textile finishing dyes and their impact on aquatic environs. Heliyon 5(11):e02711
15. Berradi M, El Harfi A (2016) Review of the water resources, pollution sources and wastewaters of the Moroccan textile industry. J Water Sci Environ Technol 1(2):80–86
16. Berradi M, Cherkaoui O, El Harfi A (2017) Comparative study the performance of ultrafiltration asymmetric membranes. Application to the bleaching of colored water with vat dyes. Appl J Environ Eng Sci 3(2):114–120
17. Berradi M, Essamri A, El Harfi A (2016) Discoloration of water loaded with vat dyes by the membrane process of ultrafiltration. J Mater Environ Sci 7(4):1098–1106
18. Bhatia D, Sharma NR, Singh J, Kanwar RS (2017) Biological methods for textile dye removal from wastewater: a review. Crit Rev Environ Sci Technol 47(19):1836–1876
19. Bilal M, Rasheed T, Iqbal HMN, Li C, Wang H, Hu H, Wang W, Zhang X (2018) Photocatalytic degradation, toxicological assessment and degradation pathway of C.I. Reactive blue 19 dye. Chem Eng Res Des 129:384–390
20. Broch-Due A, Andersen R, Opheim B (1997) Treatment of integrated newsprint mill wastewater in moving bed biofilm reactors. Water Sci Technol 35(2–3):173–180
21. Bui XT, Vo TPT, Ngo HH, Guo WS, Nguyen TT (2016) Multi-criteria assessment of advanced treatment technologies for micro pollutants removal at large-scale applications. Sci Total Environ 564:1050–1067
22. Burkinshaw SM, Son YA (2010) The dyeing of super microfiber nylon with acid and vat dyes. Dyes Pigm 87:132–138
23. Butani SA, Mane SJ (2017) Coagulation/flocculation process for cationic, anionic dye removal using water treatment residuals–a review. Int J Technol Manag 6:121–125
24. Cardoso NF, Lima EC, Calvete T, Pinto IS, Amavisca CV, Fernandes THM, Pinto RB, Alencar WS (2011) Application of aqai stalks as biosorbents for the removal of the dyes reactive black 5 and reactive orange 16 from aqueous solution. J Chem Eng Data 56(5):1857–1868
25. Cardoso NF, Lima EC, Pinto IS, Amavisca CV, Royer B, Pinto RB, Alencar WS, Pereira SFP (2011) Application of cupuassu shell as biosorbents for theremoval of textile dyes from aqueous solution. J Environ Manage 92(4):1237–1247
26. Cardoso NF, Lima EC, Royer B, Bach MV, Dotto GL, Pinto LAA, Calvete T (2012) Comparison of *Spirulina platensis* microalgae and commercial activated carbon as adsorbents for the removal of reactive red 120 dye from aqueous effluents. J Hazard Mater 241–242:146–153
27. Carneiro PA, Umbuzeiro GA, Oliveira DP, Zanoni MVB (2010) Assessment of water contamination caused by a mutagenic textile effluent/dyehouse effluent bearing disperses dyes. J Hazard Mater 174(1–3):694–699
28. Casas ME, Chhetri RK, Ooi G, Hansen KM, Litty K, Christensson M, Kragelund C, Andersen HR, Bester K (2015) Biodegradation of pharmaceuticals in hospital wastewater by staged moving bed biofilm reactors (MBBR). Water Res 83(10):293–302

29. Cao Y, Kwok BH, van Loosdrecht M, Daigger G, Png HY, Long WY, Eng OK (2018) The influence of dissolved oxygen on partial nitritation/anammox performance and microbial community of the 200,000 m3/d activated sludge process at the Changi water reclamation plant (2011 to 2016). Water Sci Techno 78(3):634–643

30. Castro FD, Bassin JP, Alves TLM, Sant'Anna GL, Dezotti M (2021) Reactive Orange 16 dye degradation in anaerobic and aerobic MBBR coupled with ozonation: addressing pathways and performance. Int J Environ Sci Technol 18:1991–2010

31. Chaari I, Feki M, Medhioub M, Bouzid J, Fakhfakh E, Jamoussi F (2009) Adsorption of a textile dye "indanthrene blue RS (C.I. vat blue 4)" from aqueous solutions onto smectite-rich clayey rock. J Hazard Mater 172:1623–1628

32. Chakraborty JN (2010) Dye-fibre interaction. Fundam Pract Colouration Text 2010:20–28

33. Chang MW, Chern JM (2010) Decolourization of peach red azo dye, HF6 by Fenton reaction: initial rate analysis. J Taiwan Inst Chem Eng 41(2):221–228

34. Chellam S, Sari MA (2016) Aluminium electrocoagulation as pre-treatment during microfiltration of surface water containing NOM: a review of fouling, NOM, DBP and virus control. J Hazard Mater 304:490–501

35. Chen S, Liu J (2006) Landfill leachate treatment by MBR: performance and molecular weight distribution of organic contaminant. China Sci Bull 51:2831–2838

36. Chen S, Sun D, Jong-Shik C (2007) Treatment of pesticide wastewater by moving bed biofilm reactor combined with Fenton-coagulation pre-treatment. J Hazard Mater 144:577–584

37. Chu L, Wang J, Quan F, Xing XH, Tang L, Zhang C (2014) Modification of polyurethane foam carriers and application in a moving bed biofilm reactor. Process Biochemistry 49(11):1979–1982

38. Collivignarelli MC, Abbà A, Miino MC, Damiani S (2019) Treatments for color removal from wastewater: state of the art. J Environ Manage 236:727–745

39. Croce R, Cin F, Lombardo A, Crispeyn G, Cappelli CI, Vian M, Maiorana S, Benfenati E, Baderna D (2017) Aquatic toxicity of several textile dye formulations: acute and chronic assays with Daphnia magna and Raphidocelis subcapitata. Ecotoxicol Environ Saf 144:79–87

40. Damodar RA, You SJ, Ou SH (2010) Coupling of membrane separation with photocatalytic slurry reactor for advanced dye wastewater treatment. Sep Purif Technol 76(1):64–71

41. Das N, Basak LG, Salam JA, Abigail EA (2012) Application of biofilms on remediation of pollutants–an overview. J Microbiol Biotechnol Res 2:783–790

42. de Aragão Umbuzeiro G, Freeman HS, Warren SH, de Oliveira DP, Terao Y, Watanabe T, Claxton LD (2005) The contribution of azo dyes to the mutagenic activity of the Cristais river. Chemosphere 60:55–64

43. De Lima ROA, Bazo AP, Salvadori DMF, Rech CM, Oliveira DP, Umbuzeiro GA (2007) Mutagenic and carcinogenic potential of a textile azo dye processing plant effluent that impacts a drinking water source. Mutat Res Genet Toxicol Environ Mutagen 626(12):53–60

44. Delnavaz M, Ayati B, Ganjidoust H (2010) Prediction of moving bed biofilm reactor (MBBR) performance for the treatment of aniline using artificial neural networks (ANN). J Hazard Mater 179:769–775

45. Deng L, Guo W, Ngo HH, Zhang X, Wang XC, Zhang Q, Chen R (2016) New functional biocarriers for enhancing the performance of a hybrid moving bed biofilm reactor-membrane bioreactor system. Bioresour Technol 208:87–93

46. Di Fabio S, Lampis S, Zanetti L, Cecchia F, Fatone F (2013) Role and characteristics of problematic biofilms within the removal and mobility of trace metals in a pilot-scale membrane bioreactor. Process Biochem 48(11):1757–1766

47. Dias NC, Alves TLM, Azevedo DA, Bassin JP, Dezotti M (2020) Metabolization of by-products formed by ozonation of the azo dye Reactive Red 239 in moving-bed biofilm reactors in series. Braz J Chem Eng 37:495–504

48. Dong B, Chen H, Yang Y, He Q, Dai X (2014) Treatment of printing and dyeing wastewater using MBBR followed by membrane separation process. Desalin Water Treat 52(22–24):4562–4567

49. dos Santos AB, Cervantes FJ, van Lier JB (2007) Review paper on current technologies for decolourisation of textile wastewaters: perspectives for anaerobic biotechnology. Bioresour Technol 98:2369–2385
50. Donlan RM, Costerton JW (2002) Biofilms: survival mechanisms of clinically relevant microorganisms. Clin Microbiol Rev 15(2):167–193
51. Dotto J, Fagundes-Klen MR, Veit MT, Palácio SM, Bergamasco R (2019) Performance of different coagulants in the coagulation/flocculation process of textile wastewater. J Clean Prod 208:656–665
52. Dulkadiroglu H, Cokgor EU, Artan N, Orhon D (2005) The effect of temperature and sludge age on COD removal and nitrification in a moving bed sequencing batch biofilm reactor. Water Sci Technol 51(11):95–103
53. El Abed S, Ibnsouda KS, Latrache H, Boutahari S (2012) Theoretical effect of cedar wood surface roughness on the adhesion of conidia from *Penicillium expansum*. Ann Microbiol 62(4):1361–1366
54. Eren Z (2012) Ultrasound as a basic and auxiliary process for dye remediation: a review. J Environ Manage 104:127–141
55. Feria-Diaz JJ, Tavera-Quiroz MJ, Vergara-Suarez O (2018) Efficiency of chitosan as a coagulant for wastewater from slaughterhouses. Indian J Sci Technol 11(3):1–12
56. Fernando E, Keshavarz T, Kyazze G (2014) Complete degradation of the azo dye Acid Orange-7 and bioelectricity generation in an integrated microbial fuel cell, aerobic two-stage bioreactor system in continuous flow mode at ambient temperature. Bioresour Technol 156:155–162
57. Feng G, Cheng Y, Wang SY, Borca-Tasciuc DA, Worobo RW, Moraru CI (2015). Bacterial attachment and biofilm formation on surfaces are reduced by smalldiameter nanoscale pores: how small is small enough? npj Biofilms and Microbiomes 1(1):1–9
58. Ferronato C, Silva B, Costa F, Tavares T (2016) Vermiculite bio-barriers for Cu and Zn remediation: an eco-friendly approach for freshwater and sediments protection. Int J Environ Sci Technol 13:1219–1228
59. Fitzpatrick CS, Fradin E, Gregory J (2004) Temperature effects on flocculation, using different coagulants. Water Sci Technol 50:171–175
60. Forrez I, Carballa M, Boon N, Verstraete W (2009) Biological removal of 17α-ethinylestradiol (EE2) in an aerated nitrifying fixed bed reactor during ammonium starvation. J Chem Technol Biotechnol 84:119–125
61. Francis A, Sosamony KJ (2016) Treatment of pre-treated textile wastewater using moving bed bio-film reactor. Procedia Technol 24:248–255
62. Franciscon E, Zille A, Fantinatti-Garboggini F, Silva IS, Cavaco-Paulo A, Durrant LR (2009) Microaerophilic–aerobic sequential decolourization/biodegradation of textile azo dyes by a facultative *Klebsiella* sp. strain VN-31. Process Biochem 44:446–452
63. Gao B, Yue Q, Miao J (2001) Evaluation of polyaluminium ferric chloride (PAFC) as a composite coagulant for water and wastewater treatment. Water Sci Technol 47(1):127–132
64. Garcia-Montano J, Torrades F, Perez-Estrada LA, Oller I, Malato S, Maldonado MI, Peral J (2008) Degradation path-ways of the commercial reactive azo dye Procion Red H-E7B undersolar-assisted photo-Fenton reaction. Environ Sci Technol 42:6663–6670
65. Gautam S, Saini G (2020) Use of natural coagulants for industrial wastewater treatment. Glob J Environ Sci Manag 6(4):553–578
66. Ghaly AE, Ananthashankar R, Alhattab M, Ramakrishnan VV (2014) Production, characterization and treatment of textile effluents: a critical review. J Chem Eng Process Technol 5:182
67. Gong XB (2016) Advanced treatment of textile dyeing wastewater through the combination of moving bed biofilm reactors and ozonation. Sep Sci Technol 51(9):1589–1597
68. Guo W, Ngo HH, Dharmawan F, Palmer CG (2010) Roles of polyurethane foam in aerobic moving and fixed bed bioreactors. Bioresour Technol 101:1435–1439
69. Hai FI, Yamamoto K, Fukushi K (2007) Hybrid treatment systems for dye wastewater. Crit Rev Environ Sci Technol 37(4):315–377

70. Hamad H, Bassyouni D, El-Ashtoukhy ES, Amin N, El-Latif MA (2017) Comparative performance of anodic oxidation and electrocoagulation as clean processes for electro-catalytic degradation of diazo dye Acid Brown 14 in aqueous medium. J Hazard Mater 335:178–187

71. Hapeshi E, Lambrianides A, Koutsoftas P, Kastanos E, Michael C, Fatta-Kassinos D (2013) Investigating the fate of iodinated X-ray contrast media iohexol and diatrizoate during microbial degradation in an MBBR system treating urban wastewater. Environ Sci Pollution Research 20(6):3592–3606

72. Hall-Stoodley L, Stoodley P (2002) Developmental regulation of microbial biofilms. Current Opinion in Biotechnology 13(3):228–233

73. Hassaan MA, El Nemr A, Madkour FF (2017) Testing the advanced oxidation processes on the degradation of direct blue 86 dye in wastewater. Egypt J Aquat Res 43(1):11–19

74. He CC, Hu CY, Lo SL (2016) Evaluation of sono-electrocoagulation for the removal of Reactive Blue 19 passive film removed by ultrasound. Sep Purif Technol 165:107–113

75. Hoh D, Watson S, Kan E (2016) Algal biofilm reactors for integrated wastewater treatment and biofuel production: a review. Chem Eng J 287:466–473

76. Holkar CR, Jadhav AJ, Pinjari DV, Mahamuni NM, Pandit AB (2016) A critical review on textile wastewater treatments: possible approaches. J Environ Manage 182:351–366

77. Hossain L, Sarker SK, Khan M (2018) Evaluation of present and future wastewater impacts of textile dyeing industries in Bangladesh. Environ Dev 26:23–33

78. Hosseini Koupaie E, Alavi Moghaddam MR, Hashemi SH (2011) Post-treatment of anaerobically degraded azo dye Acid Red 18 using aerobic moving bed biofilm process: enhanced removal of aromatic amines. J Hazard Mater 195:147–154

79. Hosseini Koupaie E, Alavi Moghaddam MR, Hashemi SH (2012) Investigation of decolorization kinetics and biodegradation of azo dye Acid Red 18 using sequential process of anaerobic sequencing batch reactor/moving bed sequencing batch biofilm reactor. Int Biodeterior Biodegrad 71:43–49

80. Hosseini Koupaie E, Alavi Moghaddam MR, Hashemi SH (2013) Evaluation of integrated anaerobic/aerobic fixed-bed sequencing batch biofilm reactor for decolorization and biodegradation of azo dye Acid Red 18: comparison of using two types of packing media. Bioresour Technol 127:415–421

81. Hu B, Shen L, Xu X, Zheng P (2011) Anaerobic ammonium oxidation (anammox) in different natural ecosystems. Biochem Soc Trans 39:1811–1816

82. Huarachi-Olivera R, Dueñas-Gonza A, Yapo-Pari U, Vega P, Romero-Ugarte M, Tapia J, Molina L, Lazarte-Rivera A, Pacheco-Salazar DG, Esparza M (2018) Bioelectrogenesis with microbial fuel cells (MFCs) using the microalga *Chlorella vulgaris* and bacterial communities. Electron J Biotechnol 31:34–43

83. Hussein FH (2013) Chemical properties of treated textile dyeing wastewater. Asian J Chem 25:9393–9400

84. Imran M, Crowley DE, Khalid A, Sabir Hussain S, Mumtaz MW, Arshad A (2014) Microbial biotechnology for decolorization of textile wastewaters. Rev Environ Sci Biotechnol 14:73–92

85. Iyim TB, Güçlü G (2009) Removal of basic dyes from aqueous solutions using natural clay. Desalination 249:1377–1379

86. Jafari N, Soudi MR, Kasra-Kermanshahi R (2014) Biodegradation perspectives of azo dyes by yeasts. Microbiology 83(5):484–497

87. Jia Y, Zhou M, Chen Y, Hu Y, Luo J (2020) Insight into short-cut of simultaneous nitrification and denitrification process in moving bed biofilm reactor: effects of carbon to nitrogen ratio. Chem Eng J 400:125905

88. Ji Q, Liu G, Jiti ZW, Jing R, Jin HL (2012) Removal of water-insoluble Sudan dyes by *Shewanella oneidensis* MR-1. Bioresour Technol 114:144–148

89. Jonstrup M, Kumar N, Murto M, Mattiasson B (2011) Sequential anaerobic–aerobic treatment of azo dyes: decolourisation and amine degradability. Desalination 280:339–346

90. Kapdan IK, Alparslan S (2005) Application of anaerobic-aerobic sequential treatment system textile wastewater for colour and COD removal. Enzyme Microb Technol 36:273–279

91. Karam J, Nicell JA (1997) Potential applications of enzymes in waste treatment. J Chem Technol Biotechnol 69:141–153
92. Karigar CS, Rao SS (2011) Role of microbial enzymes in the bioremediation of pollutants: a review. Enzyme Res 2011: 805187
93. Kargi F, Ozmıhcı S (2004) Biosorption performance of powdered activated sludge for removal of different dyestuffs. Enzyme Microb Technol 35:267–271
94. Kermani M, Bina B, Movahedian H, Amin MM, Nikaein M (2008) Application of moving bed biofilm process for biological organics and nutrients removal from municipal wastewater. Am J Environ Sci 4(6):682–689
95. Kesaano M, Sims RC (2014) Algal biofilm based technology for wastewater treatment. Algal Res 5:231–240
96. Khandegar V, Saroha AK (2013) Electrocoagulation for the treatment of textile industry effluent—a review. J Environ Manage 128:949–963
97. Khataee AR, Vatanpou V, Ghadim ARA (2009) Decolorization of CI Acid Blue 9 solution by UV/nano-TiO2, Fenton, Fenton-like, electro-Fenton and electrocoagulation processes: a comparative study. J Hazardous Materials 161:1225–1233
98. Khattab TA, Rehan M, Hamouda T (2018) Smart textile framework: photochromic and fluorescent cellulosic fabric printed by strontium aluminate pigment. Carbohydr Polym 195:143–152
99. Krivorot M, Kushmaro A, Oren Y, Gilron J (2011) Factors affecting biofilm formation and biofouling in membrane distillation of seawater. J Membr Sci 376:15–24
100. Kimura K, Hane Y, Watanabe Y (2005) Effect of precoagulation on mitigating irreversible fouling during ultrafiltration of a surface water. Water Sci techno 51(6–7):93–100
101. Kurade MB, Waghmode TR, Patil SM, Jeon BH, Govindwar SP (2017) Monitoring the gradual biodegradation of dyes in a simulated textile effluent and development of a novel triple layered fixed bed reactor using a bacterium-yeast consortium. Chem Eng J 307:1026–1036
102. Le-Clech P, Chen V, Fane TAG (2006) Fouling in membrane bioreactors used in wastewater treatment. J Membr Sci 284:17–53
103. Lewandowski Z, Boltz J (2011) Biofilms in water and wastewater treatment. In: Treatise on water science. Academic, Oxford, pp 529–570
104. Ledakowicz S, Paździor K (2021) Recent achievements in dyes removal focused on advanced oxidation processes integrated with biological methods. Molecules 26(4):870
105. Liang J, Ning X, Kong M, Liu D, Wang G, Cai H, Sun J, Zhang Y, Lu X, Yuan Y (2017) Elimination and eco-toxicity evaluation of phthalic acid esters from textile-dyeing wastewater. Environ Pollut 231:115–122
106. Liao BQ, Kraemer JT, Bagley DM (2006) Anaerobic membrane bioreactors: applications and research directions. Crit Rev Environ Sci Technol 36(6):489–530
107. Libra JA, Borchert M, Vigelahn L, Storm T (2004) Two stage biological treatment of a diazo reactive textile dye and the fate of the dye metabolites. Chemosphere 56(2):167–180
108. Lim SL, Chu WL, Phang SM (2010) Use of Chlorella vulgaris for bioremediation of textile wastewater. Bioresour Technol 101(19):7314–7322
109. Lin H, Gao W, Meng F, Liao BQ, Leung KT, Zhao L, Chen J, Hong H (2012) Membrane bioreactors for industrial wastewater treatment: a critical review. Crit Rev Environ Sci Technol 42:677–740
110. Liu C, Li J, Zhu X, Zhang L, Ye D, Brown RK, Liao Q (2013) Effects of brush lengths and fiber loadings on the performance of microbial fuel cells using graphite fiber brush anodes. Int J Hydrog Energy 38:15646–15652
111. Liu T, Yang X, Wang ZL, Yan X (2013) Enhanced chitosan beads-supported Fe(0)-nanoparticles for removal of heavy metals from electroplating wastewater in permeable reactive barriers. Water Res 47(17):6691–6700
112. Liu X, Kokare C (2017) Microbial enzymes of use in industry. In: Biotechnology of microbial enzymes. Elsevier 2017, pp 267–298
113. Liu Y, Li K, Xu W, Du B, Wei Q, Liu B, Wei D (2020) GO/PEDOT: NaPSS modified cathode as heterogeneous electro-Fenton pre-treatment and subsequently aerobic granular sludge biological degradation for dye wastewater treatment. Sci Total Environ 700:134536

114. Logan B, Rabaey K (2012) Conversion of wastes into bioelectricity and chemicals using microbial electrochemical technologies. Science 337(6095):686–690
115. Logroño W, Pérez M, Urquizo G, Kadier A, Echeverría M, Recalde C, Rákhely G (2017) Single chamber microbial fuel cell (SCMFC) with a cathodic microalgal biofilm: a preliminary assessment of the generation of bioelectricity and biodegradation of real dye textile wastewater. Chemosphere 176:378–388
116. Luo Y, Guo W, Ngo HH, Nghiem LD, Hai FI, Zhang J, Liang S, Wang XC (2014) A review on the occurrence of micropollutants in the aquatic environment and their fate and removal during wastewater treatment. Sci Total Environ 473:619–641
117. Mansour H, Boughzala O, Barillier D, Chekir-Ghedira L, Mosrati R (2011) Les colorants textiles sources de contamination de l'eau: CRIBLAGE de la toxicité et des méthodes de traitement. Revue des sciences de l'eau/Journal of Water Science 24(3):209–238
118. Martin KJ, Nerenberg R (2012) The membrane biofilm reactor (MBfR) for water and wastewater treatment: principles, applications, and recent developments. Bioresour Technol 122:83–94
119. Mahto KU, Das S (2022) Bacterial biofilm and extracellular polymeric substances in the moving bed biofilm reactor for wastewater treatment: a review. Bioresour Technol 345:126476
120. Malik SN, Ghosh PC, Vaidya AN, Mudliar SN (2020) Hybrid ozonation process for industrial wastewater treatment: principles and applications: a review. J Water Process Eng 35:101193
121. Mao Y, Quan X, Zhao H, Zhang Y, Chen S, Liu T, Quan, W (2017) Accelerated startup of moving bed biofilm process with novel electrophilic suspended biofilm carriers. Chem Eng J 315:364–372
122. Megharaj M, Ramakrishnan B, Venkateswarlu K, Sethunathan N, Naidu R (2011) Bioremediation approaches for organic pollutants: a critical perspective. Environ Int 37(8):1362–1375
123. Meng X, Liu G, Zhou J, Fu QS, Wang G (2012) azo dye decolorization by shewanella aquimarina under saline conditions. Bioresource technology, 114, 95–101
124. Mijin DZ, Avramov ML, Antonije I, Onjia E, Grgur BN (2012) Decolorization of textile dye CI basic yellow 28 with electrochemically generated active chlorine. Chem Eng J 204–206:151–157
125. Mohorcic M, Teodorovič S, Golob V, Friedrich J (2006) Fungal and enzymatic decolourisation of artificial textile dye baths. Chemosphere 63(10):1709–1717
126. Mu Y, Rabaey K, Rozendal RA, Yuan Z, Keller J (2009) Decolorization of azo dyes in bioelectrochemical systems. Environ Sci Technol 43(13):5137–5143
127. Mungasavalli DP, Viraraghavan T, Jin YC (2007) Biosorption of chromium from aqueous solutions by pre-treated *Aspergillus niger*: batch and column studies. Colloids Surf A Physicochem Eng Asp 301:214–223
128. Müller-Renno C, Buhl S, Davoudi N, Aurich JC, Ripperger S, Ulber R, Muffler K, Ziegler C (2013) Novel materials for biofilm reactors and their characterization. Productive Biofilms 207–233
129. Muzamil K, Farooq A, Irfan S, Duy-Nam P, Qamar K, Zeeshan K, Hoik L, Ick-Soo K (2017) Dyeing and characterization of regenerated cellulose nanofibers with vat dyes. Carbohydr Polym 174:443–449
130. Nath N, Mishra A, Pande PP (2021) A review natural polymeric coagulants in wastewater treatment. Materials Today: Proceedings 46(14):6113–6117
131. Nguyen TA, Fu CC, Juang RS (2016) Effective removal of sulfur dyes from water by biosorption and subsequent immobilized laccase degradation on cross-linked chitosan beads. Chem Eng J 304:313–324
132. Ochs P, Martin BD, Germain E, Wu Z, Lee P-H, Stephenson T, van Loosdrecht M, Soares A (2021) Evaluation of a full-scale suspended sludge deammonification technology coupled with an hydrocyclone to treat thermal hydrolysis dewatering liquors. Processes 9:278
133. Ødegaard H, Rusten B, Westrum T (1994) A new moving bed biofilm reactor-applications and results. Water Sci Techno 29(10–11):157
134. Ødegaard H (2016) A road-map for energy-neutral wastewater treatment plants of the future based on compact technologies (including MBBR). Front Environ Sci Eng 10:2

135. Ong C, Lee K, Chang Y (2020) Biodegradation of mono azo dye-Reactive Orange 16 by acclimatizing biomass systems under an integrated anoxic-aerobic REACT sequencing batch moving bed biofilm reactor. J Water Process Eng 36:101268
136. Pal S, Sarkar U, Dasgupta D (2010) Dynamic simulation of secondary treatment processes using trickling filters in a sewage treatment works in Howrah, west Bengal, India. Desalination 253:135–140
137. Pan Y, Zhu T, He Z (2018) Enhanced removal of azo dye by a bioelectrochemical system integrated with a membrane biofilm reactor. Ind Eng Chem Res 57(48):16433–16441
138. Park HO, Oh S, Bade R, Shin WS (2011) Application of fungal moving-bed biofilm reactors (MBBRs) and chemical coagulation for dyeing wastewater treatment. KSCE J Civ Eng 15(3):453–461
139. Pastorelli G, Canziani R, Pedrazzi L, Rozzi A (1999) Phosphorus and nitrogen removal in moving-bed sequencing batch biofilm reactors. Water Sci Technol 40(4–5):169–176
140. Petrovic M, De Alda MJL, Diaz-Cruz S, Postigo C, Radjenovic J, Gros M, Barcelo D (2009) Fate and removal of pharmaceuticals and illicit drugs in conventional and membrane bioreactor wastewater treatment plants and by riverbank filtration. Philosophical Transactions of the Royal Society A: Mathematical, Physical Eng Sci 367(1904):3979–4003
141. Pearce CI, Lioyd JR, Guthrie JT (2003) The removal of color from textile wastewater using whole bacterial cells: a review. Dyes Pigm 58:179–196
142. Pisoni DDS, Marluza PDA, Cesar LP, Rodembusch FS, Campo LF (2013) Synthesis, photo physical study and BSA association of water-insoluble squaraine dyes. J Photochem Photobiol A Chem 252:77–83
143. Plósz BG, Vogelsang C, Macrae K, Heiaas HH, Lopez A, Liltved H, Langford KH (2010) The BIOZO process—a biofilm system combined with ozonation: occurrence of xenobiotic organic micro-pollutants in and removal of polycyclic aromatic hydrocarbons and nitrogen from landfill leachate. Water Sci Technol 61:3188–3197
144. Prasad ASA, Satyanarayana VSV, Bhaskara KVR (2013) Biotransformation of direct blue 1 by a moderately *Halophilicbacterium marinobacter* sp. strain HBRA and toxicity assessment of degraded metabolites. J Hazard Mater 262:674–684
145. Prigione V, Varese GC, Casieri L, Marchisio VF (2008) Biosorption of simulated dyed effluents by inactivated fungal biomasses. Bioresour Technol 99:3559–3567
146. Puttaswamy VKN, Ninge GKN (2008) oxidation of CI acid red 27 by chloramine-T in perchloric acid medium: spectrophotometric, kinetic and mechanistic approaches, Dyes and Pigments 78(2):131–138
147. Punzi M, Nilsson F, Anbalagan A, Svensson BM, Jönsson K, Mattiasson B, Jonstrup M (2015) Combined anaerobic–ozonation process for treatment of textile wastewater: removal of acute toxicity and mutagenicity. J Hazard Mater 292:52–60
148. Quintelas C, Silva B, Figueiredo H, Tavares T (2010) Removal of organic compounds by a biofilm supported on GAC: modelling of batch and column data. Biodegradation 21:379–392
149. Quintelas C, da Silva VB, Silva B, Figueiredo H, Tavares T (2011) Optimization of production of extracellular polymeric substances by *Arthrobacter viscosus* and their interaction with a 13X zeolite for the biosorption of Cr(VI). Environ Technol 32(14):1541–1549
150. Rai HS, Bhattacharyya MS, Singh J, Bansal TK, Vats P, Banerjee UC (2005) Removal of dyes from the effluent of textile and dyestuff manufacturing industry: a review of emerging techniques with reference to biological treatment. Crit Rev Environ Sci Technol 35(3):219–238
151. Rajasimman M, Babu SV, Rajamohan N (2017) Biodegradation of textile dyeing industry wastewater using modified anaerobic sequential batch reactor–start-up, parameter optimization and performance analysis. J Taiwan Inst Chem Eng 72:171–181
152. Rauf MA, Ashraf SS (2009) Fundamental principles and application of heterogeneous photocatalytic degradation of dyes in solution. Chem Eng J 151:10–18
153. Rawat D, Sharma RS, Karmakar S, Arora LS, Mishra V (2018) Ecotoxic potential of a presumably non-toxic azo dye. Ecotoxicol Environ Saf 148:528–537

154. Riera-Torres M, Gutierrez-Bouzan C, Crespi M (2010) Combination of coagulation, floc-culation and nanofiltration techniques for dye removal and water reuse in textile effluents. Desalination 252(1–3):53–59
155. Royer B, Cardoso NF, Lima EC, Ruiz VSO, Macedo TR, Airoldi C (2009) Organo-functionalized kenyaite for dye removal from aqueous solution. J Colloid Interf Sci 336:398–405
156. Rosales E, Pazos M, Longo MA, Sanromán MA (2009) Electro-fenton decoloration of dyes in a continuous reactor: a promising technology in colored wastewater treatment. Chem Eng J 155(1–2):62–67
157. Saha P, Shinde O, Sarkar S (2016) Phytoremediation of industrial mines wastewater using water hyacinth. Int J Phytoremed 19(1):87–96
158. Saini RD (2017) Textile organic dyes: polluting effects and elimination methods from textile wastewater. Int J Chem Eng Res 9(1):121–136
159. Samsamia S, Mohamadia M, Sarrafzadeha M, Reneb ER, Firoozbahr M (2020) Recent advances in the treatment of dye-containing wastewater from textile industries: overview and perspectives. Process Saf Environ Prot 143:138–163
160. Santhi P, Moses J (2011) Reduction process of vat dye on cotton fabric assisted by ferrous sulphate. Asian J Chem 23(1):169–172
161. Sarvajith M, Reddy GKK, Nancharaiah YV (2018) Textile dye biodecolourization and ammo-nium removal over nitrite in aerobic granular sludge sequencing batch reactors. J Hazard Mater 342:536–543
162. Segura PA, Takada H, Correa JA, Saadi KE, Koike T, Onwona-Agyeman S, Ofosu-Anim J, Sabi EB, Wasonga OV, Mghalu JM, dos Santos AM, Newman B, Weerts S, Yargeau V (2015) Global occurrence of anti-infectives in contaminated surface waters: impact of income inequality between countries. Environ Int 80:89–97
163. Sehar S, Naz I (2016) Role of the biofilms in wastewater treatment. In: Microbial biofilms—importance and applications. InTech Chapt. 7, pp 121–144
164. Selvakumar S, Manivasagan R, Chinnappan K (2013) Biodegradation and decolourization of textile dye wastewater using Ganoderma lucidum. 3 Biotech 3:71–79
165. Sghaier I, Guembri M, Chouchane H, Mosbah A, Ouzari HI, Jaouani A, Cherif A, Neifar M (2019) Recent advances in textile wastewater treatment using microbial consortia. J Text Eng Fash Technol 5(3):134–146
166. Shanmugam BK, Easwaran SN, Mohanakrishnan AS, Kalyanaraman C, Mahadevan S (2019) Biodegradation of tannery dye effluent using Fenton's reagent and bacterial consortium: a biocalorimetric investigation. J Environ Manage 242:106–113
167. Shariati FP, Mehrnia MR, Salmasi BM, Heran M, Wisniewski C, Sarrafzadeh MH (2010) Membrane bioreactor for treatment of pharmaceutical wastewater containing acetaminophen. Desalination 250:798–800
168. Sharma S, Bhattacharya A (2017) Drinking water contamination and treatment techniques. Appl Water Sci 7:1043–1067
169. Shin DH, Shin WS, Kim YH, Han MH, Choi SJ (2006) Application of a combined process of moving-bed biofilm reactor (MBBR) and chemical coagulation for dyeing wastewater treatment. Water Sci Technol 54(9):181–189
170. Silva B, Figueiredo H, Quintelas C, Neves IC, Tavares T (2008) Zeolites as supports for the biorecovery of hexavalent and trivalent chromium. Micropor Mesopor Mat 116:555–560
171. Smith ML, Gourdon D, Little WC, Kubow KE, Eguiluz RA, Luna-Morris S, Vogel V (2007) Force-induced unfolding of fibronectin in the extracellular matrix of living cells. PLoS Biol 5:10
172. Solpan D, Guven O (2002) Discoloration and degradation of some textile dyes by gamma irradiation. Radiat Phys Chem 65:549–558
173. Şolpan D, Güven O, Takács E, Wojnárovits L, Dajka K (2003) High-energy irradiation treat-ment of aqueous solutions of azo dyes: steady state gamma radiolysis experiments. Radiat Phys Chem 67(3):531–534

174. Sonwani RV, Swain G, Jaiswal RP, Singh RS, Rai BN (2021) Moving bed biofilm reactor with immobilized low-density polyethylene–polypropylene for Congo red dye removal. Environ Technol Innov 23:101558
175. Spagni A, Grilli S, Casu S, Mattioli D (2010) Treatment of a simulated textile wastewater containing the azo-dye reactive orange 16 in an anaerobic-biofilm anoxic-aerobic membrane bioreactor. Int Biodeterior Biodegrad 64(7):676–681
176. Srinivasan A, Viraraghavan T (2010) Decolorization of dye wastewaters by biosorbents: a review. J Environ Manage 91(10):1915–1929
177. Stylianou SK, Szymanska K, Katsoyiannis IA, Zouboulis AI (2015) Novel water treatment processes based on hybrid membrane-ozonation systems: a novel ceramic membrane contactor for bubbleless ozonation of emerging micropollutants. J Chem 2015:214927
178. Surti A, Ansari R (2018) Characterization of dye degrading potential of suspended and nanoparticle immobilized cells of *Pseudomonas aeruginosa* AR-7. Microbiol Biotechnol Food Sci 8(2):774–780
179. Tan ZX, Lal R, Wiebe KD (2005) Global soil nutrient depletion and yield reduction. J Sustain Agric 26:123–146
180. Tang K, Ooi GT, Litty K, Sundmark K, Kaarsholm KM, Sund C, Kragelund C, Christensson M, Bester K, Andersen HR (2017) Removal of pharmaceuticals in conventionally treated wastewater by a polishing moving bed biofilm reactor (MBBR) with intermittent feeding. Bioresource Techno 236:77–86
181. Tang K, Ooi GTH, Spiliotopoulou A, Kaarsholm KMS, Sundmark K, Florian B, Kragelund C, Bester K, Andersen HR (2020) Removal of pharmaceuticals, toxicity and natural fluorescence by ozonation in biologically pre-treated municipal wastewater, in comparison to subsequent polishing biofilm reactors. Water 12:1059
182. Tarjányi-Szikora S, Oláh J, Makó M (2013) Comparison of different granular solids as biofilm carriers. Microchem J 107(1):101–107
183. Turcanu A, Bechtold T (2017) Cathodic decolourisation of reactive dyes in model effluents released from textile dyeing. J Clean Prod 142:1–9
184. Turhan K, Ozturkcan SA (2013) Decolorization and degradation of reactive dye in aqueous solution by ozonation in a semi-batch bubble column reactor. Water Air Soil Pollut 224:1353
185. Umukoro EH, Peleyeju MG, Ngila JC, Arotiba OA (2016) Photocatalytic degradation of acid blue 74 in water using Ag-Ag$_2$O-ZnO nanostructures anchored on graphene oxide. Solid State Sci 51:66–73
186. Üstün GE, Solmaz SKA, Birgül A (2007) Regeneration of industrial district wastewater using a combination of Fenton process and ion exchange—a case study. Resour Conserv Recycl 52:425–440
187. van Belkum A (2007) The biofilm mode of life. J Microbiol Methods 71:348–349
188. van Nieuwenhuijzen AF, Evenblij H, Uijterlinde CA, Schulting FL (2008) Review on the state of science on membrane bioreactors for municipal wastewater treatment. Water Sci Technol 57(7):979–986
189. van der Zee FP, Villaverde S (2005) Combined anaerobic-aerobic treatment of azo dyes—a short review of bioreactor studies. Water Res 39:1425–1440
190. Verma RK, Sankhla MS, Rathod NV, Sonone SS, Parihar K, Singh GK (2021) Eradication of fatal textile industrial dyes by wastewater treatment. Biointerface Res Appl Chem 12(1):567–587
191. Wan T, Li D, Jiao Y, Ouyang X, Chang L, Wang X (2017) Bifunctional MoS$_2$ coated melamine-formaldehyde sponges for efficient oil-water separation and water-soluble dye removal. Appl Mater Today 9:551–559
192. Wang H, Qu J (2003) Combined bio-electrochemical and sulphur autotrophic denitrification for drinking water treatment. Water Res 37:3767–3775
193. Wang X, Guan W, Pan J (2011) Preparation of magnetic imprinted polymer particles via microwave heating initiated polymerization for selective enrichment of 2-amino-4 nitrophenol from aqueous solution. Chem Eng J 178:85–92

194. Wang S, Yang X, Menga H, Zhang Y, Li X, Xu J (2019) Enhanced denitrification by nano α-Fe$_2$O$_3$ induced self-assembled hybrid biofilm on particle electrodes of three-dimensional biofilm electrode reactors. Environ Int 125:142–151
195. Webb J (2009) Differentiation and dispersal in biofilms. Bacterial Biofilm Form Adapt 1996:1–8
196. Webster S, Fu J, Padilha LA, Przhonska OA, Hagan DJ, Van Stryland EW, Bondar MV, Slominsky YL, Kachkovski AD (2008) Comparison of nonlinear absorption in three similar dyes: polymethine, squaraine and tetra one. Chem Phys 348:143–151
197. Yamagami A, Kawano K, Futaki S, Kuramochi K, Tsubaki K (2017) Syntheses and properties of second-generation V-shaped xanthene dyes with piperidino groups. Tetrahedron 73:7061–7066
198. Yaseen DA, Scholz M (2019) Textile dye wastewater characteristics and constituents of synthetic effluents: a critical review. Int J Environ Sci Technol 16:1193–1226
199. Yadav DN, Naz I, Kishore KA, Saroj D (2020) Evaluation of tire derived rubber (TDR) fixed biofilm reactor (FBR) for remediation of Methylene blue dye from wastewater. Environ Technol 1–27
200. Yusuff RO, Sonibare JA (2004) Characterization of textile industries effluents in Kaduna, Nigeria and pollution implications. Glob Nest Int J 6(1):212–221
201. Zadeh BS, Esmaeili H, Foroutan R, Mousavi SM, Hashemi SA (2020) Removal of Cd^{2+} from aqueous solution using eucalyptus sawdust as a bio-adsorbent: kinetic and equilibrium studies. J Environ Treat Tech 8(1):112–118
202. Zaroual Z, Azzi M, Saib N, Chainet E (2006) Contribution to the study of electrocoagulation mechanism in basic textile effluent. J Hazard Mater 131(B):73–78
203. Zhang L, Li J, Zhu X, Ye D, Qiang L (2013) Anodic current distribution in a liter-scale microbial fuel cell with electrode arrays. Chem Eng J 223:623–631
204. Zhao B, Xiao W, Shang Y, Zhu H, Han R (2017) Adsorption of light green anionic dye using cationic surfactant-modified peanut husk in batch mode. Arab J Chem 10:3595–3602
205. Zhang B, Dong X, Yu D He J (2012) Stabilization mechanisms of C.I. Disperses red 60 dispersions in the presence of its dye-polyether derivatives. Colloids Surf A Physicochem Eng Asp 405:65–72
206. Zhao Y, Liu D, Huang W, Yang Y, Ji M, Nghiem LD, Trinh QT, Tran NH (2019) Insights into biofilm carriers for biological wastewater treatment processes: current state-of-the-art, challenges, and opportunities. Bioresour Technol 288:121619
207. Zheng Q, Dai Y, Han X (2016) Decolorization of azo dye C.I. Reactive Black 5 by ozonation in aqueous solution: influencing factors, degradation products, reaction pathway and toxicity assessment. Water Sci Technol 73:1500–1510
208. Zhang L, Fu G, Zhang Z (2019) High-efficiency salt, sulfate and nitrogen removal and microbial community in biocathode microbial desalination cell for mustard tuber wastewater treatment. Bioresource Techno 289:121630

Biotransformation of Anthraquinone Dye by Microbial Enzymes

Tony Hadibarata, Khaloud Mohammed Alarjani, and Amal M. Al-Mohaimeed

Abstract The fabric industry consumes a significant amount of water and generates huge amounts of wastewater; one of the primary pollutants is categorized as anthraquinone dyes. Due to the structure composition, anthraquinone dyes are more resistant to physico-chemical treatment than natural dyes, making them more difficult to decolorize. Currently, it is important to develop efficient, eco-friendly, and cheap methods for remediation of anthraquinone dye in wastewater. Microbial dye degradation is an intriguing method of dyes from wastewater because it provides numerous benefits over common methods. Microbial degradation is a cheap and eco-friendly biotechnological method based on the utilization of their enzymes. This chapter assesses the most recent advances in the biotransformation of synthetic dyes from wastewater, especially anthraquinone-typed dyes. The utilization of various microorganisms such as bacteria, fungi, yeast/filamentous fungi, as well as their extracellular enzymes for the transformation of dye are discussed in detail. This chapter will discuss the role of dioxygenase, monooxygenase, peroxidase, laccase, hydrolases, and transferase enzymes in the transformation of anthraquinone dyes. The study will give a significant contribution to the development of advanced bioprocess technology to eliminate hazardous materials in the aquatic environment.

Keywords Anthraquinone dye · Enzyme · Biodegradation · Metabolites

T. Hadibarata (✉)
Environmental Engineering Program, Faculty of Engineering and Science, Curtin University Malaysia, CDT 250, 98009 Miri, Malaysia
e-mail: hadibarata@curtin.edu.my

K. M. Alarjani
Department of Botany and Microbiology, College of Science, King Saud University, Riyadh 11451, Saudi Arabia

A. M. Al-Mohaimeed
Department of Chemistry, College of Science, King Saud University, P.O. Box 22452, Riyadh 11495, Saudi Arabia

1 Introduction

Contamination of drinking water is a major concern in emerging societies. Rapid industrial and agricultural expansion have resulted in an increase in the generation of dangerous pollutants such as plastic, dye, heavy metals, phenolic compounds, and polycyclic aromatic hydrocarbons [21, 24, 30, 31, 39, 45]. Synthetic dyes are widely implemented in various manufacturing, including clothing, plastics, cosmetics, apparel, food and beverage, and printing. Most of them are categorized as textile dyes due to massive utilization in coloring cloth, wool, cotton, etc. Synthetic dye contamination in aquatic ecosystems caused by the textile industry is a major environmental concern [11]. Not only dyes are found in textile industry wastewater, but also various hazardous and carcinogenic substances that block sunlight, impair photosynthesis, oxygenation processes, and increase biological oxygen demand (BOD), chemical oxygen demand (COD), total solids (TS), and other elements [16]. Synthetic dyes are categorized based on their structure (anthraquinone, azo, indigo, nitro, phthalein, and triphenylmethane) and application (acid, direct, disperse, ingrain, moderate, reactive, simple, and vat) [25]. Table 1 summarizes the typical and classification of synthetic dyes. Given synthetic dyes' environmental impact, it is critical to develop and evaluate a viable treatment for decolorizing and detoxifying dyes before realizing wastewater to the environment. As a result, substantial research has been performed to create unique and effective strategies for dye wastewater cleanup. Numerous physical and chemical methods have been utilized to treat textile effluents, including membrane filtration, advanced oxidation, adsorption, flocculation, coagulation, precipitation, ion exchange, and biodegradation [25, 51]. However, most conventional treatment showed a lot of limits, such as ineffective, expensive, complicated procedures, significant sludge production, and low commercial viability [27].

Due to the numerous advantages that bioremediation has over conventional treatment techniques, it has been demonstrated to be a viable alternative treatment method for textile effluents. Many microbes, either a pure culture or a mixed microbial culture have the ability to degrade and remediate various types of dyes in wastewater. A previous study showed that a consortium microbe can efficiently degrade dyes due to synergistic metabolic actions. Bioremediation is primarily based on microorganisms that enzymatically degrade hazardous contaminants and transform them into metabolites. However, because bioremediation effectiveness is contingent on the presence of favorable environmental conditions for microbial growth and activity, it is frequently used in conjunction with environmental parameter manipulation to accelerate microbial growth and degradation. While the majority of bioremediation process function in aerobic environments, anaerobic condition might also allow microbes to degrade recalcitrant molecules. To remediate recalcitrant lignin and organopollutants, these microorganisms rely on the involvement of distinct intracellular and extracellular enzymes [2, 7, 22, 54]. This review was compiled in an effort to assess the general extent of scientific knowledge regarding the properties of various synthetic dyes especially anthraquinone-type dye as well as remediation

Table 1 Typical and classification of synthetic dyes

Based on chemical nature

Class	Description	Example	Application
Acid	Anionic and soluble in water	Acid Black 1, Acid Yellow 36, Acid Orange 3	Food industry, textile, printing, cosmetic, leather
Basic	Cationic and soluble in water	Methylene Blue, Crystal Violet, Basic Blue 9	Textile, paper, ink, medicine, polyester
Direct	Anionic and soluble in water; retain sulphonic acid group structure	Congo Red, Brilliant Yellow, Direct Black 38	Textile, leather, cellulosic fabric, paper
Disperse	Non-ionic in nature and partially soluble in water; better diffusion to a higher temperature	Disperse Orange 1, Disperse Red 9	Printing, Synthetic fabrics, plastic, polyester
Reactive	Anionic and water soluble; good fastness properties	Procion Red, Remazol Brilliant Blue R, Reactive Black 5	Food industry, textile, pulp and paper, additive to detergent
Sulfur	Insoluble in water; produced from thionation or sulphuration	Sulfur Black 1, Sulfur Brilliant Green, Sulfur Blue	Fibers, textile, rayon, wood, paper, leather
Vat	Insoluble in water; high color fastness; require a reducing agent to solubilize	Vat Blue 1, Vat Orange 3, Vat Yellow 1	Fibers, rayon, leather, wool, polyesters

Based on chemical structure

Class	Structure	Example
Anthraquinone		Acid Green 25, Acid Blue 129, Disperse Blue 3, Disperse Red 3B, Pigment Yellow 24, Procion Navy H-EXL, Reactive Blue 4, Reactive Blue 19, Remazol Brilliant Blue R, Sulphonyl Blue TLE
Azo	R——N═N——R	Congo Red, Maxilon Blue, Procion Yellow H-EXL, Reactive Black 5, Remazol Black N, Sulphonyl Blue TLE, Sulphonyl Green BLE, Sulphonyl Scarlet

(continued)

technique for dyes contamination. The role of microbial enzymes in the transformation of anthraquinone dyes is extensively discussed. The study will give a significant contribution to the development of advanced bioprocess technology to eliminate hazardous materials in the aquatic environment.

Table 1 (continued)

Based on chemical structure

Class	Structure	Example
Heterocyclic		Basic Violet 10, Disperse Red CBN, Fluorol Yellow 88
Indigoid		Tyrian Purple, Thioindigo, Indigo Carmine, Direct Blue 71, C.I. Vat Blue 35, Fuchsin Basic
Nitro	R—N═O	Acid Yellow 1, Palatine Orange, Fast Green O, Pigment Green 8
Nitroso		Acid Yellow 24, Disperse Yellow 14, Naphthol Yellow, Fast Yellow GG
Phthalein		o-Cresolphthalein, Dixylenolphthalein, Guaiacolphthalein, α-Naphtholphthalein, Phenolphthalein, Phenolsulfonphthalein

2 Anthraquinone Dye

Anthraquinones are commonly found in the most common natural red dyes that were widely used in fabric industries from thousands of years ago to the late nineteenth century. These type of dyes are one of preferred synthetic dyes in industry after azo dyes because they are cheap, shows best performance, and ease of access. However, anthraquinones dyes are significantly more toxic than azo dyes to humans and microorganisms due to the existence of chromophore groups in the stable and intricate structure, which consist of a benzene ring bonded with two carbonyl groups on both sides. Natural red pigments are frequently composed of anthraquinones, a chemical compound that is broadly implemented in a lot of industries including textile and printing. Due to their discovery thousands of years ago and their use in wrapping mummies, anthraquinone dyes have been dubbed the world's oldest dyes. The major class of naturally occurring quinoids is anthraquinones. Anthraquinones are naturally occurring compounds produced by a variety of scale insects and plant roots. The Rubiaceae family plants such as chai root, madder, and Indian mulberry, as well as scale insects such as cochineal, kermes, and lac, produced stunning color palettes of red hues on many variety fibers. The palette pigments were determined by the metallic salt used as a mordant, which is available in a limited range of hues including purple, brown, and orange. There are four distinct classes of anthraquinone dyes: heterocyclic anthraquinone dyes, heterocyclic anthrone dyes, anthraquinone

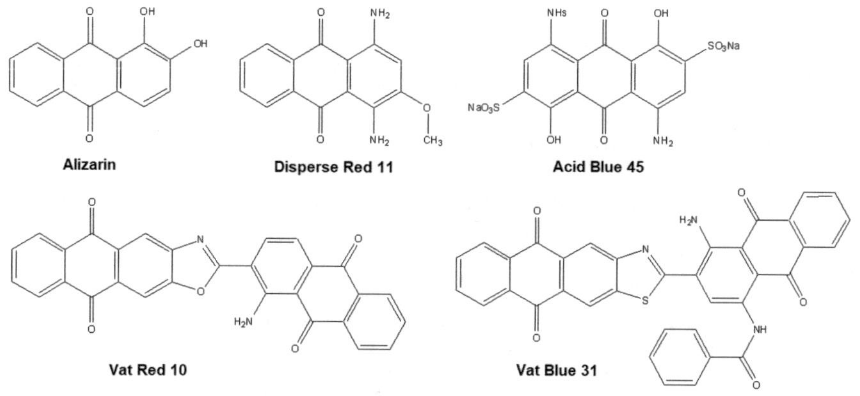

Fig. 1 Example of anthraquinone dyes

derivatives, and fused ring anthrone dyes [32]. Acid Blue 45, Alizarin, Disperse Red 11, Vat Red 10, and Vat Blue 31 are all examples of anthraquinone dyes (Fig. 1).

3 Remediation Technique for Dyes Contamination

The three primary classes of remediation techniques for dye contamination are physical, chemical, and biological. Some conventional methods are widely implemented to treat dyes wastewater such as membrane filtration, adsorption, flocculation, coagulation, precipitation, ozonation, bleaching, and ion exchange. However, there are many limitations in these conventional methods including inefficiency, expensive, complicated procedures, and hazardous by-products. Apart from these well-established techniques, bioremediation has recently received significant consideration as a relatively inexpensive and effective method of treating textile wastewater. The following section summarizes the physical, chemical, and biological methods for dye remediation as shown in Table 2. While some treatments are effective, their prohibitively high investment and operating costs, as well as the generation of secondary waste, preclude their widespread use. Nano-filtration, reverse osmosis, and ultrafiltration have all been used for remediation of dye effluent before it was released to the environment [17, 27, 51].

Membrane-based dye treatment has several advantages, including no chemical requirement, high flexibility, and effectiveness for water recovery. However, membrane-based dye treatment also has some limits, including the inefficient low solubility and large quantity of contaminant, concentrated residue production, and membrane clogging [1]. Advanced Oxidation Processes (AOPs) have been demonstrated to be significantly more effective at treating wastewater than other methods. When energy consumption is kept to a minimum, AOP systems are considered environmentally sustainable [48]. In a homogeneous solution, the Fenton reaction

Table 2 Chemical, physical, and biological technologies used for synthetic dye removal

Treatment	Advantages	Disadvantages	References
Chemical methods			
Electrokinetic	Competitive in cost, effective method	Limited by solubility of contaminant	[48]
Electrooxidation	No addition of chemicals, simple preparation, high efficiency, operation at room temperature and pressure	Expensive	[48]
Fenton's oxidation	High performance and simplicity	Expensive, high generation of sludge, long reaction time, work only on low pH	[14]
Ozonation	Fast process, very effective for decolorization, no sludge generation, no increase of wastewater volume fast reaction	Expensive treatment, short half-life, toxic by-products Generation, Unstable system	[49]
Ion exchange	Produce high quality water, can be regenerated, no loss of adsorbent during regeneration	High operational cost, only ionic analytes, less efficient column	[43]
Photochemical	Very effective for dye removal, No foul odor and sludge generation	High operational cost, generation of by-products	[29]
Ultraviolet irradiation	No addition of chemicals, no sludge generation, reduction in foul odors	Impact to human health due to UV radiation	[12]
Physical methods			
Membrane filtration	No requirement of chemical, high flexibility, effective for water recovery	Equipment cost can be high, membrane can crack easily, applicable for only liquid	[50]
Irradiation	Effective at laboratory	Expensive, high amount of dissolved oxygen	[12]
Adsorption (activated carbon)	Highly porous material; large surface area, effective removal of wide ranges of dyes	Not recommended for low molecular weight and high purity	[53]
Adsorption (activated charcoal)	Low cost, easy to prepare, good efficiency	Low surface area of adsorption	[38]
Adsorption (agricultural waste)	low cost, abundance, affective adsorption capacity, and recyclability	Slow process, require long retention time	[27]

(continued)

Table 2 (continued)

Treatment	Advantages	Disadvantages	References
Nano-filtration & ultra-filtration	Low cost of operation, low energy cost, removal of virus, bacteria and pesticide, very effective for decolorization	Require high pressure, constant blockage of membrane pores, short life span	[1]
Reverse osmosis	Produce high quality water, Effective desalination and decolourization of many dyes	Costly, high energy consumption	[1]
Coagulation & Flocculation	Easy operation, low cost, high efficiency, effective removal of various types of dyes	High sludge production, increase cost of sludge treatment,	[17]
Biological methods			
Biosorption	No sludge generation, low cost, easy operation, high efficiency	Disposal of biosorbent at the end of life, time consuming	[33]
Fungal decolourization	Less toxic by-product generation, easy operation, flexible method	Slow process, nutrient, and oxygen supply	[7]
Consortium cultures	Fast process, eliminating limitation of single culture	Require high stability, nutrient, and oxygen to culture	[26]
Aerobic-anaerobic remediation	Low cost, wide range of dyes, effective removal, bioenergy recovery	Low rates, require large land area; deal with some inhibitors, time consuming	[18]
Algae degradation	Cheap in operational, Greater production of biomass, high ability to accumulate	Time consuming, slow process, Unstable system	[35]
Enzymatic process	Fast compared to microbial culture, non-toxic by-product, easily reproduction	Maintain the stability of enzyme	[20]

occurs. The Fe^{2+} ions generate hydroxyl radicals (OH*), which are extremely reactive and capable of completely mineralizing complex organic molecules, particularly non-biodegradable dyes [14].

4   Biological Methods

Bioremediation is a biotechnological process that focuses on rehabilitating recalcitrant environments and managing pollutants via detoxification and mineralization.
Organic pollutant remediation using microbial enzymes is considered to be environmentally friendly, cost effective, innovative, and promising [23]. However, bioremediation does have some drawbacks. It is a lengthy process, and only a few microorganisms are capable of producing specific enzymes capable of degrading pollutants.
As a result, genetically engineered microorganisms are preferred for bioremediation
because they produce a high concentration of desired enzymes under optimal conditions. Previous study showed the ability of microorganisms to catalyze or metabo-
lize xenobiotic pollutants in order to obtain energy and biomass. These pollutants
may include aromatic hydrocarbons, halogenated chemicals, microplastics, and other
emerging pollutants. Bioremediation is a process that transforms toxic substances
into nontoxic and, in some cases, novel products. Bioremediation may offer a number
of advantages over other psyco-chemical methods because this method requires less
cost, easy operation, flexible procedure, high efficiency, and less toxic by-product
generation. On-site bioremediation is possible without interfering with normal operations. The process ensures that toxic substances are completely degraded without
the use of hazardous chemicals and that contaminants are removed from the environment without being transferred to another type of materials or environmental media
[16]. While there are undoubtedly a number of advantages to bioremediation, there
are also a number of disadvantages. For example, the process is mostly ineffective to
inorganic contaminants, slow processes, require the supply of nutrients, oxygen, and
stability of culture condition [10]. Thus, development of bioremediation techniques
is required to eliminate these limitations.

4.1   Degradation of Dyes by Fungal Strains

It has been demonstrated that fungal culture is effective at degrading and transforming
a wide variety of textile dyes due to the high adaptability of fungal strain metabolism
in changing environmental conditions [2, 4]. Enzymes mechanisms contribute to
degrading and converting various dyes found in textile wastewater. The degradation
of anthraquinone dyes by various fungal strains, as well as the metabolic product
is shown in Table 3. Extracellular enzymes found in fungi's hyphae initiate dye
degradation by adsorption on the surface and subsequent dissociation of chemical
bonds within dye structure. Fungi are the most dominant microbes for decolorization
and degradation of dye due to their oxidoreductase enzyme [20, 36]. However, dye
degradation by fungi has a number of inherent disadvantages, including inconsistent
enzyme production, slow growth, requiring a large reactor and nitrogen-restricted
condition. Furthermore, the degradation of pollutants by the single fungal strain
was not suggested due to instability of the system/substrate for a relatively brief

Table 3 Degradation of anthraquinone dye by fungal strain

Fungal strain	Antraquinone dye	Removal (%)	Metabolic product	References
Polyporus spp.	RBBR	90	Sodium 1-amino-9,10-dioxo-9,10-dihydroanthracene-2-sulfonate; sodium 2-((3-aminophenyl)sulfonyl)ethyl sulfate	[23]
Cerrena unicolor	Reactive Blue 4	61	2-formylbenzoic acid; 1,2,4,5-tetrahydroxy-3-benzoic acid; 2,3,4-trihydroxybenzenesulfonic acid; 1,2,3,4-pentahydroxybenzene	[46]
Trametes hirsuta	RBBR	95	salicylic acid, hydroxyisophthalic acid	[8]
Trametes hirsuta	Acid Blue 129	96	salicylic acid, hydroxyisophthalic acid	[8]
Trametes hirsuta	Reactive Blue 4	90	Cyanuric acid, salicylic acid, 4-phenolsulfonic acid	[8]
Thermomucor indicae-seudaticae	RBBR	79	–	[44]
Pleurotus ostreatus	RBBR	85	2-amino-4-((2-hydroxyethyl)sulfonyl)phenol, 2-(4-oxopentanoyl)benzoic acid	[54]
Armillaria spp.	RBBR	79	–	[21]

period of time. This is because bacteria will grow and fungi will lose their ability to control the system and degrade the dyes [7, 16]. The effectiveness of fungal strains to degrade synthetic dyes were mostly depended on culture conditions including supply of nutrients and oxygen, temperature, and acidity [20, 37]. Hadibarata and Kristanti have explored the effect of environmental conditions on the decolorization of Remazol brilliant blue r, including pH, agitation, carbon and nitrogen sources, metal ion, salinity, and phenolic compound. The results indicated that decolorization occurs more rapidly at pH 5 and with agitation. Glucose and ammonium tartrate were found to be the most suitable carbon and nitrogen sources, respectively, when compared to other carbon and nitrogen sources tested. Iron ion was found to be the most inhibitive at a concentration of 10 mM, whereas copper ion was found to be the most suitable for decolorization [19].

4.2 Role of Bacterial Strains

The degradation of anthraquinone dyes by various bacterial strains is shown in Table 4. Bacteria have demonstrated tremendous decolorization potential in both pure cultures and consortia. However, bacterial consortiums were typically more efficient than pure isolates at removing dye [34]. Numerous dyes have been degraded using various bacteria [10, 13, 16]. Bacteria typically decolorize anthraquinone dyes by first deaminating them to form quinone compounds and then oxidizing them to form aromatic acid. On the other hand, deamination and hydroxylation processes were used to transform the dye structure into aromatic phenol and aromatic acid [16]. Numerous studies have demonstrated that dyes degrade significantly when bacteria such as *Bacillus cereus, Hortaea spp., Proteus mirabilis*, and *Pseudomonas aeruginosa* are used. Certain bacteria exhibited a high degree of decolorization in the presence of metal ions such as Cu^{2+}, owing to the strain's laccase activity. Additionally, bacteria consortiums demonstrated efficient decolorization of anthraquinone dye when carbon and nitrogen sources were added. Extracts from agricultural wastes have been determined as superior additives for increasing dye decolorization [10, 13, 16, 34].

4.3 Role of Yeast and Filamentous Fungi

When it comes to degrading anthraquinone dyes, yeasts have a number of advantages over filamentous fungi and bacteria. For example, they grow rapidly and are resistant to adverse environmental conditions [52]. Previous studies have reported the decolorization and degradation of anthraquinones dyes by yeasts or filamentous fungi [2, 5, 6, 9]. Adnan et al. investigated the effect of nutrients on *Trichoderma lixii* F21 degradation of alizarin red s and quinizanine green ss. It was concluded that

Table 4 Degradation of anthraquinone dye by bacterial strain

Bacteria	Antraquinone dye	Removal (%)	Metabolic product	References
Hortaea spp.	Solvent Green 3	100	P-cresol, phthalic acid, 4-hydroxybenzoic acid	[16]
Consortium of *Bacillus flexus, Proteus mirabilis, and Pseudomonas aeruginosa*	Indanthrene Blue RS	80	1,2-diaminoanthracene-9,10-dione, anthracene-9,10-dione, 1,2-dihydroxyanthracene-9,10-dione, phthalic acid, 1-hydroxyanthracene-9,10-dione, benzoic acid	[34]
Aerobic bacterial granules	Reactive Blue 4	72	4-amino-9,10- dihydro-9,10-dioxoanthracene-2-sulfonic acid, 2-(4,6-dichloro1,3,5-triazin-2-ylamino)-4-aminophenol (b), 1- aminoanthracene-9, 10-dione	[10]
Bacillus cereus	Acid Blue 25	90	4,40 -bis(dimethylamino)benzophenone	[13]

glucose and yeast extract was fit nutrients for enhancing biosorption and biodegradation processes [2]. Pan et al. determined that the anthraquinone dye cleavage occurred under microaerophilic conditions. The degradation pathways for Disperse Blue 2BLN by yeast were formed as a result of the anthraquinone chromophore being broken at various positions, including C_{10}–C_{11}, C_{13}–C_{14} and C_{10}-C_{11}, C_{12}–C_{13} [41]. The degradation of anthraquinone dyes using yeasts and filamentous fungi is summarized in Table 5.

4.4 Role of Microbial Enzymes

Enzymes are biological catalysts that play role in substrates conversion to form products by lowering the reaction's activation energy. In recent years, biocatalysts have gained popularity for a variety of industrial applications. Enzymatic process of dye wastewater offers several advantages including ease of access, eco-friendly, substrate transformation, low cost, and high efficiency [40]. Additionally, enzymatic degradation processes are advantageous due to their non-toxic and environmentally friendly by-product, as well as their reusability. However, a significant impediment to enzymatic degradation is the biocatalyst's deactivation as a result of denaturation [28]. The enzymatic decolorization of anthraquinone dyes is shown in Table 6.

Monooxygenase, dioxygenase, peroxidase, hydrolase, dehydrogenase, laccases, and transferase appear to be the most promising enzymes in the enzymatic process of anthraquinone dyes. The potential benefits of using enzymes rather than fungal cultures include a shorter treatment time, the ability to operate at any concentration of substrate, the absence of delays associated with the lag phase of biomass, a low production of sludge, and simple maintenance and process. Additionally, the production of enzymes by microbes was impacted by physiological factors such as pH and effluent composition variation. The enzyme-based treatment may be sufficient on its own, particularly when the enzymes convert toxic compounds to less harmful products. Complete degradation of the contaminants may not be necessary in these instances. However, some practical problems might occur during implementation of enzymes such as stability of enzyme, high cost production, and long process for isolation and purification [34, 54]. In the latter case, immobilization of the enzyme has been shown to increase its stability. Enzyme immobilization provides some benefits such as higher enzyme stability, reducing enzyme cost production, and recovery for utilization, which is crucial for commercial application. However, the concept of enzyme reutilization implies that the enzyme's stability must be sufficient. As a result, the immobilized enzyme must be extremely stable in order to develop a process that is suitable [8, 9].

Monooxygenases are oxygenases that contain heme and are found in single-celled organisms such as bacteria and archaea, and prokaryotic organisms such as fungi, plant, and animal. This enzyme is a large group of enzymes that catalyze the oxidative reactions of a broad range of substrates, from single-bonded carbon functional groups to complex aromatic and heterocyclic hydrocarbons. Monooxygenases play a role as

Table 5 Degradation of anthraquinone dye by yeast and filamentous fungi

Yeast strain	Antraquinone dye	Removal (%)	Metabolic product	References
Trichoderma lixii F21	Quinizarine Green SS	98	3-nitro-phthalic acid; 1-ethyl-4-methyl-benzene	[2]
Trichoderma lixii F21	Alizarin Red s	78	(3-hydroxy-phenyl)-acetic acid, 4-hydroxy-benzoic acid, Succinic acid	[2]
Aspergillus spp.	Disperse Blue 2BLN	93	2-(*N*-methyl-*p*-phenylenediamine) ethanol, dibutyl phthalate, methyl 3-(3,5-di-*tert*-butyl-4-hydroxyphenyl) propionate, ethyl propionate, hydroquinone, 2,4-di-*tert*-butylphenol, 1,3-indanone	[41]
Aspergillus flavus	Drimarene blue K2R	71	Phthalic acid, benzoic acid, 1,4-dihydroxyanthraquinone, 2,3-dihydro-9,10-dihydroxy-1,4-anthracenedione, and catechol	[9]
Candida tropicalis	Remazol Blue	90	–	[5]

Table 6 Type of microbial enzyme for degradation of anthraquinone dye

Enzyme name	Mechanism	Dye	References
Dioxygenase	Oxidation and reduction of substrate	RBBR, Solvent Green 3, Alizarin Red S	[2, 16, 21, 41]
Monooxygenase	Incorporate one hydroxyl group into substrate	RBBR	[3]
Lignin peroxidase	Catalyze a reaction using hydrogen peroxide and mediator such as veratryl alcohol	RBBR, Solvent Green 3, Disperse Blue 2BLN	[16, 22, 41]
Manganese peroxidase	Catalyze oxidation using hydrogen peroxide and ion Mn^{2+}	RBBR, Solvent Green 3, Disperse Blue 2BLN	[16, 23, 41]
Versaile peroxidase	Catalyzes the transport of electron from organic pollutants	RBBR	[36]
Laccase	Multicopper oxidation, crosslinking	RBBR, Solvent Green 3, Alizarin Red S, Disperse Blue 2BLN, Reactive Blue 4	[2, 8, 16, 20, 21, 41]
Dehydrogenase	Catalyzes the removal of hydrogen atoms from a particular molecule	RBBR, Poly R-478	[3, 15]
Hydrolase	Enyzme that hydrolase of substrate	RBBR	[47]
Transferase	Transfer of functional groups such as methyl and hydroxymethyl	RBBR	[3, 42]

biocatalysts in degradation and biotransformation due to their high region selectivity and stereoselectivity over a wide variety of organic pollutants. Dioxygenase is an important enzyme in environmental remediation applications that catalyzes organic pollutants through oxygenation reaction. The enzyme lignin peroxidase catalyzes a reaction involving hydrogen peroxide and a mediator such as veratryl alcohol. Lignin peroxidase is capable of oxidizing aromatic compounds with greater than 1.4 V redox potentials. Manganese peroxidase catalyzes the oxidation reaction by utilizing hydrogen peroxide and the ion Mn^{2+}. Versatile peroxidase catalyzes the transfer of electrons from an oxidizable substrate to a non-oxidizable substrate. Laccase enzymes catalyze the oxidation of a broad range of phenolic and aromatic hydrocarbons while simultaneously reducing molecular oxygen to water. Dehydrogenase catalyzes the hydrogen atoms being removed from a particular molecule. Hydrolases play a role in condensation and alcoholysis reactions of various organic pollutants. The function of transferase is to facilitate the transfer of functional groups such as methyl and hydroxymethyl [23, 28].

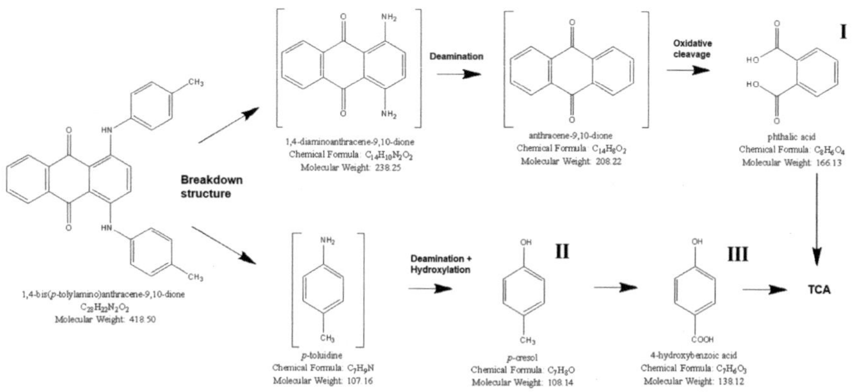

Fig. 2 The pathway for Solvent Green 3 degradation by *Hortaea* sp [7]

Al Farraj et al. reported that a halophilic bacteria Hortaea degraded the solvent green 3 structure during the initial transformation process, resulting in the formation of two distinct pathways, 1,4-diaminoanthracene-9,10-dione and *p*-toluidine. Following that, deamination of 1,4-diaminoanthracene-9,10-dione to form 9,10-anthraquinone was performed, which was likely catalyzed by laccase. Then the dioxygenase enzyme initiates the oxidative cleavage to form phthalic acid. The phenyl group is one of the metabolites formed during the degradation of solvent green 3. The phenyl group was converted and mineralized into the citric acid cycle, H_2O, and CO_2. The main metabolites of solvent green 3 degradation by Hortaea sp. were determined to be phthalic acid, 4-methylphenol, and 4-hydroxybenzoic acid (Fig. 2).

The role of laccase and dioxygenase is demonstrated in *Trichoderma lixii* F21's biodegradation of quinizarine green and alizarin s. The pathway demonstrated the ability of laccase and 1,2-dioxygenase to convert anthraquinone dyes to form carboxylic acid and phenolic compounds without producing any toxic metabolites, making it an eco-friendly treatment. Desulphonation and oxidative cleavage was the primary transformation of alizarin red s to convert carbonyl group to form 3-hydroxyphenylacetic acid and 4-hydroxybenzoic acid. The decarboxylation and ring fission reaction were catalyzed by laccase resulting in the formation of succinic acid and the subsequent entry into the tricarboxylic acid cycle (Fig. 3). Laccase enzymes catalyzed the majority of the oxidative cleavage, decarboxylation, and hydroxylation reactions in biotransformation of quanizarine ss. Prolonged treatment with the dye catalyzed the C–C bond cleavage to form 3-nitrophthalic acid. The decarboxylation of phthalic acid to phenylacetic acid was the first step in the mechanism. Phenylacetic acid may also be decarboxylated and hydroxylated to form catechol. (Fig. 4). The biodegradation of quinizarine green and alizarin s revealed the oxidative reaction catalyzed by the oxidoreductase laccase and 1,2-dioxygenase [2].

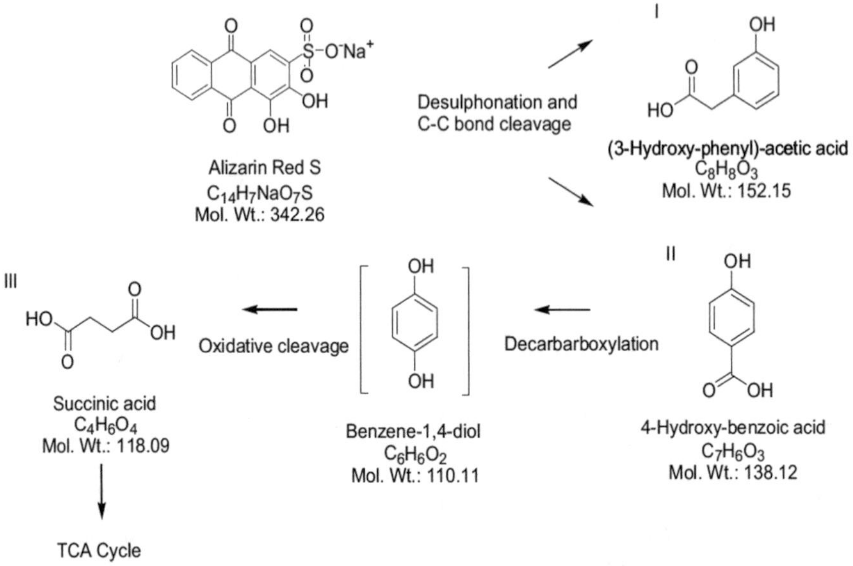

Fig. 3 The pathway for Alizarin Red S degradation by *Trichoderma lixii* F21 [2]

5 Conclusion

Microorganisms have demonstrated tremendous decolorization potential in both pure cultures and consortia. The application of genetically modified organisms and immobilization is advantageous in order to increase the efficiency of the degradation process. Global research and interest in the development and application of enzymes in the dye wastewater are receiving great attention. This is demonstrated by the high volume of scientific papers and patents published on this subject, which continues to grow year after year. Monooxygenase, dioxygenase, peroxidase, hydrolase, dehydrogenase, laccases, and transferase appear to be the most promising enzymes in the enzymatic process of anthraquinone dyes. The potential benefits of using enzymes rather than microbial cultures include a shorter treatment time, the ability to operate at any concentration of substrate, the absence of delays associated with the lag phase of biomass, a low production of sludge, and simple maintenance and process. The study will make a significant contribution to the development of advanced bioprocess technology in order to eliminate hazardous materials in aquatic environments.

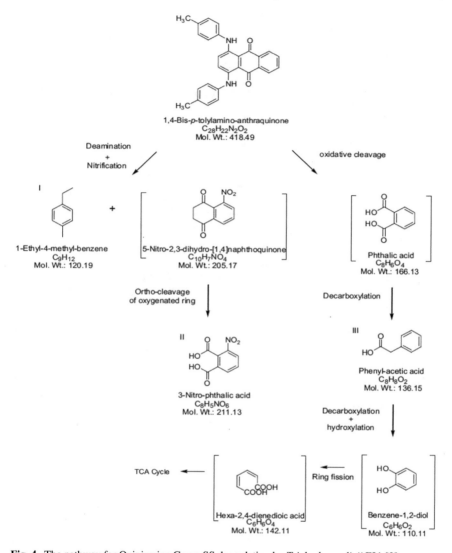

Fig. 4 The pathway for Quinizarine Green SS degradation by *Trichoderma lixii* F21 [2]

References

1. Abid MF, Zablouk MA, Abid-Alameer AM (2012) Experimental study of dye removal from industrial wastewater by membrane technologies of reverse osmosis and nanofiltration. Iran J Environ Health Sci Eng 9:17. https://doi.org/10.1186/1735-2746-9-17
2. Adnan LA, Sathishkumar P, Yusoff AR, Hadibarata T, Ameen F (2017) Rapid bioremediation of Alizarin Red S and Quinizarine Green SS dyes using Trichoderma lixii F21 mediated by biosorption and enzymatic processes. Bioprocess Biosyst Eng 40:85–97. https://doi.org/10.1007/s00449-016-1677-7
3. Aksu O, Yildirim NC, Danabas D, Yildirim N (2017) Biochemical impacts of the textile dyes

Remazol Brillant Blue R and Congo Red on the crayfish Astacus leptodactylus (Decapoda, Astacidae). Crustaceana 90:1563–1574. https://doi.org/10.1163/15685403-00003738

4. Aksu Z (2003) Reactive dye bioaccumulation by *Saccharomyces cerevisiae*. Process Biochem 38:1437–1444. https://doi.org/10.1016/S0032-9592(03)00034-7

5. Aksu Z, Dönmez G (2003) A comparative study on the biosorption characteristics of some yeasts for Remazol Blue reactive dye. Chemosphere 50:1075–1083. https://doi.org/10.1016/s0045-6535(02)00623-9

6. Aksu Z, Dönmez G (2005) Combined effects of molasses sucrose and reactive dye on the growth and dye bioaccumulation properties of *Candida tropicalis*. Process Biochem 40:2443–2454. https://doi.org/10.1016/j.procbio.2004.09.013

7. Al Farraj DA, Elshikh MS, Al Khulaifi MM, Hadibarata T, Yuniarto A, Syafiuddin A (2019) Biotransformation and detoxification of Antraquione dye green 3 using halophilic Hortaea sp. Int Biodeter Biodegr 140:72–77. https://doi.org/10.1016/j.ibiod.2019.03.011

8. Alam R, Ardiati FC, Solihat NN, Alam MB, Lee SH, Yanto DHY, Watanabe T, Kim S (2021) Biodegradation and metabolic pathway of anthraquinone dyes by Trametes hirsuta D7 immobilized in light expanded clay aggregate and cytotoxicity assessment. J Hazard Mat 405:124176. https://doi.org/10.1016/j.jhazmat.2020.124176

9. Andleeb S, Atiq N, Robson GD, Ahmed S (2012) An investigation of anthraquinone dye biodegradation by immobilized Aspergillus flavus in fluidized bed bioreactor. Environ Sci Pollut Res 19:1728–1737. https://doi.org/10.1007/s11356-011-0687-x

10. Chaudhari AU, Paul D, Dhotre D, Kodam KM (2017) Effective biotransformation and detoxification of anthraquinone dye reactive blue 4 by using aerobic bacterial granules. Water Res 122:603–613. https://doi.org/10.1016/j.watres.2017.06.005

11. Dalalibera A, Vilela PB, Vieira T, Becegato VA, Paulino AT (2020) Removal and selective separation of synthetic dyes from water using a polyacrylic acid-based hydrogel: characterization, isotherm, kinetic, and thermodynamic data. J Environ Chem Eng 8:104465. https://doi.org/10.1016/j.jece.2020.104465

12. Dehghani MH, Mahdavi P (2015) Removal of acid 4092 dye from aqueous solution by zinc oxide nanoparticles and ultraviolet irradiation. Des Water Treat 54:3464–3469. https://doi.org/10.1080/19443994.2014.913267

13. Deng D, Guo J, Zeng G, Sun G (2008) Decolorization of anthraquinone, triphenylmethane and azo dyes by a new isolated Bacillus cereus strain DC11. Int Biodeter Biodegr 62:263–269. https://doi.org/10.1016/j.ibiod.2008.01.017

14. Ebrahiem EE, Al-Maghrabi MN, Mobarki AR (2017) Removal of organic pollutants from industrial wastewater by applying photo-Fenton oxidation technology. Arab J Chemi 10:S1674–S1679. https://doi.org/10.1016/j.arabjc.2013.06.012

15. Enayatzamir K, Tabandeh F, Yakhchali B, Alikhani HA, Rodríguez Couto S (2009) Assessment of the joint effect of laccase and cellobiose dehydrogenase on the decolouration of different synthetic dyes. J Hazard Mat 169:176–181. https://doi.org/10.1016/j.jhazmat.2009.03.088

16. Farraj D, Elsheikh M, Khulaifi M, Hadibarata T, Yuniarto A, Syafiuddin A (2019) Biotransformation and detoxification of antraquione dye green 3 using halophilic Hortaea sp. Int Biodeter Biodegr 140:72–77. https://doi.org/10.1016/j.ibiod.2019.03.011

17. Gadekar MR, Ahammed MM (2016) Coagulation/flocculation process for dye removal using water treatment residuals: modelling through artificial neural networks. Des Water Treat 57:26392–26400. https://doi.org/10.1080/19443994.2016.1165150

18. Gadow SI, Li Y-Y (2020) Development of an integrated anaerobic/aerobic bioreactor for biodegradation of recalcitrant azo dye and bioenergy recovery: HRT effects and functional resilience. Biores Technol Rep 9:100388. https://doi.org/10.1016/j.biteb.2020.100388

19. Hadibarata T, Kristanti RA (2012) Effect of environmental factors in the decolorization of remazol brilliant blue R by polyporus SP. S133. J ChilChem Soc 57:1095–1098. https://doi.org/10.4067/S0717-97072012000200007

20. Hadibarata T, Nor NM (2014) Decolorization and degradation mechanism of Amaranth by Polyporus sp. S133. Bioproc and Biosys Eng 37:1879–1885. https://doi.org/10.1007/s00449-014-1162-0

21. Hadibarata T, Yusoff ARM, Aris A, Salmiati HT, Kristanti RA (2012) Decolorization of azo, triphenylmethane and anthraquinone dyes by laccase of a newly isolated Armillaria sp. F022. Water Air Soil Pollut 223:1045–1054. https://doi.org/10.1007/s11270-011-0922-6

22. Hadibarata T, Yusoff ARM, Kristanti RA (2012) Acceleration of anthraquinone-type dye removal by white-rot fungus under optimized environmental conditions. Water Air Soil Pollut 223:4669–4677. https://doi.org/10.1007/s11270-012-1177-6

23. Hadibarata T, Yusoff ARM, Kristanti RA (2012) Decolorization and metabolism of anthraquionone-type dye by laccase of white-rot fungi Polyporus sp. S133. Water Air Soil Pollut 223:933–941. https://doi.org/10.1007/s11270-011-0914-6

24. Hii HT (2021) Adsorption isotherm and kinetic models for removal of methyl orange and Remazol brilliant Blue R by coconut shell activated Carbon. Trop Aqua Soil Pollut 1:1–10. https://doi.org/10.53623/tasp.v1i1.4

25. Ihsanullah I, Jamal A, Ilyas M, Zubair M, Khan G, Atieh MA (2020) Bioremediation of dyes: current status and prospects. J Water Proc Eng 38:101680. https://doi.org/10.1016/j.jwpe.2020.101680

26. Jiménez S, Velásquez C, Mejía F, Arias M, Hormaza A (2019) Comparative studies of pure cultures and a consortium of white-rot fungi to degrade a binary mixture of dyes by solid-state fermentation and performance at different scales. Intl Biodeter Biodegr 145:104772. https://doi.org/10.1016/j.ibiod.2019.104772

27. Kadhom M, Albayati N, Alalwan H, Al-Furaiji M (2020) Removal of dyes by agricultural waste. Sus Chem Pharm 16:100259. https://doi.org/10.1016/j.scp.2020.100259

28. Karigar CS, Rao SS (2011) Role of microbial enzymes in the bioremediation of pollutants: a review. Enzyme Res 2011:805187. https://doi.org/10.4061/2011/805187

29. Kostyukov AA, Egorov AE, Mestergazi MG, Shmykova AM, Podrugina TA, Borissevitch IE, Shtil AA, Kuzmin VA (2020) Photochemical properties of new bis-cyanine dye as a promising agent for in vivo imaging. Mendeleev Comm 30:442–444. https://doi.org/10.1016/j.mencom.2020.07.012

30. Kristanti RA, Liong RMY, Hadibarata T (2021) Soil remediation applications of nanotechnology. Trop Aqua Soil Pollut 1:35–45. https://doi.org/10.53623/tasp.v1i1.12

31. Lai HJ (2021) Adsorption of Remazol Brilliant Violet 5R (RBV-5R) and Remazol Brilliant Blue R (RBBR) from aqueous solution by using agriculture waste. Trop Aqua Soil Pollut 1:11–23. https://doi.org/10.53623/tasp.v1i1.10

32. Li HH, Wang YT, Wang Y, Wang HX, Sun KK, Lu ZM (2019) Bacterial degradation of anthraquinone dyes. J Zhejiang Univ Sci B 20:528–540. https://doi.org/10.1631/jzus.B1900165

33. Maurya NS, Mittal AK, Cornel P, Rother E (2006) Biosorption of dyes using dead macro fungi: effect of dye structure, ionic strength and pH. Biores Technol 97:512–521. https://doi.org/10.1016/j.biortech.2005.02.045

34. Mohanty SS, Kumar A (2021) Enhanced degradation of anthraquinone dyes by microbial monoculture and developed consortium through the production of specific enzymes. Scientif Rep 11:7678. https://doi.org/10.1038/s41598-021-87227-6

35. Mokhtar N, Aziz EA, Aris A, Ishak WFW, Mohd Ali NS (2017) Biosorption of azo-dye using marine macro-alga of Euchema Spinosum. J Environ Chem Eng 5:5721–5731. https://doi.org/10.1016/j.jece.2017.10.043

36. Moreira PR, Bouillenne F, Almeida-Vara E, Xavier Malcata F, Frère JM, Duarte JC (2006) Purification, kinetics and spectral characterisation of a new versatile peroxidase from a Bjerkandera sp. isolate. Enzyme Microbial Technol 38:28–33. https://doi.org/10.1016/j.enzmictec.2004.12.035

37. Mostafa AA-F, Elshikh MS, Al-Askar AA, Hadibarata T, Yuniarto A, Syafiuddin A (2019) Decolorization and biotransformation pathway of textile dye by Cylindrocephalum aurelium. Bioproc Biosys Eng 42:1483–1494. https://doi.org/10.1007/s00449-019-02144-3

38. Nayak SS, Mirgane NA, Shivankar VS, Pathade KB, Wadhawa GC (2021) Adsorption of methylene blue dye over activated charcoal from the fruit peel of plant hydnocarpus pentandra. Mater Today: Proc 37:2302–2305. https://doi.org/10.1016/j.matpr.2020.07.728

39. Ngieng HY, Hadibarata T, Rubiyatno (2021) Utilization of construction and demolition waster and environmental management practice in South East Asian Countries. Trop Aqua Soil Pollut 1:46–61. https://doi.org/10.53623/tasp.v1i1.13
40. Novotný Č, Svobodová K, Kasinath A, Erbanová P (2004) Biodegradation of synthetic dyes by Irpex lacteus under various growth conditions. Int Biodeter Biodegr 54:215–223. https://doi.org/10.1016/j.ibiod.2004.06.003
41. Pan H, Xu X, Wen Z, Kang Y, Wang X, Ren Y, Huang D (2017) Decolorization pathways of anthraquinone dye Disperse Blue 2BLN by Aspergillus sp. XJ-2 CGMCC12963. Bioengineered 8:630–641. https://doi.org/10.1080/21655979.2017.1300728
42. Shu Z, Wu H, Lin H, Li T, Liu Y, Ye F, Mu X, Li X, Jiang X, Huang J (2016) Decolorization of Remazol Brilliant Blue R using a novel acyltransferase-ISCO (in situ chemical oxidation) coupled system. Biochem Eng J 115:56–63. https://doi.org/10.1016/j.bej.2016.08.008
43. Suteu D, Bilba D, Coseri S (2014) Macroporous polymeric ion exchangers as adsorbents for the removal of cationic dye basic blue 9 from aqueous solutions. J Appl Polym Sci 131. https://doi.org/10.1002/app.39620
44. Taha M, Adetutu EM, Shahsavari E, Smith AT, Ball AS (2014) Azo and anthraquinone dye mixture decolourization at elevated temperature and concentration by a newly isolated thermophilic fungus, Thermomucor indicae-seudaticae. J Environ Chem Eng 2:415–423. https://doi.org/10.1016/j.jece.2014.01.015
45. Tang KHD (2021) Interactions of microplastics with persistent organic pollutants and the ecotoxicological effects: a review. Trop Aqua Soil Pollut 1:24–34. https://doi.org/10.53623/tasp.v1i1.11
46. Verma AK, Raghukumar C, Parvatkar RR, Naik CG (2012) A rapid two-step bioremediation of the anthraquinone dye, reactive Blue 4 by a marine-derived fungus. Water Air Soil Pollut 223:3499–3509. https://doi.org/10.1007/s11270-012-1127-3
47. Vieira GAL, Cabral L, Otero IVR, Ferro M, Faria AU, Oliveira VM, Bacci M, Sette LD (2021) Marine associated microbial consortium applied to RBBR textile dye detoxification and decolorization: combined approach and metatranscriptomic analysis. Chemosphere 267:129190. https://doi.org/10.1016/j.chemosphere.2020.129190
48. Wang A, Qu J, Liu H, Ge J (2004) Degradation of azo dye Acid Red 14 in aqueous solution by electrokinetic and electrooxidation process. Chemosphere 55:1189–1196. https://doi.org/10.1016/j.chemosphere.2004.01.024
49. Wijannarong S, Aroonsrimorakot S, Thavipoke P, Kumsopa C, Sangjan S (2013) Removal of reactive dyes from textile dyeing industrial effluent by ozonation process. APCBEE Procedia 5:279–282. https://doi.org/10.1016/j.apcbee.2013.05.048
50. Xu R, Jia M, Li F, Wang H, Zhang B, Qiao J (2012) Preparation of mesoporous poly (acrylic acid)/SiO_2 composite nanofiber membranes having adsorption capacity for indigo carmine dye. Appl Phys A 106:747–755. https://doi.org/10.1007/s00339-011-6697-1
51. Yagub MT, Sen TK, Afroze S, Ang HM (2014) Dye and its removal from aqueous solution by adsorption: a review. Adv Coll Interf Sci 209:172–184. https://doi.org/10.1016/j.cis.2014.04.002
52. Yu Z, Wen X (2005) Screening and identification of yeasts for decolorizing synthetic dyes in industrial wastewater. Int Biodeter Biodegr 56:109–114. https://doi.org/10.1016/j.ibiod.2005.05.006
53. Zhang J, Zhou Q, Ou L (2016) Removal of indigo carmine from aqueous solution by microwave-treated activated carbon from peanut shell. Des Water Treat 57:718–727. https://doi.org/10.1080/19443994.2014.967729
54. Zhuo R, Zhang J, Yu H, Ma F, Zhang X (2019) The roles of Pleurotus ostreatus HAUCC 162 laccase isoenzymes in decolorization of synthetic dyes and the transformation pathways. Chemosphere 234:733–745. https://doi.org/10.1016/j.chemosphere.2019.06.113

Microalgae-Based Remediation Approaches in Textile Dye Removal

Jucélia T. Ferreira, Kyria C. de A. Bortoleti, Laysla dos S. Motta, Sávia Gavazza, Ana C. Brasileiro-Vidal, and Raquel P. Bezerra

Abstract Synthetic azo dyes are among the main constituents of textile effluents, whose improper treatment and discharge have been responsible for environmental degradation and contamination. This fact is associated with the stability and low aerobic biodegradability of these aromatic compounds, which provide acute and chronic effects to the aquatic biota and human health, because of their toxic, geno-toxic, mutagenic, and carcinogenic properties. Given the inefficiency of dye removal by the conventional physical and chemical treatments of textile effluents, biore-mediation with microalgae (phycoremediation) is considered an efficient and alter-native tool for dye removal, due to the high adsorption capacity, quick operation, cost-effectiveness, and eco-friendly technology. Phycoremediation involves biosorp-tion, bioaccumulation, and/or bioconversion processes. It may occur by the interac-tion between the functional chemical groups present in the cell wall of microalgae and the chemical structure of the dyes and/or by the activation of the azoreductase enzymes, which degrade the azo bonds into more simple or intermediate compounds.

Jucélia T. Ferreira, Kyria C. de A. Bortoleti—These authors contributed equally to this work.

J. T. Ferreira · S. Gavazza
Laboratório de Saneamento Ambiental, Universidade Federal de Pernambuco, Cidade Universitária, Rua Acdo. Hélio Ramos, s/n°, Recife, Pernambuco CEP 50740-530, Brazil
e-mail: savia@ufpe.br

K. C. de A. Bortoleti
Laboratório de Citogenética, Campus Ciências Agrárias, Universidade Federal Do Vale Do São Francisco, Rod. BR 407 Km 12 Lote 543 Projeto de Irrigação Senador Nilo Coelho, s/n° - C1, Petrolina, Pernambuco CEP 56.300-990, Brazil
e-mail: kyria.bortoleti@univasf.edu.br

Laysla dos S. Motta · A. C. Brasileiro-Vidal
Laboratório de Genética e Biotecnologia Vegetal, Departamento de Genética, Universidade Federal de Pernambuco, Av. Da Engenharia, s/n°, Recife, Pernambuco CEP 50740-600, Brazil
e-mail: ana.vidal@ufpe.br

R. P. Bezerra (✉)
Laboratório de Tecnologia de Bioativos, Universidade Federal Rural de Pernambuco, Rua Manuel de Medeiros, s/n°, Dois Irmãos, Recife, Pernambuco CEP 52171-900, Brazil
e-mail: Raquel.pbezerra@ufrpe.br

In this sense, this chapter highlights phycoremediation as an emerging and efficient technology in azo dye remediation, summarizes the remediation strategies used by microalgae, and discusses the advantages of the use of the generated algal biomass in the acquisition of products with added value, such as biodiesel, biofertilizers, and others.

Keywords Phycoremediation · Microalgae · Azo dyes · Biosorption · Bioaccumulation · Bioconversion · Azoreductase enzymes · Algal biomass · Textile wastewater · Environmental pollution

Abbreviations

$-CH_3$	Methyl group
$-Cl$	Chlorine
$-N=N-$	Azo group
$-NH_2$	Amino group
$-NO_2$	Nitro group
$-OH$	Hydroxyl group
BOD	Biological Oxygen Demand
CAGR	Compound Annual Growth Rate
COD	Chemical Oxygen Demand
CO_2	Carbon dioxide
FAME	Fatty Acid Methyl Ester
GHS	Globally Harmonized System of Classification and Labelling of Chemicals
H_2O	Water
OECD	Organization for Economic Co-operation and Development
$ZnCl_2$	Zinc chloride

1 Introduction

The textile sector consists of a production chain that integrates the production of natural, synthetic, and artificial fibers; spinning, weaving and knitting; processing, finishing and manufacture of the textile article. In 2020, considering the COVID-2019 pandemic, the global textile industry had a turnover of around US$920 billion, projecting a CAGR (Compound Annual Growth Rate) growth of 2.7% between 2020 and 2027, reaching a revised size of US$1.1 trillion in 2027 [90]. Worldwide, this sector employs from 60 to 75 million people [103], being China, the European Union, the United States, and India the main producing and exporting countries [90].

On the other hand, the textile sector faces challenges to raise productivity with sustainable solutions, since this anthropogenic activity is identified as one of the main causes of environmental pollution, resulting in acute effects to environmental and human health. The global production of textile fibers and clothing reaches more than 110 million tons per year, generating a large amount of waste, including residual waters, pre-consumption, and post-consumption solid waste, whose volume represents around 5% of the total waste produced worldwide [110]. Therefore, the importance of strategies which promote a balanced relationship between industrial production and the sustainable exploration of natural resources is emphasized [100, 104].

The production and consumption of textile products trigger environmental modifications in two ways: use of natural resources and pollution with the generated waste. The discard of textile effluents into the water bodies without proper treatment, especially after dyeing, printing, and finishing activities, has been responsible for around 20% of the environmental contamination of water bodies and soils [13]. This environmental impact is associated to heterogeneous features and a complex mixture of various pollutants in the effluents, containing large amounts of dyes (dispersive dyes, mordant dyes, reactive dyes, and solvent dyes), disinfectants, surfactants, heavy metals, and other recalcitrant organic substances ([62, 86]). These pollutants present difficult decomposition and are responsible for water pollution, loss of environmental balance, rise in Chemical Oxygen Demand (COD) and Biological Oxygen Demand (BOD) in water, alteration in pH and in the organic and inorganic chemical content of water bodies, obstruction of light penetration and inhibition of oxygen transfer, causing an effect of acute toxicity to the aquatic ecosystem and, consequently, its biota [6, 33, 81].

Organic dyes are soluble compounds, being classified into natural or synthetic [44]. These latter are predominant in the textile industry and are divided into different classes, based on the chemical structure of the chromophore group [3], with around 20–30 different classes of dyes [106]. Among them, the most applied dyes in the textile industry belong to the classes azo, anthraquinone, phthalocyanine, and triarylmethane [91, 120].

The class of azo dyes represents around 70% of the dyes used in the textile industry [89]. They are easily-synthesized compounds, present a wide variety of colors and good ability to impart color to a given substrate because of the presence of the chromophoric groups. They present an excellent property of fixation and permanence in the fibers because of the polar-polar interaction between them and the auxotrophic groups compared to natural dyes [42, 97, 111, 114]. Given these properties, these dyes are very versatile, being used for the dyeing of polyester, nylon, cellulose diacetate and triacetate, and acrylic fibers, as well as for additives in products derived from petroleum and in the dyeing of leather, paints, plastics, paper, wood, oils, cosmetics, pharmaceutical products, metals, and foods [55, 105].

On the other hand, some azo dyes and their degradation products are potentially hazardous to living organisms due to their possible toxicity and carcinogenicity [95]. Several countries have adopted an environmental legislation to restrict the use of dangerous chemical products in the textile production chain. In 1994, the German government devised the law Second Amendments to the Consumer Protection Act, one of the best known laws worldwide, prohibiting the use of azo dyes for their allergenic (Disperse Yellow 1/3, Disperse Orange 3/37/76; Disperse Red 1; Disperse Blue 1/35/106/124) and carcinogenic (Acid Red 26, Basic Red 9, Basic Violet 14, Direct Black 38, Direct Red 28, Direct Blue 6, Disperse Yellow 3, Disperse Orange 11, Disperse Blue 1) actions [85]. On the other hand, other European countries, Sweden, France, and Denmark, formulated their own legislation for the use of these dyes [85]. In Portugal, the Decree-Law n° 208/2003 was published, establishing that the azo dyes that produce aromatic amines at detectable concentrations (superior to 30 ppm), after reductive cleavage, must not be used on textile and leather articles that are susceptible to direct contact with the human skin or the oral cavity [50]. Nevertheless, in 2014, the European Union allowed the use of azo dyes in the textile industry, excluding them from the EU Ecolabel Textile Restricted Substances List [23].

Textile industries produce 200 billion liters of colored effluents per year. The large quantity of industrial effluents containing dyes (above 50%) is directed to conventional waste treatment plants, where they are subjected to primary, secondary, and tertiary treatments based on physical, chemical, and biological processes, or directly discarded into the aquatic ecosystem [51, 120]. According to Tkackyk et al. [108], dyes have been found in several types of environmental samples, such as water, suspended particulate matter, sediment, and fish, thus being considered micropollutants of aquatic ecosystems. Environmental persistence allows these pollutants to reach all trophic levels of a food chain, providing biomagnification [92], so that the organisms at high trophic levels show high levels of contamination compared to their preys [74].

In the conventional wastewater treatment plants, which generally apply aerobic activated sludge technology, the processes are low efficient in degrading some undesirable chemical compounds, especially synthetic organic dyes [106, 108]. Many chemicals used in the treatments, degradation products, metabolites, and refractory contaminants at low concentrations but with high toxicity are still maintained after treatment, which poses serious adverse effects to natural water bodies [7, 63]. In this sense, the decolorization and degradation of dyes using bacteria in biological treatment by sequential combination of anaerobic and aerobic processes have been receiving special attention because of their high efficiency, simple operation, inexpensive nature, and eco-friendliness [4, 17]. Furthermore, several organisms can degrade the target-compounds, eliminating the problem of pollutant transfer [67], including fungi, yeasts, and microalgae [47, 119].

The use of microalgae for dye degradation is more recent and emerges from the biosorption, bioaccumulation, and/or bioconversion processes. Known as Phycoremediation, it has been proposed as an alternative to mitigate the toxic and recalcitrant pollutants in the effluents. Microalgae, with emphasis to the green microalgae, have the capacity to remove different types of contaminants [59], because of the presence of different functional groups in their cell wall, with the example of amino, carboxyl, hydroxyl, and phosphate groups, which are responsible for the dye removal process [8]. Despite the progressive development of phycoremediation, some limitations still exist, demanding research studies which contribute to their effective application on a large scale.

2 Azo Dyes

Azo dyes represent around 2/3 of the synthetic aromatic dyes [40], being the most used class of dyes in the textile sector, with a representativeness from 60 to 70% [27, 94]. They have a chemical structure represented by the backbone, the auxochrome groups, the chromophore groups, and the solubilizing groups, which can form covalent bonds with the textile substrates [12].

These dyes are composed of one or more azo bonds, R_1–$N=N$–R_2, associated with one or more aromatic rings and sulfonic groups (SO^{3-}) [45], which increase dye solubility in water, rendering them non-biodegradable [60]. Aromatic compounds are complex and present a significant structural diversity, with properties that are altered to provide high degree of chemical, biological and photocatalytic stability [93], offering the dyes high photolytic stability and resistance to the main oxidizing agents [20, 38, 83]. Usually, they are bound to phenyl and naphthyl radicals, which are substituted by some functional groups, as the example of amino (–NH_2), chlorine (–Cl), hydroxyl (–OH), methyl (–CH_3), nitro (–NO_2), and sulfonates [11]. Thus, the synthetic nature, the stable and strong color, the high organic content, and the complex aromatic structure of azo dyes make them stable and difficult to degrade [43].

Considering hydrophobicity, azo dyes are classified into two types: (1) hydrophobic azo dyes, which are absorbed and reduced inside the microorganism cell; (2) hydrophilic azo dyes, which are reduced outside the microorganism cell [94]. According to the number of azo groups (–$N=N$–), they are classified as monoazo, disazo, trisazo, polyazo, or azoic [12]. For a certain time, it was believed that the stability and degradation rate of these dyes was associated to the increase in the number of azo bonds and, consequently, their high molecular weights [12]. Nonetheless, the stability of these dyes is not associated to the number of existing azo bonds, but to their respective structural symmetries, as observed in a study on the removal efficiency and speed of degradation of different azo dyes, including Reactive Orange 4 (monoazo), Reactive Black 5 (disazo), Direct Blue 71 (trisazo), and Direct Black 22 (polyazo) (Fig. 1). The dye Reactive Black 5, which is a disazo, was the molecule with the highest symmetry and which presented the lowest removal efficiency and degradation speed, which suggests the highest toxic potential for the referred molecule [101].

The toxic effects of the dyes have already been tested and proven in aquatic organisms from different trophic levels [108]. Considering the guidelines of OECD (Organisation for Economic Co-operation and Development) [76] and GHS (Globally Harmonized System of Classification and Labelling of Chemicals) [39], azo dyes were toxic for different bioindicators, as the example of *Vibrio fischeri* [41, 75], *Tetrahymena pyriformis* [75], *Chlorella vulgaris* [113], *Daphnia magna* [37],

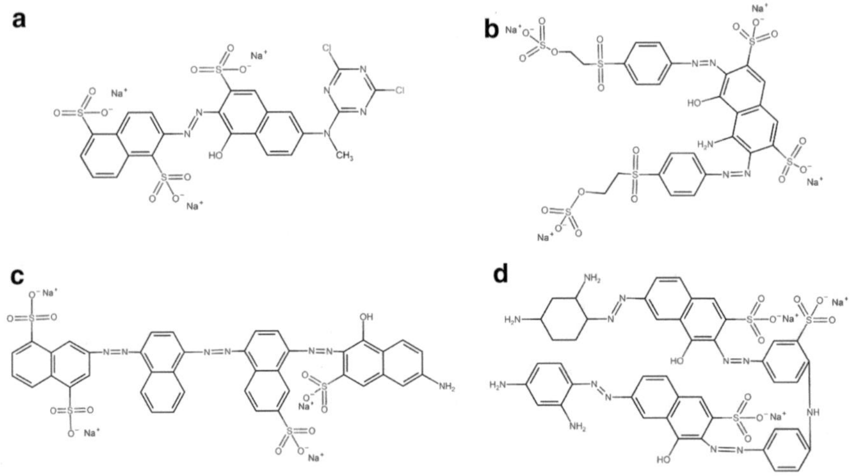

Fig. 1 Chemical structure of the textile dyes: **a** Reactive Orange 4 (monoazo), **b** Reactive Black 5 (diazo), **c** Direct Blue 71 (triazo), and **d** Direct Black 22 (polyazo)

Poecilia reticulata [99], among others. Additionally, several authors have reported the toxic, cytogenotoxic, mutagenic, and carcinogenic potential of azo dyes [14, 16, 36, 112]. These effects may result from the direct action of these compounds in the organisms, or associated to the formation of aromatic amines, by-products resulting from the reductive cleavage of the azo bonds present in the molecules. Depending on the metabolic activation of the amino group of the aromatic amines, chemical compounds are generated, causing damage to the DNA and proteins [73], the reason why amines present carcinogenic and mutagenic potential for the exposed organisms, including human beings [84].

3   Use of Microalgae in the Removal of Azo Dyes from Textile Effluents

3.1   Processes for the Treatment of Textile Effluents by Microalgae

The bioremediation processes have been considered promising alternatives, given the limitations of the physical and chemical treatments of dye-contaminated wastewater. They are eco-friendly methods, in which biological systems are used to direct the degradation or transformation of toxic chemical compounds into less toxic or non-toxic substances [8, 98]. Among the bioremediation methods, Phycoremediation addresses the use of macroalgae, microalgae, and cyanobacteria for the removal or biotransformation of contaminants, including nutrients, synthetic dyes, heavy metals, and xenobiotics of dye-contaminated wastewater [8], associated to the simultaneous biomass production. These organisms are widely available at natural resources, and are able to perform eco-friendly, fast, and low-cost bioremediation processes [109].

Microalgae are unicellular photoautotrophic microorganisms, usually in the size range of 1–400 µm, visible using a microscope [2, 70]. These organisms live in groups or colonies and are able to grow under different environmental conditions, such as soil, freshwater, and marine habitat, as well as in domestic and industrial effluent dump sites [66, 70]. In terms of biomass, microalgae are among the largest primary producers, and responsible for at least 32% of the global photosynthesis and oxygen production, as well as the best-known CO_2 sequestration [54].

In the past few years, studies which prove the efficiency of removal of textile pollutants by the microalgae have been published, emphasizing the possibility of use of this technology from a renewable and biodegradable source (Table 1). Nonetheless, it is known that this efficiency is variable and dependent on the molecular structure of the dye, the microalgal species and their respective strategies for dye degradation and other factors, such as pH, temperature, initial concentration of dyes, agitation speed, and biomass dosage [8, 24]. Figure 2 shows for instance Direct Black 22 dye decolorization after *Tetradesmus obliquus* cultivation.

The mechanisms of algal decolorization may occur by the processes of biosorption, bioaccumulation, and/or bioconversion during the microalgae cultivation [8, 61]. Briefly, biosorption consists in removing dye molecules from a liquid phase to a solid phase (the bioadsorbent), in an extracellular process. The complexation and electrostatic force of attraction between dye molecules and the solid surface of algal cells play an important role during this process. The biosorption capacity of microalgae is mainly determined by the presence of lipid compounds and heteropolysaccharide in the cell wall, since these compounds have many functional groups, such as amino, carboxyl, hydroxyl, and phosphate groups, which provide a strong attractive force between the dyes and the cell wall [35, 72, 102].

On the other hand, bioaccumulation comprehends the removal of contaminants from wastewater in an intracellular process. Some compounds can cross algal cell membranes and can be accumulated inside the cells [61]. However, in high concentrations, the accumulated dyes can induce the production of toxic metabolites, causing damage to cells or eventually death [116]. In some cases, the algal cell can neutralize the dyes, and the bioaccumulation becomes the initial step for the bioconversion process [61]. In the process of bioconversion, also known as biodegradation or enzymatic degradation, the activity of microalgal azoreductase enzymes promote the reductive cleavage of azo bond (–N=N–), converting dyes into simpler or intermediate compounds, such as aromatic amines (–NH_2), which can be further degraded under aerobic conditions, generating CO_2 and H_2O, at the final process [48, 95]. Thereby, bioaccumulation and biosorption can occur sequentially during the cultivation of microalgae for the bioremediation of textile wastewater [22].

Among the most used microalgal species for the removal of pollutants from textile wastewater, species of the green algal genera *Chlorella* sp. [57, 59, 71] and *Scenedesmus* sp. [59], belonging to the class of Chlorophycea [77, 82] can be cited,

Table 1 Microalgae used in the removal of textile dyes, specifying the mechanism of treatment, the dye evaluated, percentage of dye removal, removal time, the type of effluent, and the respective references

Microalgae	Mechanism	Dye	Color removal (%)	Removal time	Type of effluent	References
Anabaena sp. Bory de Saint-Vincent ex Bornet & Flahault 1886	Bioconversion	Indigo (ANIL)	71.2	14 days	Synthetic	Dellamatrice et al. [26]
Aphanocapsa elachista West & G. S. West 1894	Bioconversion	Orange 2RL	49.1	7 days	Synthetic	El-Sheekh et al. [31]
Chlorella sp. M.Beijerinck, 1890	Bioconversion	Orange 2RL	55.2	7 days	Synthetic	El-Sheekh et al. [31]
Chlorella vulgaris Beijerinck 1890	Bioconversion	G-Red (FN-3G)	59.1	7 days	Synthetic	El-Sheekh et al. [29]
C. vulgaris	Biosorption	NU	75.6	15 days	Real	El-Kassas and Mohamed [28]
C. vulgaris	Bioconversion	Congo Red	83.0	15 days	Synthetic	Hernández-Zamora et al. [46]
C. vulgaris	Biosorption	Congo Red	90.0	15 days	Synthetic	Mahalakshmi et al. [65]
C. vulgaris	Biosorption, bioaccumulation and bioconversion	Methylene Blue	98.5	8 days	Synthetic	Fazal et al. [35]
Chlorella pyrenoidosa Chick 1903	Bioconversion	Direct Red-31	78.6	180 min	Real	Sinha et al. [102]
C. pyrenoidosa	Biosorption	Methylene Blue	98.2	2 h	Synthetic	Lebron et al. [57]
C. pyrenoidosa	Biosorption	Methylene Blue	99.7[a]	4 h	Synthetic	Lebron et al. [58]

(continued)

Table 1 (continued)

Microalgae	Mechanism	Dye	Color removal (%)	Removal time	Type of effluent	References
Desmosdesmus sp. (Chodat) S. S. An, T. Friedl & E. Hegewald	Biosorption	Direct Red 31	36.0	4 h	Real	Behl et al. [10]
Microcystis aeruginosa Kützing 1846	Bioconversion	Reactive Black NN	55.12	7 days	Synthetic	El-Sheekh et al. [30]
M. aeruginosa	Bioconversion	Orange 2RL	65.07	7 days	Synthetic	El-Sheekh et al. [30]
Phormidium sp. Kützing ex Gomont 1892	Bioconversion	Indigo (ANIL)	91.2	14 days	Synthetic	Dellamatrice et al. [26]
Pseudoanabaena sp. Anagnostidis & Komárek 1974	Bioconversion	Reactive yellow 3RN	58.4	7 days	Synthetic	El-Sheekh et al. [30]
Pseudoanabaena sp.	Bioconversion	Tracid red BS	78.44	7 days	Synthetic	El-Sheekh et al. [30]
Spirogyra sp. C. G. Nees 1820	Biosorption	Reactive Yellow 22	92.0	120 h	Synthetic	Mohan et al. [69]
Spirulina platensis Sitzenberger ex Gomont 1892	Biosorption	Reactive Red 120	97.1	NI[2]	Synthetic	Cardoso et al. [15]
Spirulina maxima Sitzenberger ex Gomont 1892	Biosorption	Methylene Blue	94.1	2 h	Synthetic	Lebron et al. [57]
S. maxima	Biosorption	Methylene Blue	41.6[a]	4 h	Synthetic	Lebron et al. [58]

[a]Percentage of removal by the unmodified algal sample

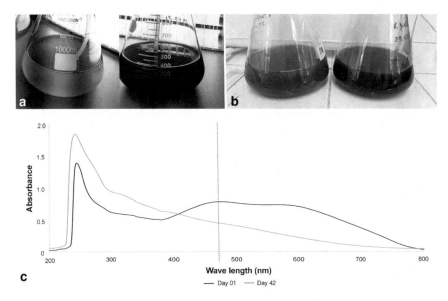

Fig. 2 Cultivation of the species *Tetradesmus obliquus* Kützing in the medium BG-11 without addition of the dye (left) and with addition of the dye Direct Black 22 (right) on the first day of experiment (**a**) and 42 **days** after the beginning of the experiment (**b**). UV–vis spectrophotometric scan of the cultivation of the species *T. obliquus* growing in the medium BG-11 with addition of the dye Direct Black 22, on the first day of cultivation, and after the period of 42 days, being 473 nm the dye peak absorbance (**c**)

as well as filamentous blue-green cyanobacterial genera, such as *Spirulina* sp. [57]. Figure 3 illustrates some important species for bioremediation.

Chlorella pyrenoidosa and *S. maxima*, for instance, were able to remove Methylene Blue, a cationic dye, by biosorption from a commercial dye sample and a synthetic textile effluent, in a rate varying from 92.20% to 99.70%, and from 41.64

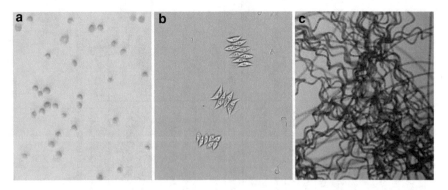

Fig. 3 Microphotographs of photosynthetic microorganisms: **a** *Chlorella vulgaris*, **b** *Tetradesmus obliquous*, and **c** *Arthrospira (Spirulina) platensis*

to 94.19%, respectively [57, 58]. The physic biosorption (physisorption) mechanism was directed by electrostatic interactions between Methylene Blue and the functional groups, especially hydrogen bonds, present on the surface of both microalgae. According to Lebron et al. [57], this high efficiency rate was associated to the pH, which was above 6, a condition in which Methylene Blue appears in its cationic form (MB^+) and with higher affinity to the negative charge of the adsorbent (algae), favoring dye biosorption, supporting the information that biosorption occurs because of the strong binding affinity of the negatively charged surface of the algal cells with the cationic azo dye [35, 57].

Additionally, modified *C. pyrenoidosa* and *S. maxima* samples using zinc chloride ($ZnCl_2$) improved the Methylene Blue biosorption efficiency to 99.7% for *S. maxima*, while a removal rate of 96% was maintained for the modified *C. pyrenoidosa* [58]. On the other hand, the esterified biomass of *Arthospira* (*Spirulina*) *platensis* presented a decrease in Methylene Blue biosorption capacity of 25.5% compared to the biosorption capacity of the untreated biomass due to the blockage of the surface carboxyl groups, which indicates a chemisorption process, with the participation of carboxyl groups in Methylene Blue binding by the untreated biomass [68]. The cell wall composition in these microalgae may be associated to the different dye removal rates, as well as to the removal rates that diverge by the unmodified and modified *S. maxima*. Cyanobacteria contain a thick structural layer of peptidoglycan and an extended layer of glycoproteins and polysaccharides, which are the main sources of reactive carboxyl groups on the biosorbent surface [34].

Different research works have been evaluating the phycoremediation potential of *Chlorella vulgaris*, emphasizing that this microalga acts by both the biosorption and bioconversion mechanisms. Fazal et al. [35] analyzed the employment of *C. vulgaris* in the process of biosorption of undiluted and diluted textile waster, observing a reduction of 99.7% and 98.5% of the dye Methylene Blue after eight days. Bioremediation by biosorption or bioconversion by *C. vulgaris* also promoted the removal of the dye Congo Red, also known as Direct Red 28, presenting a removal rate between 83 and 90%, after 96 h and 15 days, respectively [46, 65]. The biosorption ability of *C. vulgaris* was inversely proportional to the rise in Congo Red concentration [46], a fact also observed for the dyes Tartrazine [79] and Yellow dye Tectilon G [1], emphasizing that the microalgal capacity of azo dye biosorption depends on the structure and concentration of the dye, as well as the experimental conditions.

Regarding an inactive biomass of *C. vulgaris*, Hernández-Zamora et al. [46] observed a rapid decrease in the concentration of the dye Congo Red during the first 10 h of treatment, with a subsequent slow decrease in the residual concentration. Nevertheless, a decrease in biosorption efficiency was observed with the rise in dye concentration, resulting in a higher amount of non-adsorbed Congo Red dye molecules because of the saturation of biosorbent binding sites. It is suggested that Congo Red adsorption is performed by the cell wall of *C. vulgaris*, by the interaction of the dye molecule and the functional groups, such as amino, carboxyl, hydroxyl, sulfate, and other charged groups, to which dyes can bind [46]. In addition, an activity of the enzyme azo reductase of *C. vulgaris* was noticed from the initial concentrations of Congo Red, reaching a maximum value for the highest concentration of the dye and indicating the occurrence of biodegradation [46].

The bioconversion process has been studied for different azo dyes [Methyl red, Basic cationic, Basic Fuschin, Orange II, and G-Red (FN-3G)] for three, five, and seven days of cultivation using different microalgae [29]. Among them, the species *C. vulgaris* demonstrated an efficiency in degrading the dye G-Red (FN-3G) varying between 56.5% (three days) and 59.1% (seven days), with a noticeable induction of 72.25% in the activity of the enzyme azo reductase and, subsequently, the formation of aromatic amines. The addition of p-amino azo benzene increased the activity of this enzyme in *C. vulgaris* [29], since this compound acts as an electron donor for azo reductase, breaking the azo bond [49].

Evaluating the removal of two azo dyes (Disp Orange 2L and Reactive Yellow 3RN) by *C. vulgaris* and *Aphanocapsa elachita*, bioconversion was related to the molecular structure of the dye and the algal species used. After seven days of incubation, a degradation of 55.2% and 31.3% of Disp Orange 2L and Reactive Yellow 3RN was observed, respectively, for the treatment with *C. vulgaris*, and of 26.89% and 49.1%, respectively, for the treatment with *A. elachita*. The concentrations of both dyes induced the activation of the enzyme azoreductase, whose percentage was proportional to the incubation time in the culture of *A. elachita*. Additionally, the degradation of Disp Orange 2RL by the activation of the enzyme azoreductase of *C. vulgaris* was confirmed by Gas chromatography—mass spectrometry, observing a damage in the primary chromophore and the formation of intermediate compounds [31].

Additionally, the bioconversion of the dye Indigo (ANIL) by the microalga *Phormidium* sp. was analyzed, with the observation that the dye was biotransformed into two by-products, less toxic: anthranilic acid and isatin, after 17 days of treatment [26]. As observed, one of the advantages of the use of microalgae as an alternative for the treatment of textile effluents is associated with the bioconversion of the azo dyes into less toxic or non-toxic secondary compounds [26, 102].

In short, as presented, microalgae have three unique systems for dye removal: (1) dye adsorption by the algae cell wall; (2) dye intracellular accumulation, and (3) dye biodegradation, through the activation of the azo reductase enzyme, reductive cleavage of azo-bonds, and formation and mineralization of aromatic amines [32, 61].

3.2   Use of the Microalgal Biomass Generated in the Treatment of Textile Effluents

Although few studies have investigated the use of the microalgal biomass produced in the treatment of textile wastewater, this biomass is a commercial value-added product. Microalgae growing in textile wastewater store the excess energy in the form of lipids, carbohydrates, and proteins, which are the source of valuable bioproducts, such as animal food and feed, health supplements, biopharmaceuticals, cosmetics, biofertilizers, bioplastic, bioelectricity, as well as potential renewable energy as the feedstock for biofuel production, adding value to the bioproduct generated in the treatment of textile effluents [5, 25, 64, 87, 88]. Depending on the application of the biomass produced in the textile wastewater treatment, it can cover the cost of a microalgal production and/or can make the process cyclical when biomass is used as an energy source at the feeding process. Furthermore, this process can lower the environmental impact using wastewater instead of freshwater as well as CO_2 mitigation through microalgae growth during photosynthesis.

So far, microalgae-based textile wastewater treatment has been evaluated regarding energy-related questions, mainly as a biofuel source which are classified as solid (biochar), liquid (ethanol, vegetable oil, and biodiesel), and gaseous (biogas, biosyngas, and biohydrogen) fuels that can compete with fossil fuels. Biochar is produced through pyrolysis; bio-oil through thermochemical conversion; biomethane, biohydrogen, and bioalcohols through fermentation; and biodiesel, the most studied, through transesterification ([53, 117, 118]). Wu et al. [115] reported that the biomass of *Chlorella* sp. G23 accumulates fatty acids when cultivated at several textile wastewater concentrations, and the addition of extra phosphate and nitrogen sources could enhance pollutant removal efficiency, as well as the production of fatty acids. Similarly, the biomass of *C. vulgaris* and *Coelastrella* sp. produced in the treatment of textile effluents showed a high content of long-chain fatty acids, such as C16, C18, and C22, which are essential to biodiesel production [9, 35, 56]. Additionally, *C. vulgaris* cultivated in diluted synthetic textile wastewater showed high content of palmitic acid (C16:0, saturated fatty acid, 10.64 mg g^{-1}) and palmitoleic acid (C16:1, monounsaturated fatty acid, 0.43 mg g^{-1}). On the other hand, microalgae from undiluted synthetic textile wastewater presented higher content of linolenoic acid (C18:3, polyunsaturated fatty acid) when compared to 13-docosenic acid (C22:1, monounsaturated fatty acid). Additionally, *Coelastrella sp.* cultivated in the culture medium BG11 with 100 mg L^{-1} of Rhodamine B produced higher contents of linoleic (33.83%), heptadecanoic (32.1%), and pentadecanoic (23.18%) acids when compared to the cultivation in the standard culture medium. These results show that the dye potentially interferes in the microalgal metabolism to store energy and impact fatty acid yield [9, 35].

Another possible application of the microalgal biomass cultivated in wastewater is as biofertilizers, providing, mainly, nitrogen and/or growth regulators to plants. *Aulosira* sp., *Anabaena* sp., *Nostoc* sp., *Tolypothrix* sp., *Scytonema* sp., and *Plectonema* sp. are nitrogen-fixing cyanobacteria often used as biofertilizers, and, similarly to microalgae, can be used as a source of substances that promote germination, stem or leaf growth, and the flowering of plants ([18, 21, 52, 78]).

Furthermore, the biomass generated can be used to produce value-added products, such as polyunsaturated fatty acids, sterols, proteins, enzymes, vitamins, and pigments with commercial applications in the food industry, health sector, and cosmetic industry. Shah [96] described astaxanthin production by *Haematococcus pluvialis* in a wastewater treatment plant since it provides water and the necessary nutrients for algal cultivation. Astaxanthin plays an important role in the food, cosmetics, nutraceutical, and aquaculture industries.

The application of microalgal biomass depends on the species and the wastewater characteristics. Therefore, the integration of algal biomass with wastewater treatment may provide higher overall economics for the wastewater treatment process.

4   Concluding Remarks

Among the different methods for the treatment of textile effluents, phycoremediation with microalgae has been prominent in the current scenario, given the efficiency of azo dye remediation, reaching up to 94%. Microalgae can remove dyes by biosorption, bioaccumulation, and/or bioconversion processes, thus degrading or transforming it into less toxic chemical compounds. It is worth highlighting that the efficiency depends on the microalgal species employed, on the chemical structure of the dye and on the abiotic factors, such as pH.

The efficiency of microalgae is associated with the presence of functional chemical groups in the cell wall, such as amino, carboxyl, hydroxyl, and phosphate groups, which interact with azo dye molecules. Furthermore, microalgae have the capacity to accumulate the nutrients available in the textile effluents, such as organic carbon, nitrogen, phosphorus and trace metals, promoting their growth and their transformation into value-added bioproducts. Some studies include the use of the residual biomass as substrate for bioethanol, biodiesel, biogas, food source for fish and birds, and fertilizers. Therefore, the use of the residual biomass is a way to maximize the energy production obtained by the microalga and to reduce the total costs of the processes and treatment of waste.

Considering the promising and effective action of phycoremediation in dye removal, recent works have been applying biological treatments, associating microalgae, macroalgae, and bacteria to the treatment of textile effluents, as well as the use of nanoparticles, with promising results. Hence, microalgae present an effective bioremediation capability for the treatment of textile effluents, an advantageous treatment for requiring little energy, presenting inexpensive nature, forming little sludge, being easy to apply and environmentally-friendly, promoting economic and environmental sustainability.

Acknowledgements   We thank FACEPE (Fundação de Amparo à Ciência e Tecnologia do Estado de Pernambuco - grant number APQ-0456-3.07/20) and CNPq (Conselho Nacional de Desenvolvimento Científico e Tecnológico - grant number 304862/2018-5) for the financial support and scholarships (FACEPE grant numbers IBPG-1585-3.07/18 and BCT-0231-2.02/21).

References

1. Acuner E, Dilek FB (2004) Treatment of tectilon yellow 2G by *Chlorella vulgaris*. Process Biochem 39:623–631. https://doi.org/10.1016/S0032-9592(03)00138-9
2. Aiyar P, Schaeme D, García-Altares M, Flores DC, Dathe H, Hertweck C, Mittag M (2017) Antagonistic bacteria disrupt calcium homeostasis and immobilize algal cells. Nat Commun 8:1–13. https://doi.org/10.1038/s41467-017-01547-8
3. Almeida EJR, Mazzeo DEC, Sommaggio LRD, Marin-Morales MA, de Andrade AR, Corso CR (2019) Azo dyes degradation and mutagenicity evaluation with a combination of microbiological and oxidative discoloration treatments. Ecotoxicol Environ Saf 183:109484. https://doi.org/10.1016/j.ecoenv.2019.109484
4. Amaral FM, Florêncio L, Kato MT, Santa-Cruz PA, Gavazza S (2017) Hydraulic retention time influence on azo dye and sulfate removal during the sequential anaerobic–aerobic treatment of real textile wastewater. Water Sci Technol 76(12):3319–3327. https://doi.org/10.2166/wst.2017.378
5. Ángeles R, Rodero R, Carvajal A, Muñoz R, Lebrero R (2019) Potential of microalgae for wastewater treatment and its valorization into added value products. Application of microalgae in wastewater treatment, pp 281–315. https://doi.org/10.1007/978-3-030-13909-4_13
6. Asad S, Amoozegar MA, Pourbabaee A, Sarbolouki MN, Dastgheib SMM (2007) Decolorization of textile azo dyes by newly isolated halophilic and halotolerant bacteria. Bioresour Technol 98:2082–2088. https://doi.org/10.1016/j.biortech.2006.08.020
7. Asghar A, Raman AAA, Daud WMAW (2015) Advanced oxidation processes for in-situ production of hydrogen peroxide/hydroxyl radical for textile wastewater treatment: a review. J Clean Prod 87:826–838. https://doi.org/10.1016/j.jclepro.2014.09.010
8. Ayele A, Getachew D, Kamaraj M, Suresh A (2021) Phycoremediation of synthetic dyes: an effective and eco-friendly algal technology for the dye abatement. J Chem. https://doi.org/10.1155/2021/9923643
9. Baldev E, MubarakAli D, Ilavarasi A, Pandiaraj D, Ishack KSS, Thajuddin N (2013) Degradation of synthetic dye, Rhodamine B to environmentally non-toxic products using microalgae. Colloids Surf B Biointerfaces 105:207–214. https://doi.org/10.1016/j.colsurfb.2013.01.008
10. Behl K, Joshi M, Sharma M, Tandon S, Chaurasia AK, Bhatnagar A, Nigam S (2019a) Performance evaluation of isolated electrogenic microalga coupled with graphene oxide for decolorization of textile dye wastewater and subsequent lipid production. Chem Eng J 375:121950. https://doi.org/10.1016/j.cej.2019.121950
11. Bell J, Plumb JJ, Buckley CA, Stuckey DC (2000) Treatment and decolorization of dyes in an anaerobic baffled reactor. J Environ Eng 126:1026–1032
12. Benkhaya S, M'rabet S, El Harfi A (2020) Classifications, properties, recent synthesis and applications of azo dyes. Heliyon 6:e03271. https://doi.org/10.1016/j.heliyon.2020.e03271
13. Bhatia SC, Devraj S (2017) Pollution control in textile industry. Woodhea Publishing India Textiles. https://doi.org/10.1201/9781315148588
14. Brambilla CMCC, Garcia ALH, Silva FR, Taffarel SR, Grivicich I, Picada JN, Scotti A, Dalberto D, Misik M, Knasmuller S, Silva J (2019) Amido Black 10B a widely azo dye causes DNA damage in pro- and eukaryotic indicator cells. Chemosphere 217:430–436. https://doi.org/10.1016/j.chemosphere.2018.11.026
15. Cardoso NF, Lima EC, Royer B, Bach MV, Dotto GL, Pinto LA, Calvete T (2012) Comparison of *Spirulina platensis* microalgae and commercial activated carbon as adsorbents for the removal of Reactive Red 120 dye from aqueous effluents. J Hazard Mater 241:146–153. https://doi.org/10.1016/j.jhazmat.2012.09.026
16. Caritá R, Marin-Morales MA (2008) Induction of chromosome aberrations in the *Allium cepa* test system caused by the exposure of seeds to industrial effluents contaminated with azo dyes. Chemosphere 72:722–725. https://doi.org/10.1016/j.chemosphere.2008.03.056
17. Carvalho JRS; Amaral FM; Florencio L; Kato MT; Delforno TP; Gavazza S (2020) Microaerated UASB reactor treating textile wastewater: the coremicrobiome and removal of azo

dye Direct Black 22. Chemosphere 242:125157. https://doi.org/10.1016/j.chemosphere.2019. 125157

18. Castro JS, Calijuri ML, Ferreira J, Assemany PP, Ribeiro, VJ (2020) Microalgae based biofertilizer: a life cycle approach. Sci Total Environ 724:138138. https://doi.org/10.1016/j.scitot env.2020.138138

19. Chai WS, Tan WG, Munawaroh HSH, Gupta VK, Ho SH, Show PL (2020) Multifaceted roles of microalgae in the application of wastewater biotreatment: a review. Environ Pollut 269:116236. https://doi.org/10.1016/j.envpol.2020.116236

20. Chequer FMD, Lizier TM, de Felício R, Zanoni MVB, Debonsi HM, Lopes NP, de Oliveira DP (2015) The azo dye Disperse Red 13 and its oxidation and reduction products showed mutagenic potential. Toxicol in Vitro 29:1906–1915. https://doi.org/10.1016/j.tiv.2015.08.001

21. Chittora D, Meena M, Barupal T, Swapnil P, Sharma K (2020) Cyanobacteria as a source of biofertilizers for sustainable agriculture. Biochem Biophys Rep 22:100737. https://doi.org/ 10.1016/j.bbrep.2020.100737

22. Chu WL, See YC, Phang SM (2009) Use of immobilised *Chlorella vulgaris* for the removal of colour from textile dyes. J Appl Phycol 21:641–648

23. Commission Decision of 5 June 2014 establishing the ecological criteria for the award of the EU Ecolabel for textile products (2014) Off J Eur Union 174:45–83

24. Daneshvar N, Khataee AR, Rasoulifard MH, Pourhassan M (2007) Biodegradation of dye solution containing Malachite Green: optimization of effective parameters using Taguchi method. J Hazard Mater 143:214–219. https://doi.org/10.1016/j.jhazmat.2006.09.016

25. da Silva MROB, Moura YAS, Converti A, Porto ALF, Marques DDAV, Bezerra RP (2021) Assessment of the potential of Dunaliella microalgae for different biotechnological applications: a systematic review. Algal Res 58:102396. https://doi.org/10.1016/j.algal.2021. 102396

26. Dellamatrice PM, Silva-Stenico ME, Moraes LABD, Fiore MF, Monteiro RTR (2017) Degradation of textile dyes by cyanobacteria. Braz J Microbiol 48:25–31. https://doi.org/10.1016/ j.bjm.2016.09.012

27. Dos Santos AB, Bisschops IA, Cervantes FJ, van Lier JB (2004) Effect of different redox mediators during thermophilic azo dye reduction by anaerobic granular sludge and comparative study between mesophilic (30 C) and thermophilic (55 C) treatments for decolourisation of textile wastewaters. Chemosphere 55:1149–1157. https://doi.org/10.1016/j.chemosphere. 2004.01.031

28. El-Kassas HY, Mohamed LA (2014) Bioremediation of the textile waste effluent by *Chlorella vulgaris*. Egypt J Aquat Res 40(301):308. https://doi.org/10.1016/j.ejar.2014.08.003

29. El-Sheekh MM, Gharieb MM, Abou-El-Souod GW (2009) Biodegradation of dyes by some green algae and cyanobacteria. Int Biodeter Biodegr 63:699–704. https://doi.org/10.1016/j. ibiod.2009.04.010

30. El-Sheekh MM, Abou-El-Souod GW, El Asrag HA (2017) Biodegradation of some dyes by the cyanobacteria species *Pseudoanabaena sp.* and *Microcystis aeruginosa Kützing*. Egypt J Exp Biol (Bot.) 13:233–243. https://doi.org/10.5455egyjebb.20170628083523

31. El-Sheekh MM, Abou-El-Souod GW, El Asrag HA (2018) Biodegradation of some dyes by the green alga *Chlorella vulgaris* and The Cyanobacterium *Aphanocapsa elachista*. Egypt J Bot 58:311–320. https://doi.org/10.21608/EJBO.2018.2675.1145

32. El-Sheekh MM, El-Shanshoury AR, Abou-El-Souod GW, Gharieb DY, El Shafay SM (2021) Decolorization of dyestuffs by some species of green algae and cyanobacteria and its consortium. Int J Environ Sci Technol. https://doi.org/10.1007/s13762-020-03108-x

33. Fang H, Wenrong H, Yuezhong L (2004) Biodegradation mechanisms and kinetics of azo dye 4BS by a microbial consortium. Chemosphere 57:293–301. https://doi.org/10.1016/j.chemos phere.2004.06.036

34. Fang L, Zhou C, Cai P, Chen W, Rong X, Dai K, Huang Q (2011) Binding characteristics of copper and cadmium by cyanobacterium *Spirulina platensis*. J Hazard Mater 190:810–815. https://doi.org/10.1016/j.jhazmat.2011.03.122

35. Fazal T, Rehman MSU, Javed F, Akhtar M, Mushtaq A, Hafeez A, Rehman F (2021) Integrating bioremediation of textile wastewater with biodiesel production using microalgae (*Chlorella vulgaris*). Chemosphere 281:130758. https://doi.org/10.1016/j.chemosphere.2021.130758

36. Fernandes FH, Bustos-Obregon E, Salvadori DMF (2015) Disperse Red 1 (textile dye) induces cytotoxic and genotoxic effects in mouse germ cells. Reprod Toxicol 53:75e81. https://doi.org/10.1016/j.reprotox.2015.04.002

37. Ferraz ERA, Umbuzeiro GA, De Almeida G, Caloto Oliveira A, Chequer FMD, Zanoni MVB, Oliveira DP (2011) Differential toxicity of Disperse Red 1 and Disperse Red 13 in the Ames test, HepG2 cytotoxicity assay, and *Daphnia acute* toxicity test. Environ Toxicol 26:489–497. https://doi.org/10.1002/tox.20576

38. Franciscon E, Mendonca D, Seber S, Morales DA, Zocolo GJ, Zanoni MB, Umbuzeiro GA (2015) Potential of a bacterial consortium to degrade azo dye Disperse Red 1 in a pilot scale anaerobic–aerobic reactor. Process Biochem 50:816–825. https://doi.org/10.1016/j.procbio.2015.01.022

39. GHS (2011) Globally harmonized system of classification and labelling of chemicals (GHS). United Nations, New York and Geneva, pp 1–568

40. Gičević A, Hindija L, Karačić A (2020) Toxicity of azo dyes in pharmaceutical industry. In: International conference on medical and biological engineering. Springer, Cham, pp 581–587. https://doi.org/10.1007/978-3-020-17971-7_88

41. Gottlieb A, Shaw C, Smith A, Wheatley A, Forsythe S (2003) The toxicity of textile reactive azo dyes after hydrolysis and decolourisation. J Biotechnol 101:49–56. https://doi.org/10.1016/S0168-1656(02)00302-4

42. Griffiths J (1984) Developments in the chemistry and technology of organic

43. Gupta VK, Suhas (2009) Application of low cost adsorbents for dye removal—a review. J Environ Manage 90:2313–2342. https://doi.org/10.1016/j.jenvman.2008.11.017

44. Gürses A, Açıkyıldız M, Güneş K, Gürses MS (2016) Dyes and pigments: their structure and properties. Dyes and pigments. Springer, Cham, pp 13–29. https://doi.org/10.1007/978-3-319-33892-7_2

45. Halaburgi V, Karegoudar T (2016) Biocatalysis and Biotransformation 34(6):265–271. https://doi.org/10.1080/10242422.2016.1247828

46. Hernández-Zamora M, Cristiani-Urbina E, Martínez-Jerónimo F, Perales-Vela HV, Ponce-Noyola T, del Carmen M-H, Cañizares-Villanueva RO (2015) Bioremoval of the azo dye Congo Red by the microalga *Chlorella vulgaris*. Environ Sci Pollut Res 22:10811–10823. https://doi.org/10.1007/s11356-015-4277-1

47. Jadhav I, Vasniwal R, Shrivastav D, Jadhav K (2016) Microorganism-based treatment of azo dyes. J Environ Sci Technol 9:188–197. https://doi.org/10.3923/jest.2016.188.197

48. Jamee R, Siddique R (2019) Biodegradation of synthetic dyes of textile effluent by microorganisms: an environmentally and economically sustainable approach. Eur J Microbiol Immunol 9:114–118. https://doi.org/10.1556/1886.2019.00018

49. Jingi L, Houtian L (1992) Defradation of azo dyes by algae. Environ Pollut 75:273–278. https://doi.org/10.1016/0269-7491(92)90127-V

50. Journal of the Portuguese Diary of September 15, 2003. 1ª A, n° 213 (2003) Min of Economy, Portugal, pp 6014–6016

51. Kant R (2012) Textile dyeing industry an environmental hazard. Nat Sci 4:22–26. https://doi.org/10.4236/ns.2012.41004

52. Khan SA, Sharma GK, Malla FA, Kumar A, Gupta N (2019) Microalgae based biofertilizers: A biorefinery approach to phycoremediate wastewater and harvest biodiesel and manure. J Cleaner Production 211:1412–1419. https://doi.org/10.1016/j.jclepro.2018.11.281

53. Klinthong W, Yang YH, Huang CH, Tan CS (2015) A review: microalgae and their applications in CO2 capture and renewable energy. Aerosol Air Qual Res 15:712–742. https://doi.org/10.4209/aaqr.2014.11.0299

54. Kumar KS, Dahms HU, Won EJ, Lee JS, Shin KH (2015) Microalgae–a promising tool for heavy metal remediation. Ecotoxicol Environ Saf 113:329–352. https://doi.org/10.1016/j.ecoenv.2014.12.019

55. Kunz A, Peralta-Zamora P, Moraes SG, Durán N (2002) Degradation of reactive dyes by the system metallic iron/hydrogen peroxide. Degradação de corantes reativos pelo sistema ferro metálico/peróxido de hidrogênio. Quím Nova 25:78–82

56. Lam MK, Lee KT (2012) Immobilization as a feasible method to simplify the separation of microalgae from water for biodiesel production. Chem Eng J 191:263–268. https://doi.org/10.1016/j.cej.2012.03.013

57. Lebron YAR, Moreira VR, Santos LVS, Jacob RS (2018) Remediation of methylene blue from aqueous solution by *Chlorella pyrenoidosa* and *Spirulina maxima* biosorption: equilibrium, kinetics, thermodynamics and optimization studies. J Environ Chem Eng 6:6680–6690. https://doi.org/10.1016/j.jece.2018.10.025

58. Lebron YAR, Moreira VR, Santos LVS (2019) Studies on dye biosorption enhancement by chemically modified *Fucus vesiculosus*, *Spirulina maxima* and *Chlorella pyrenoidosa* algae. J Clean Prod 240:118197. https://doi.org/10.1016/j.jclepro.2019.118197

59. Lekshmi B, Joseph RS, Jose A, Abinandan S, Shanthakumar S (2015) Studies on reduction of inorganic pollutants from wastewater by *Chlorella pyrenoidosa* and *Scenedesmus abundans*. Alex Eng J 54:1291–1296. https://doi.org/10.1016/j.aej.2015.09.013

60. Lellis B, Fávaro-Polonio CZ, Pamphile JA, Polonio JC (2019) Effects of textile dyes on health and the environment and bioremediation potential of living organisms. Biotechnol Res Innov 3:275–290. https://doi.org/10.1016/j.biori.2019.09.001

61. Leng L, Wei L, Xiong Q, Xu S, Li W, Lv S, Zhou W (2020) Use of microalgae based technology for the removal of antibiotics from wastewater: a review. Chemosphere 238:124680. https://doi.org/10.1016/j.chemosphere.2019.124680

62. Liang J, Ning XA, Kong M, Liu D, Wang G, Cai H, Sun J, Zhang Y, Lu X, Yuan Y (2017) Elimination and ecotoxicity evaluation of phthalic acid esters from textile-dyeing wastewater. Environ Pollut 231:115–122. https://doi.org/10.1016/j.envpol.2017.08.006

63. Liang J, Ning XA, Sun J, Song J, Hong Y, Cai H (2018) An integrated permanganate and ozone process for the treatment of textile dyeing wastewater: efficiency and mechanism. J Clean Prod 204:12–19. https://doi.org/10.1016/j.jclepro.2018.08.112

64. Logroño W, Pérez M, Urquizo G, Kadier A, Echeverría M, Recalde C, Rákhely G (2017) Single chamber microbial fuel cell (SCMFC) with a cathodic microalgal biofilm: a preliminary assessment of the generation of bioelectricity and biodegradation of real dye textile wastewater. Chemosphere 176:378–388. https://doi.org/10.1016/j.chemosphere.2017.02.099

65. Mahalakshmi S, Lakshmi D, Menaga U (2015) Biodegradation of different concentration of dye (Congo red dye) by using green and blue green algae. Int J Environ Res 9:735–744

66. Menezes M, Bicudo CE, Moura CW, Alves AM, Santos AA, Pedrini ADG, Silva WJD (2015) Update of the Brazilian floristic list of Algae and Cyanobacteria. Rodriguésia 66:1047–1062. https://doi.org/10.1590/2175-7860201566408

67. Menezes O, Brito R, Hallwass F, Florêncio L, Kato MT, Gavazza S (2019) Coupling intermittent micro-aeration to anaerobic digestion improves tetra-azo dye Direct Black 22 treatment in sequencing batch reactors. Chem Eng Res Des 146:369–378. https://doi.org/10.1016/j.cherd.2019.04.020

68. Mitrogiannis D, Markou G, Çelekli A, Bozkurt H (2015) Biosorption of methylene blue onto *Arthrospira platensis* biomass: kinetic, equilibrium and thermodynamic studies. Int J Environ Chem Eng 3:670–680. https://doi.org/10.1016/j.jece.2015.02.008

69. Mohan SV, Rao NC, Prasad KK, Karthikeyan J (2002) Treatment of simulated Reactive Yellow 22 (Azo) dye effluents using Spirogyra species. Waste Manage 22:575–582. https://doi.org/10.1016/S0956-053X(02)00030-2

70. Monteiro CM, Castro PM, Malcata FX (2012) Metal uptake by microalgae: underlying mechanisms and practical applications. Biotechnol Prog 28:299–311. https://doi.org/10.1002/btpr.1504

71. Moreira VR, Lebron YAR, Freire SJ, Santos LVS, Palladino F, Jacob RS (2019) Biosorption of copper ions from aqueous solution using *Chlorella pyrenoidosa*: optimization, equilibrium and kinetics studies. Microchem J 145:119–129. https://doi.org/10.1016/j.microc.2018.10.027

72. Murugesan S, Sridharan MC, Yoganandam M (2016) Biological decolorization and removal of metal from dye industry effluent by microalgae. Biosci Biotechnol Res Asia 6:111–120
73. Neumann HG (2010) Aromatic amines: mechanisms of carcinogenesis and implications for risk assessment. Front Biosci 15:1119–1130
74. Newman MC (2015) Fundamentals of ecotoxicology: the science of pollution. CRC Press, Boca Raton
75. Novotný Č, Dias N, Kapanen A, Malachová K, Vándrovcová M, Itävaara M, Lima N (2006) Comparative use of bacterial, algal and protozoan tests to study toxicity of azo-and anthraquinone dyes. Chemosphere 63:1436–1442. https://doi.org/10.1016/j.chemosphere.2005.10.002
76. OECD (2001) Guidance document on the use of the harmonised system for the classification of chemicals which are hazardous for the aquatic environment. OECD Publishing, Paris
77. Oliva-Martínez MG, Godínez-Ortega JL, Zuñiga-Ramos CA (2014) Biodiversity of inland water phytoplankton in Mexico. Rev Mex Biodivers 85:S54–S61. https://doi.org/10.7550/rmb.32706
78. Osman MEH, El-Sheekh MM, El-Naggar AH, Gheda SF (2010) Effect of two species of cyanobacteria as biofertilizers on some metabolic activities, growth, and yield of pea plant. Biol Fertil Soils 46:861–875. https://doi.org/10.1016/j.bbrep.2020.100737
79. Omar HH (2008) Algal decolorization and degradation of monoazo and diazo dyes. Pak J Biol Sci 11:1310–1316. https://doi.org/10.3923/pjbs.2008.1310.13161
80. Oyebamiji OO, Boeing WJ, Holguin FO, Ilori O, Amund O (2019) Green microalgae cultured in textile wastewater for biomass generation and biodetoxification of heavy metals and chromogenic substances. Bioresour Technol Rep 7:100247. https://doi.org/10.1016/j.biteb.2019.100247
81. Pandey A, Singh P, Iyengar L (2007) Bacterial decolorization and degradation of azo dyes. Int Int Biodeter Biodegr 59:73–84. https://doi.org/10.1016/j.ibiod.2006.08.006
82. Pathak VV, Kothari R, Chopra AK, Singh DP (2015) Experimental and kinetic studies for phycoremediation and dye removal by *Chlorella pyrenoidosa* from textile wastewater. J Environ Manag 163:270–277. https://doi.org/10.1016/j.jenvman.2015.08.041
83. Pérez-Ibarbia L, Majdanski T, Schubert S, Windhab N, Schubert US (2016) Safety and regulatory review of dyes commonly used as excipients in pharmaceutical and nutraceutical applications. Eur J Pharm Sci 93:264–273. https://doi.org/10.1016/j.ejps.2016.08.026
84. Pira E, Piolatto G, Negri E, Romano C, Boffetta P, Lipworth L (2010) Bladder cancer mortality of workers exposed to aromatic amines: a58-year follow-up. J Natl Cancer Inst 102:1096–1099. https://doi.org/10.1093/jnci/djq214
85. PRCEE, Policy Research Center for Environment and Economy (1999) The 3rd meeting of the 2nd phase of CCICED impacts of environmental standards and Requirements in EU countries on China's textile industry
86. Punzi M, Anbalagan A, Börner RA, Svensson BM, Jonstrup M, Mattiasson B (2015) Degradation of a textile azo dye using biological treatment followed by photo-Fenton oxidation: evaluation of toxicity and microbial community structure. Chem Eng J 270:290–299. https://doi.org/10.1016/j.cej.2015.02.042
87. Raheem A, Prinsen P, Vuppaladadiyam AK, Zhao M, Luque R (2018) A review on sustainable microalgae based biofuel and bioenergy production: recent developments. J Clean Prod 181:42–59. https://doi.org/10.1016/j.jclepro.2018.01.125
88. Rashid N, Rehman MSU, Han JI (2013) Recycling and reuse of spent microalgal biomass for sustainable biofuels. Biochem Eng J 75:101–107. https://doi.org/10.1016/j.bej.2013.04.001
89. Rawat D, Sharma RS, Karmakar S, Arora LS, Mishra V (2018) Ecotoxic potential of a presumably non-toxic azo dye. Ecotoxicol Environ Saf 148:528–537. https://doi.org/10.1016/j.ecoenv.2017.10.049
90. Reportlinker (2021). Indústria têxtil global. Pesquisa de impacto no mercado - covid-19 & recessão de looming. https://www.reportlinker.com/p05961230/?utm_source=GNW. Accessed June 2021

91. Saleh SMAA (2005) HPLC determination of four textile dyes and studying their degradation using spectrophotometric technique (Doctoral dissertation)
92. Sandhya S (2010) Biodegradation of azo dyes under anaerobic condition: role of azoreductase. Biodegradation of azo dyes. Handb Environ Chem 9:39–57
93. Saratale RG, Saratale GD, Chang JS, Govindwar SP (2011) Bacterial decolorization and degradation of azo dyes: a review. J Taiwan Inst Chem Eng 42:138–157. https://doi.org/10.1016/j.jtice.2010.06.006
94. Sarkar S, Banerjee A, Halder U, Biswas R, Bandopadhyay R (2017) Degradation of synthetic azo dyes of textile industry: a sustainable approach using microbial enzymes. Water Conserv Sci Eng 2:121–131. https://doi.org/10.1007/s41101-017-0031-5
95. Selvaraj V, Swarna Karthika T, Mansiya C, Alagar M (2021) An over review on recently developed techniques, mechanisms and intermediate involved in the advanced azo dye degradation for industrial applications. J. Mol. Struc. 1224:129195. https://doi.org/10.1016/j.molstruc.2020.129195
96. Shah MMR (2019) Astaxanthin production by microalgae *Haematococcus pluvialis* through wastewater treatment: waste to resource. In: Gupta S, Bux F (eds) Application of microalgae in wastewater treatment. Springer, Cham. https://doi.org/10.1007/978-3-030-13909-4_2
97. Shamey R, Zhao X (2014) Modelling, simulation and control of the dyeing process. Published by Woodhead Publishing Limited in association with The Textile Institute. Elsevier
98. Sharma H, Shirkot P (2019) Bioremediation of azo dyes using biogenic iron nanoparticles. J Microbiol Exp 7:12–15. https://doi.org/10.15406/jmen.2019.07.00232
99. Sharma S, Sharma S, Sharma KP (2006) Identification of a sensitive index during fish bioassay of an azo dye methyl red (untreated and treated). J Environ Biol 27:551–555
100. Sillanpää M, Ncibi C (2019) The circular economy: case studies about the transition from the linear economy. Academic Press. https://doi.org/10.1016/B978-0-12-815267-6.00005-0
101. Silva RDB (2018) Comportamento cinético da degradação de corantes azo e aminas aromáticas sob diferentes condições redox (Master's thesis, Universidade Federal de Pernambuco). https://repositorio.ufpe.br/handle/123456789/32129
102. Sinha S, Singh R, Chaurasia AK, Nigam S (2016) Self-sustainable Chlorella pyrenoidosa strain NCIM 2738 based photobioreactor for removal of Direct Red-31 dye along with other industrial pollutants to improve the water-quality. J Hazard Mater 306:386–394. https://doi.org/10.1016/j.jhazmat.2015.12.011
103. Solidarity Center (2019) Global garment and textile industries: workers, rights and working conditions [Internet]. Disponível online em https://www.solidaritycenter.org/wp-content/uploads/2019/08/Garment-Textile-Industry-Fact-Sheet.8.2019.pdf. Acessado Mai 2021
104. Stanescu MD (2021) State of the art of post-consumer textile waste upcycling to reach the zero waste milestone. Environ Sci Pollut Res 1–18. https://doi.org/10.1007/s11356-021-12416-9
105. Stolz A (2001) Basic and applied aspects in the microbial degradation of azo dyes. Appl Microbiol Biotechnol 56:69–80. https://doi.org/10.1007/s002530100686
106. Sudha M, Saranya A, Selvakumar N (2014) Microbial degradation of azo dyes: a review. Int J Curr Microbiol Appl Sci 3:670–690
107. Terpou A, Kornaros M (2020) Microalgae-based remediation of wastewaters. Elsevier Inc., Microalgae Cultivation for Biofuels Production
108. Tkaczyk A, Mitrowska K, Posyniak A (2020) Synthetic organic dyes as contaminants of the aquatic environment and their implications for ecosystems: a review. Sci Total Environ 717:137222. https://doi.org/10.1016/j.scitotenv.2020.137222
109. Tsai WT, Chen HR (2010) Removal of malachite green from aqueous solution using low-cost chlorella-based biomass. J Hazard Mater 175:844–849. https://doi.org/10.1016/j.jhazmat.2009.10.087
110. Ütebay B, Çelik P, Çay A (2020) Textile wastes: status and perspectives. Waste in textile and leather sectors. IntechOpen. https://doi.org/10.5772/intechopen.92234
111. Ventura-Camargo BDC, Maltempi PP, Marin-Morales MA (2011) The use of the cytogenetic to identify mechanisms of action of an azo dye in *Allium cepa* meristematic cells. J Environ Anal Toxicol 1:1–12. https://doi.org/10.4172/2161-0525.1000109

112. Ventura-Camargo BDC, Angelis DF, Marin-Morales MA (2016) Assessment of the cytotoxic, genotoxic and mutagenic effects of the commercial black dye in *Allium cepa* cells before and after bacterial biodegradation treatment. Chemosphere 161:325–332. https://doi.org/10.4172/2161-0525.1000109

113. Vinitnantharat S, Chartthe W, Pinisakul A (2008) Toxicity of reactive red 141 and basic red 14 to algae and waterfleas. Water Sci Technol 58:1193–1198. https://doi.org/10.2166/wst.2008.476

114. Wardman RH (2017) An introduction to textile coloration: principles and practice. Society of Dyers and Colourists-John Wiley & Series

115. Wu JY, Lay CH, Chen CC, Wu SY (2017) Lipid accumulating microalgae cultivation in textile wastewater: Environmental parameters optimization. J Taiwan Inst Chem Eng 79:1–6. https://doi.org/10.1016/j.jtice.2017.02.017

116. Xiong JQ, Kim SJ, Kurade MB, Govindwar S, Abou-Shanab RAI, Kim JR, Roh HS, Khan MA, Jeon BH (2018) Combined effects of sulfamethazine and sulfamethoxazole on a freshwater microalga, *Scenedesmus obliquus*: toxicity, biodegradation, and metabolic fate. J Hazard Mater 370:138–146. https://doi.org/10.1016/j.jhazmat.2018.07.049

117. Xuan J, Leung MKH, Leung DYC (2009) A review of biomass-derived fuel processors for fuel cell systems. Renew Sustain Energy Rev 13:1301–1313. https://doi.org/10.1016/j.rser.2008.09.027

118. Yu KL, Show PL, Ong HC, Ling TC, Lan JCW, Chen WH, Chang JS (2017) Microalgae from wastewater treatment to biochar—feedstock preparation and conversion technologies. Energy Convers Manag 150:1–13. https://doi.org/10.1016/j.enconman.2017.07.060

119. Zabłocka-Godlewska E, Przystaś W, Grabińska-Sota E (2018) Possibilities of obtaining from highly polluted environments: new bacterial strains with a significant decolorization potential of different synthetic dyes. Water Air Soil Pollut 229:1–13. https://doi.org/10.1007/s11270-018-3829-7

120. Zaharia C, Suteu D (2012) Organic pollutants ten years after the Stockholm convention-environmental and analytical update. In: Puzyn T (ed), InTech, pp 55–86. https://doi.org/10.5772/32373

Plant–Microbe-Based Remediation Approaches in Dye Removal

Priti Panwar, Pooja Mahajan, and Jyotsna Kaushal

Abstract In the modern world, one of the significant problems being faced is environmental pollution. Rapid industrialization and increasing demand have facilitated the use of synthetic dyes in textile industry. Textile effluents is a concern as it has increased concentration of dyes which is detrimental not only to living organisms but is also a threat to aquatic life, ultimately arising the need for its proper treatment before discharge into the environment. Many chemicals, physical, and biological techniques are already in use, but their limited efficiencies have necessitated the need to explore some other way for the remediation of dyes. The plant–microbe-based remediation has emerged as an inventive approach by utilizing both plant, and microbe synergistically for the decolorization, and degradation of dyes into less toxic or non-toxic metabolites. This chapter will focus on the plant–microbe-assisted remediation of dye, textile effluents and its mixture, the potential of synergistic approach. It will also highlight the selection of bacteria for enhancing the plant performance in remediation of dye, the analysis of metabolites and degraded products and their toxicity evaluation.

Keywords Synthetic dyes · Textile effluent · Decolorization potential · Bioremediation · Phytoremediation · Microorganisms

1 Introduction

Environmental pollution is a severe socioeconomic and health issue in the current situation [68]. The pollution of the water bodies is becoming a serious global issue as it directly affects the health of biotic community of the environment. The various industries such as pharmaceutical, textiles, paper, sugar, chemical, agricultural, metal

P. Panwar
Chitkara School of Health Sciences, Chitkara University, Rajpura 140401, Punjab, India

P. Mahajan (✉) · J. Kaushal
Centre for Water Sciences, Chitkara University Institute of Engineering and Technology, Chitkara University, Rajpura 140401, Punjab, India
e-mail: pooja.mahajan@chitkara.edu.in

refining, electroplating introduce several pollutants such as inorganic and organic pollutants into water bodies [118]. Moreover, sewage units also directly discharge their water resulted in contamination of water bodies. Inorganic pollutants constitute various noxious heavy metals such as arsenic, cadmium, lead, and chromium while in contrast phenols, endocrine-disrupting chemicals, polyaromatic hydrocarbons, chlorinated phenols, pesticides, dyes, polychlorinated biphenyls are referred to as organic pollutants [18]. Synthetic dye effluent is the most prominent environmental contaminants even their presence in very low concentration (ppm level) in aquatic ecosystems as color is one of the most apparent indicators of water pollution [106]. The textile, paper, and tannery industries utilize the large percentage of synthetic organic dyes to furnish various products such as fabrics, colored papers and files, carpets, bedding, leathers, and soft toys, etc. [76]. There are approximately 10,000 different textile dyes.

A considerable amount of synthetic dyes (approximately 10–15%) are released to the environment during dyeing and synthesis of textile fabrics without any pre-treatment [118]. Since a large quantity of water also gets disbursed for several operational and finishing processes in textile industry and the dyes that are water soluble end up in either adjacent agricultural farms or land areas and in waterbodies. Textile industries discharge approximately 10–20% of soluble dyes and other chemicals into the wastewater stream [122]. Nearly 1–10% of pigments utilized in the leather and paper and industries are also get wasted into the environment. As a result, tons of dyes are released into the environment every day as wastewater [8]. This wastewater is thus highly contaminated and has the potential to enhance biological oxygen demand (BOD) and Chemical oxygen demand (COD) of water bodies if discharged as such [15].

Various approaches followed by industries for removal of dye effluent, but the cost of those approaches is at par. Therefore, industries reluctant to use these approaches on regular basis and released the effluent as such. Hence, water and soil resources contamination resulting from dyehouse effluents dumped relentlessly and extensively has prompted the researchers to explore more cost-effective approaches for eliminating these aromatic compounds before their release in environment [46]. Biological method with utilization of plants known as phytoremediation and microbial bioremediation can effectively transform and remove such pollutants and claimed as economical and eco-friendly approaches in comparison to other techniques. A synergetic approach of use of plant with microbes for removal of dyes have dual benefit as plant-microbial associations are well known for influencing the availability and mobility of dye molecules to plants and assist plants in the quick detoxification of dye molecules. This chapter reviews the plant-microbial synergism research reports along with the mechanism adopted by plants and microbes for remediation of dye. This chapter also highlights the criteria to select the bacteria for plant-microbial synergism.

2 Dyes and Its Impact on Environment and Human Health

Synthetic dyes represent as one of the crucial categories of xenobiotic compounds utilized in various industries such as textiles, leather tanning, printing, ceramics, cosmetics, food and beverage, and used for staining in medical trials [25, 48]. Most of the synthetic dyes have complex structure which tolerates both acidic and alkaline medium, withstand color fading on exposure to sunlight and ultraviolet (UV) radiations, and high resistance against soaps and bleaching agents [17]. In addition, they are soluble as well as impart color with acidic properties in water [108]. Dyes are classified as natural and synthetic dyes, respectively, based on their origin. Synthetic dyes are further classified into anionic, cationic, and non-ionic dyes based upon their nature and further categorized as shown in Fig. 1. Based on their chemical structure, dyes can also be categorized into basic, disperse, acidic, azo, phthalocyanine, diazo, anthraquinone, indigoid, sulfur, and triphenylmethane [17]. The dye color is actual a synergetic effect of both chromophoric and auxochromic groups. These electron-withdrawing or donating groups of auxochrome not only engaged for color enhancement but also responsible for reactivity and solubility of dyes. In non-ionic and anionic dyes, the chromophores are generally azo classes. Toxic amines are produced due to the reductive cleavage of the azo group. Azo-based chromophores consist of reactive dyes such as chlorotriazine, di-fluorochloro pyrimidine, tri-chloro pyrimidine, and vinyl sulfone [5, 12]. Colored dyes are extremely noxious and constitute a significant threat to the environment. Dye wastewater is one of the most often organic effluent that is released either into the water bodies or discharged directly to agricultural farms (Fig. 2). The application of certain noxious dyes and their impact on human body is also presented in Table 1.

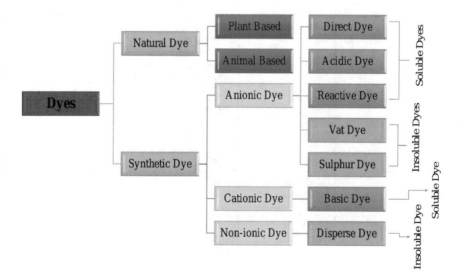

Fig. 1 Classification of dyes [5, 12, 17]

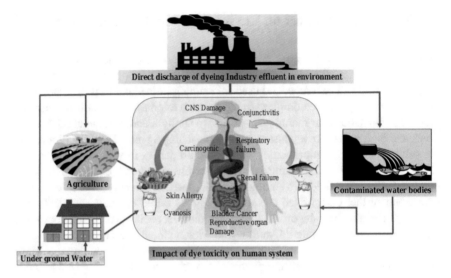

Fig. 2 Biomagnification of dyes and its toxicity impact on various organs of human body

Table 1 Application and toxicity impact of certain dyes

Name (type) of dye	Application	Impact on human	References
Methyl Orange (Acidic)	For dyeing silk, leather, paper, and wood	Tumorigenic and carcinogenic	Akansha et al. [6]
Disperse Red Disperse Orange (Disperse)	For dyeing nylon, polyester, and acrylic fibers	Mutagenic, cytotoxic, skin allergy	Lellis et al. [72]
Reactive Red 120, 198 (Anionic)	For dyeing cotton, wool, and nylon	Dermatitis, conjunctivitis, mutagenic, carcinogenic, and teratogenic, damages central nervous system and Reproductive organ	Mishra and Maiti [80]
Crystal Violet/Methylene Blue (Cationic)	Used in inks, paper, silk, and polyester dyeing	Carcinogenic, chemical cystitis, skin irritation, respiratory and renal failure, cyanosis, vomiting	Ismail et al. [45]
Indigo Vat Blue (Vat)	For dyeing denims, cotton, rayon, and cellulosic fibers	Eye and respiratory tract damage	Chowdhury et al. [27] Katheresan et al. [52]
Congo Red (Direct)	Used for dyeing cotton, paper	Carcinogenic Mutagenic, bladder cancer	Zhou et al. [133]

The noxious pollutants present in the dye wastewater get biomagnified in different trophic levels and have a negative impact on human health, the environment, and the aqua ecosystem. In various research studies, it has been proved that dyes are harmful, carcinogenic, and poisonous, and its degraded compounds in wastewater also cause damage to several organs, the reproductive system, and the central nervous system in humans (Fig. 2) [132, 133].

3 Different Approaches for the Remediation of Dyes from the Industrial Wastewater

Dyes are highly visible chemicals, and even small amounts emitted into the water bodies can cause environmental pollution [30]. Furthermore, the textile industry is among the major water consumers. Dye wastewater discharged from these industries along with various types of chemicals and colorants, needs adequate and appropriate treatment prior to their release in surrounding. Due to its highly varied composition, wastewaters are exceedingly challenging to handle. With a goal of reducing or decontaminating industrial effluents in the land and water, many physical, chemical methods, use of microorganisms, plants, and plant–microbe synergistic approaches have been exercised which are enlisted in Fig. 3 [91].

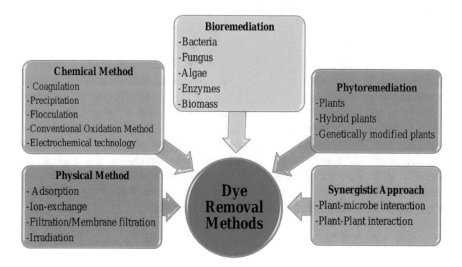

Fig. 3 Various approaches for the remediation of dye from the industrial wastewater

3.1 Physical Methods

In the adsorption method, the high adsorption capacity materials (adsorbents) such as activated carbon adsorb the dye molecules. In contrast, in the Ion-exchange method, the ions present in wastewater containing dye swap with same ions which are attached to a fixed solid surface. The irradiation treatment employs radiations typically obtained from monochromatic UV lamps operating at a wavelength of 253.7 nm. It's an effective method for eliminating organic pollutants and disinfecting harmful microbes, but it demands a large amount of dissolved oxygen [83]. Organic pollutants and bacteria in wastewater are analyzed using filtration technologies. Reverse-osmosis, micro-filtration, nano-filtration, and ultra-filtration are the most common membrane filtration types. However, the biggest issue with the physical technique is that it produces too much sludge. As a result, solid adsorbent disposal becomes a significant concern. Physical approaches, on the other hand, are helpful when the volume of effluents is minimal [83].

3.2 Chemical Methods

Chemical methods commonly used to treat industrial wastewater are coagulation and flocculation methods. They destabilize and create flocs by removing the colloidal particles, very thin solid suspensions, and some soluble substances earlier present in the wastewater [35]. The pollutants are also converted into less harmful compounds through chemical oxidation by using chemical oxidizing agents such as chlorines and Fenton reagents [100]. Electrochemical technologies have received considerable attention in recent times because of their unique environmental compatibility, safety, and versatility. Electrochemical technology competes with other conventional methods to provide solutions as per the need of different industries. In general, this technology is better than physical and membrane technologies, particularly in removing color from wastewater, because they utilize an electron as a reagent and do not produce solid residues. The method relies on direct dye degradation on the anode using chloride as an electrolyte and indirect oxidation of dyes using the formed species [64]. The major disadvantage of chemical methods is that they are expensive, which limits them from being used in the textile industry.

3.3 Biological Methods

Bioremediation uses microbes such as bacteria, fungus, algae, yeast, their enzymes, and biomass for remediating textile wastewater. Compared to other approaches, bioremediation seems to be more cost-effective, environmentally friendly, and produces less sludge [66]. In addition, because of the breakage of the link (i.e., chromophoric group), synthetic dyes degrade into a considerably less hazardous inorganic product, which helps in decolorization.

Some previous studies have shown that *Bacillus* sp. [39], *Pseudomonas aeruginosa*, *Pseudomonas putida* [83], and *Klebsiella pneumonia* [100] bacterial strains were able to remediate different textile dyes effectively. *Aspergillus niger* [102], *Phanerochaete chrysosporium* [100], *Trichoderma* sp. [40] fungus have also shown high decolorization and remediation activities of textile dyes and wastewater effluents. Other studies have reported that algae sp. such as *Chlorella vulgaris,* Cyanobacterium *Aphanocapsa elachista* [32], *Spirulina* sp., *Anabena oryzae* [1] are potential for bioremediation of dyes from textile wastewater.

Dye decolorization by microbes occurs either through biochemical reaction, adsorption on the cell surface, or by their combination. The degradation mechanisms in bacteria can be anaerobic, aerobic, or both. The treatment technique is chosen based on the dye's binding affinity, reactivity, and mechanism of binding with microbe. Application of microbial enzyme is another effective way to remediate dyes from textile effluent. The dye degradation process has been associated with various bacterial oxidative enzymes such as laccase, lignin peroxidase, and reductive enzymes like nonspecific reductases and azoreductase. Azoreductases breaks azo bonds ($-N=N-$) present in azo dyes and is responsible for its degradation and decolorization [44]. Previous studies have also reported that the enzyme such as manganese peroxidase, tyrosinanse also have the potential of remediating dye solutions [25]. Various factors affecting bioremediation have been enlisted in Fig. 4 [86].

4   Microbial Bioremediation of Dyes

Many microorganisms like fungus, bacteria, and algae are potential of remediating dye effluent. It can be achieved by bacterial species either by aerobic process or anaerobic process and sometimes it can be the combination of both the processes. Azo dyes are utilized as a source of nitrogen and carbon by various aerobic bacterial species for their metabolism and in contrast some reduces the azo dye through the enzymes. The enzymatic capability of microbes can be employed for the breakdown of organic pollutants into less environmentally hazardous substances [94]. There are several bacterial enzymes which stimulates the decolorization of dye such as Azo reductase, laccase, lignin peroxidase, tyrosinase, dichlorophenol indophenol (DCIP) reductase, and manganese peroxidase.

The azo dyes in the textile effluent are reduced into aromatic amines through the azo-reductase enzyme which cleaves the azo bonds [16] and later the amines are reduced in CO_2 and H_2O, hence decolorizing the textile dye. For breaking azo bonds ($-N=N$), this enzyme requires the cofactors Nicotinamide adenine dinucleotide phosphate hydrogen (NADPH), Flavin Adenine Dinucleotide Hydrogen (FADH), or Nicotinamide adenine dinucleotide hydride (NADH) as an electron donor. The electrons as reducing equivalents are transferred to azo dye which acts as an electron

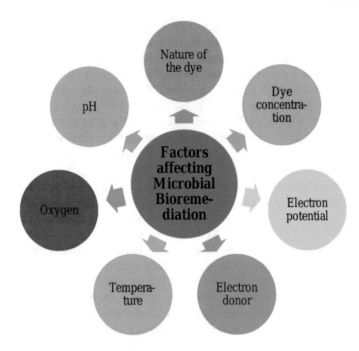

Fig. 4 Factors affecting the bioremediation

receptor during this process. Azo reductase enzyme can be either found in cyto-plasm or on the cell membrane of bacteria. A redox mediator is utilized by cell membrane-bound azoreductase as an electron shuttle in anaerobic conditions and mediators such as anthraquinonesulfonates, riboflavin is utilized. Redox mediator-dependent mechanism is showed by membrane-bound azoreductase that differs from cytoplasmic azoreductase [110]. While azoreductase is an oxygen-sensitive enzyme, in anaerobic conditions the degradation is more effective than aerobic conditions. As a result, when an enzyme interacts with oxygen, there is reduction in redox mediator rather than the azo dye in aerobic conditions [110].

To date, many bacterial species have shown promising results for the decoloriza-tion and remediation of dye and textile wastewater. Some of them are enlisted in Table 2. *Bacillus cereus* was found to decolorize 97% of Reactive Red195 [81]. *Pseudomonas entomophila* decolorized 93% of Reactive Black 5 [57]. Black WNN dye from dye effluent was 96.4% decolorized by *Paenibacillus alvei* within 48 h [92].

Staphylococcus sp. showed 100% decolorization potential for Remazol Brilliant Blue R within 12 h [123]. *Tenacibaculum* sp. was found to decolorize 99% malachite green dye within 12 h [94]. It was reported that Azo dye Procion Red—H3B was 100% decolorized by *Pseudomonas stutzeri* within 20 h [16]. It has been revealed that not only the single bacterial culture, but the bacterial consortia exhibited higher rate of decolorization potential. It is reported that bacterial consortia of *Bacillus* sp., *Bacillus*

Table 2 Bacterial species showing potential for the decolorization and remediation dye and textile effluent

Synthetic dye/dye effluent	Bacterial species	Decolorization potential	References
Reactive Orange 16 (RO16)	*Pseudomonas* sp. SUK1, LBC2 and LBC3	NA	Jadhav et al. [47]
Reactive Red195	*Bacillus cereus*	97%	Modi et al. [81]
Eriochrome Black T and Metanil Yellow Light Green and Carmine Red	*Bacillus* sp., *Bacillus cereus, Bacillus subtilis, Pseudomonas* sp., *Bacillus mycoides*, and *Micrococcus* sp.	85% 84%	Mahmood et al. [79]
Amaranth dye	*Lysinibacillus* sp. *and Bacillus* sp.	99%	Anjaneya et al. [9]
Reactive Black 5, Disperse Red 78, Reactive Orange 16, and Direct Red 81	*Pseudomonas* sp. and *Providencia rettgeri*	98–99%	Lade et al. [71]
Disperse Red 1	*Microbacterium* sp., *Leucobacter albus, Klebsiella* sp. and *Staphylococcus arlettae*	80%	Franciscon et al. [36]
Reactive Red 198	bacterial-yeast consortium (*Brevibacillus laterosporus* and *Galactomyces geotrichum*)	92%	Kurade et al. [70]
Reactive Red 198	*Alcaligenes* sp.	90%	Pandey et al. [87]
Reactive Red M8B	*Bacillus subtilis*	97%	Arulazhagan [11]
Reactive Black 5	*Pseudomonas entomophila*	93%	Khan and Malik [57]
Reactive Black 5	*Pseudomonas* sp.	83%	Mohamed [82]
Black WNN dye from dye effluent	*Paenibacillus alvei*	96%	Pokharia and Ahluwalia [92]
Congo Red	*Brevibacillus parabrevis*	95.71%	Talha et al. [113]
Remazol Brilliant Blue R	*Staphylococcus* sp.	100%	Velayutham et al. [123]
Malachite Green	*Tenacibaculum* sp.	99%	Qu et al. [94]
Metanil Yellow	Bacterial consortium *Pseudomonas, Lysinibacillus, Lactococcus, and Dysgonomonas*	96.25%	Guo et al. [38]
Reactive Black 5	Bacterial consorti *Stenotrophomonas* sp. *Sphingomonas* sp.	34% in 5 days After providing carbon, N source 100% in 5 days	Eskandari et al. [33]

(continued)

Table 2 (continued)

Synthetic dye/dye effluent	Bacterial species	Decolorization potential	References
Methyl orange dye	*Micrococcus yunnaenensis*	90%	Carolin et al. [20]
Congo red dye	*Bacillus cohnni*	99%	Kishor et al. [65]
Azo dye Procion Red—H3B	*Pseudomonas stutzeri*	100%	Bera and Tank [16]

cereus, Bacillus subtilis, Pseudomonas sp., *Bacillus mycoides*, and *Micrococcus* sp. decolorized Carmine Red and Light Green textile dye by 84 and showed 85% decolorization potential for Eriochrome Black T and Metanil Yellow dye [79]. Bacterial consortium (*Pseudomonas, Lysinibacillus, Lactococcus*, and *Dysgonomonas*) was able to decolorize 96.25% of Metanil yellow dye within 6 h [38]. Bacterial consortium of *Stenotrophomonas* sp. and *Sphingomonas* sp. was able to exhibit 34% decolorization rate for Reactive Black-5 dye in 5 days but after providing carbon and nitrogen source it showed 100% decolorization of dye within 5 days [33].

It has been reported that when nutrients such as carbon, nitrogen, sulphur starts depleting, then the fungal species starts utilizing the aromatic compounds as a source of nutrition and energy for promoting their growth. Enzymes such as lignin peroxidase, manganese peroxidase, and laccase produced by fungal species plays a key role in the remediation of aromatic compounds because of its nonspecific enzyme activity. Various fungal species which are summarized in Table 3 have been reported exhibiting high decolorization potential of dye. *Galactomyces geotrichum* reported 96% decolorization potential for Remazol Red dye [124]. *Aspergillus niger* and *Nigrospora* sp. were able to decolorize 86 and 90% Synozol red dye, respectively [42]. *Armillaria* sp., which is a white-rot fungus was found to decolorize 97.17% of Acid Red 27 dye [3]. Within 72 h, *Trichoderma tomentosum* fungus was able to decolorize 94.9% dye effluent and 99.2% Acid Red 3 R dye [40]. *Aspergillus nidulans* was found to decolorize more than 90% Synozol Red HF–6BN and Synozol Black B dye in 120 *h* whereas *Aspergillus fumigatus* was able to decolorize 89% of Synozol black B and 90% of Synozol red HF–6BN [56]. Unlike fungus and bacteria, algae don't utilize the aromatic compounds as a source of nutrients. Whereas it uses the chromophore group of dye for the production of its biomass, while conversion of color to colorless molecules it produces CO_2 and H_2O and fungal biomass also adsorbs the chromophore. Like bacteria, the algae also degraded the azo dyes by using oxidative enzyme like laccase, peroxidase, and reductive enzyme, azoreductase. Recent studies have shown the use of algae and their efficiency in the remediation of dyes (Table 4).

It has been reported that 93% malachite green dye was decolorized by *Chara vulgaris* [61]. *Cladophora* and *Spirogyra* sp. showed 93 and 95% decolorization potential, respectively, for Reactive Blue dye [127]. Cyanobacterium *Phormidium* was able to show 91% decolorization rate for indigo dye [29]. *Chara vulgaris* exhibited 90% decolorization potential for Methyl Red dye [77] and has potential for remediation of textile wastewater also [78], whereas it was also reported that *Chlorella*

Table 3 Fungal species showing potential for the decolorization and remediation dye and textile effluent

Synthetic dye/dye effluent	Fungal species	Decolorization potential (%)	References
Remazol Red	*Galactomyces geotrichum*	96	Waghmode et al. [124]
Congo Red	*Alternaria alternata*	99.99	Chakraborty et al. [22]
Synozol Red	*Aspergillus niger and Nigrospora sp.*	86 90	Ilyas and Rehman [42]
Acid Red 27	*Armillaria* sp.	97.17	Adnan et al. [3]
Congo Red	*Aspergillus niger*	More than 98.5	Lu et al. [74]
Reactive Red 31	*Aspergillus bombycis*	94.73	Khan and Fulekar [55]
Congo Red	*Ceriporia lacerata*	More than 90	Wang et al. [126]
Cotton Blue Crystal Violet	*Coriolopsis* sp.	79.6 85.1	Munck et al. [85]
Congo Red Methyl Red Reactive Blue	*Dichotomomyces cejpii* and *Phoma tropica*	NA	Krishnamoorthy et al. [67]
Dye effluent Acid Red 3 R	*Trichoderma tomentosum*	94.9 99.2	He et al. [40]
Dye effluent	*Bjerkandera* sp.	69	Gaviria-Arroyave et al. [37]
Scarlet RR dye Dye mixture Textile industry-dye effluent	*Peyronellaea prosopidis*	90 84 85	Bankole et al. [14]
Direct Blue-1	*Aspergillus terreus*	98.40	Singh and Dwivedi [109]
Synozol Black B Synozol Red HF–6BN	*Aspergillus nidulans Aspergillus fumigatus*	More than 90 (both dyes) 89 Synozol black B and 90 Synozol red HF–6BN	Khan et al. [56]
Congo Red	*Aspergillus flavus*	96.92	Chatterjee et al. [24]

vulgaris decolorized 93.55% Malachite green dye and 62.98% Crystal Violet dye [101].

Table 4 Algal species showing potential for the decolorization and remediation dye and textile effluent

Synthetic dye/dye effluent	Algal species	Decolorization potential (%)	References
Dye effluent	*Chlorella vulgaris*	44	Chu et al. [28]
Malachite Green	*Chara vulgaris*	90	Khataee et al. [61]
Acid Red 66 dye	*Acutodesmus obliquus*	NA	Sarwa et al. [104]
Dye effluent and Methylene Blue	*Chlorella pyrenoidosa*	NA	Pathak et al. [88]
Reactive Blue	*Cladophora* sp. *Spirogyra* sp.	93 95	Waqas et al. [127]
Indigo dye	*Cyanobacterium Phormidium*	91	Dellamatrice et al. [29]
Aniline Blue dye	*Chlorella vulgaris*	NA	Arteaga et al. [10]
Disperse Red 3B	*Chlorella sorokiniana and Aspergillus* sp. (Consortium)	93	Tang et al. [115]
Dye effluent	*Chara vulgaris*	78 (COD)	Mahajan et al. [78]
Methylene Blue	*Chlorella vulgaris*	83.04	Chin et al. [26]
Methyl Red	*Chara vulgaris*	90	Mahajan and Kaushal [77]
Malachite Green Crystal Violet	*Chlorella vulgaris*	93.55 62.98	Salem et al. [101]

5 Phytoremediation

Phytoremediation is an in-situ, environment-friendly method of removing pollutants. Plants are most widely used to reduce and eliminate textile dyes and heavy metals from the environment [131]. This approach has the significant benefit of preventing pollution of nearby land, water, and air. Some of the factors affecting phytoremediation have been mentioned in Fig. 5 [75, 86]. Phytoextraction, phytotransformation, phytostabilization, and phytovolatilization are some of the techniques for remediating textile wastewater containing dye as shown in Fig. 6. These techniques have a significant impact on pollutant mobility, volume, and toxicity. As a result, it can be used to remediate a wide range of pollutants, both organic and inorganic, and phytoremediation can be accomplished efficiently using either plants in their natural habitats or artificially constructed wetlands. Previous studies have reported plants such as *Petunia grandiflora, Brassica juncea, Portulaca grandiflora* have shown highly effective dye remediation [97].

Removal of contaminants and xenobiotic compounds from the environment using plants species is termed as phytoremediation. It has emerged as an effective and profitable green approach for the detoxification and remediation of textile wastewater containing pollutants/dye as it is solar energy-driven process. Plants utilizes its roots,

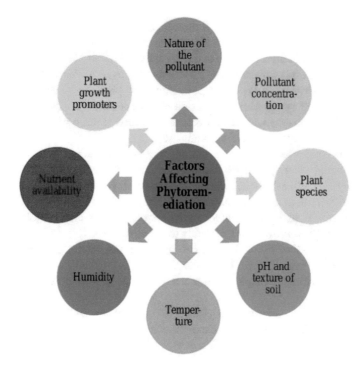

Fig. 5 Factors affecting the phytoremediation

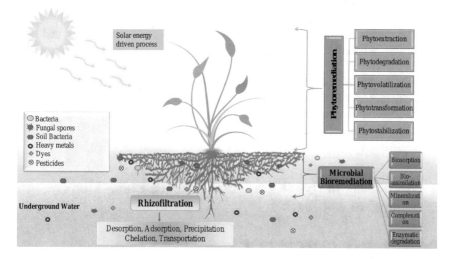

Fig. 6 Plant-soil and microbes' interaction in remediation of pollutants

root hair, callus, tissues, enzymes, and rhizosphere microflora for this process. Root hair increases the surface area for the absorption and accumulation of the nutrients as well as the pollutants present in the soil. Phytotechnology uses several mechanisms for decolorization and remediation of pollutants such as phytodegradation, phytoextraction, phytotransformation phytovolatilization, and phytostabilization [86]. Plants also have certain enzymes such as peroxidases, laccase, nitroreductase, cytochrome P450 (CYP) monooxygenase, dehalogenase, and dioxygenase which plays a very crucial role in the degradation of pollutants [69].

There are evidence of plants exhibiting high decolorization potential of dyes (Table 5). *Glandularia pulchella* had the potential of decolorizing, Rubine GFL, Navy Blue 2R, Brilliant Blue R, Scarlet RR, and Red HE3B at the rate of 100, 60, 85,100, and 55%, respectively, within 72–96 h [50]. Brilliant Blue G dye mixture and textile effluent was 86% phyto-remediated by *Petunia grandiflora* [128]. It was found that *Alternanthera philoxeroides was* able to decolorize 89, 100, 88, 93, 80, 72, 40, and 64% of Navy Blue HE2R, Remazol Red, Scarlet RR, Rubin GFL, Brown 2GR, Green HE4B, Reactive Orange HER, and Disperse Yellow 4G dye, respectively, in 72 h [96]. *Nasturtium officinale* decolorized more than 90% of malachite green dye in 7 days [119].

Sesuvium portulacastrum was found to decolorize 98% Reactive Green 19A-HE4BD dye in 5 days [73]. In one of the studies, it was seen that *Lactuca sativaa* and *Vigna radiata* were able to decolorize Methylene Blue up to 96 and 85%, respectively, and they also had a potential to decolorize Malachite Green up to 87 and 96%, respectively [2]. It was found that *Lemna minor* had decolorize Methylene Blue up to 80.56% [43] and it was also able to decolorize 94% of Acid Bordeaux B dye in 7 days [120]. *Eichhornia crassipes* exhibited 90.8, 84.8, 87.2, 79, and 87.4% decolorization potential for Methylene Blue, Rhodamine B, Crystal Violet, Auramine O, and Rose Bengal dye, respectively [107]. *Lemna minor* decolorized 98% of Methylene Blue dye [19].

Even after the high efficiency of remediation of dyes, the phytotechnology alone is not efficient for complete elimination of dyes, as the concentration of dye can halt the remediation process. Hence, it arises the need to increase the phytoremediation potential of plants using microbes.

6   Mechanism of Dye Degradation by Using Plant–Microbe Synergism Approach

Microbe-assisted phytoremediation has developed as a cost-effective, long-term soil cleanup approach with minimal soil disturbance and upkeep [13]. Plants and microorganisms dwell together and collaborate with one another. A plant gives supplements and space to the microorganisms and consequently, microorganisms improve the bioavailability, mineralization, and detoxification of different pollutants present in the soil [13].

Table 5 Plant species showing potential for the decolorization and remediation dye and textile effluent

Synthetic dye/dye effluent	Plant species	Decolorization potential	References
Direct Red 5B Red HE4B Malachite Green Reactive Red 2 Methyl Orange	*Blumea malcolmii*	42.07 76.50 96.61 80.20 87.86	Kagalkar et al. [51]
Remazol Red (mixture of dyes and a textile effluent)	*Aster amellus*	96	Khandare et al. [59]
Scarlet RR Rubine GFL Brilliant Blue R Navy Blue 2R Red HE3B	*Glandularia pulchella*	100 100 85 60 55	Kabra et al. [50]
Acid Blue 92	*Lemna minor*	77	Khataee et al. [62]
Brilliant Blue G dye (dye mixture and textile effluent)	*Petunia grandiflora*	86	Watharkar et al. [128]
Reactive Blue 160	*Tagetes patula*	90	Patil and Jadhav [89]
Basic Red 46	*Lemna minor*	68.50	Movafeghi et al. [84]
Acid Blue 92	*Azolla filiculoide*	80.00	Khataee et al. [63]
Scarlet RR dye	*Ipomoea hederifolia*	90–96	Rane et al. [95]
Remazol Red Rubin GFL Reactive Orange HER Navy Blue HE2R Disperse Yellow 4G Scarlet RR Green HE4B Brown 2GR	*Alternanthera philoxeroides*	100 93 40 89 64 88 72 80	Rane et al. [96]
Malachite Green	*Salvinia molesta*	99	Kaushal and Mahajan [54]
Reactive Green 19A-HE4BD	*Sesuvium portulacastrum*	98	Lokhande et al. [73]
Methylene Blue Methyl Orange	*Eichhornia crassipes*	98.42 66.80	Tan et al. [114]
Rubin GFL	*Salvinia molesta*	97	Chandanshive et al. [23]
Malachite Green	*Nasturtium officinale*	More than 90	Torbati [119]

(continued)

Table 5 (continued)

Synthetic dye/dye effluent	Plant species	Decolorization potential	References
Methylene Blue (MB) and Malachite Green (MG)	*Lactuca sativa* *Vigna radiata*	MB: 96 MG: 87 MB: 85 MG: 96	Adams [2]
Methylene Blue	*Lemna minor*	80.56	Imron et al. [43]
Acid Bordeaux B	*Lemna minor*	94	Torbati [120]
Methyl Orange	*Salvinia molesta*	42	Al-Baldawi et al. [7]
Methylene Blue	*Ceratophyllum demersum*	96	Ewadh [34]
EBT	*Salvinia nantas*	71	Rápó et al. [98]
Azo dyes	*Bacopa monnieri*	90	Shanmugam et al. [105]
Dyeing effluent	*Pistia stratiotes L, Salvinia dnate, Hydrilla verticillata*	73–86	Ahila et al. [4]
CI Direct Blue	*Eichhornia crassipes* *Pistia stratiotes*	100 100	Ekanayake et al. [31]
Methylene Blue Rhodamine B Crystal Violet Auramine O Rose Bengal	*Eichhornia crassipes*	90.8 84.8 87.2 79 87.4	Sharma et al. [107]
Methylene Blue	*Lemna minor*	98	Can-Terzi et al. [19]
Direct Red 28	*Salvinia molesta*	90	Kaushal and Mahajan [53]

In one of the previous studies, it was reported that the consortia of *Petunia grandiflora* and *Bacillus pumilus* potentially decolorized 96.86% of Navy Blue RX dye within 36 h. The root exudates of *P. grandiflora* provided nutrients to the *B. pumilus* for its growth and at the same time enhanced the decolorization of dye. The bacterial species showed significantly increased the enzymatic activities of NADH-DCIP reductase (15%), laccase (689%), riboflavin reductase (528%), azo reductase (10%), and Tyrosinase (660%) whereas the enzyme activity of laccase (161%), lignin peroxidase (207%), riboflavin reductase (78%) and tyrosinase (133%) in the plant roots were also increased. Through the enzymatic actions of plant and microbe, the intermediates formed may have acted as redox mediators or substrate for other enzymes of both resulting in enhanced remediation of dyes in synergistic approach [129].

Likewise, under in vitro conditions the decolorization of dye mixture and Scarlet RR dye was also studied using plant-bacterial synergistic approach. *Glandularia pulchella* plant and *Pseudomonas monteilii* consortium was used. The consortium showed decolorization of Scarlet RR dye, Textile effluent A, Textile effluent B, Dye Mixture by 100, 95, and 93% within 48, 60, and 24 h, respectively. The enzyme activity of DCIP reductase (842%) and lignin peroxidase (438%) in the roots of *Glandularia pulchella* was significantly increased. The induction in the bacterial enzyme laccase (612%) and DCIP reductase (260%) was also observed during the decolorization of Scarlet RR dye [49].

Whereas the consortium of plant *Ipomoea hederifolia* (adventitious roots) and its endophytic fungus *Cladosporium cladosporioides* decolorized Navy Blue HE2R by 97% within 36 h. The endophytic fungus colonized adventitious roots exposed to dye also revealed inductions in the enzyme's activities like laccase (59%), superoxide dismutase (151%), DCIP reductase (52%), veratryl alcohol oxidase (145%), lignin peroxidase (161%), tyrosinase (84%), and catalase (218%). Dye exposed adventitious roots—endophytic fungus showed inductions in the enzyme activities of lignin peroxidase, DCIP reductase, laccase, veratryl alcohol oxidase, and tyrosinase by 500, 103, 106, 363, and 116%, respectively [90]. A pilot-scale vertical flow constructed wetland (VFCWs) system involving *Brachiaria mutica* combined with the endophytic bacteria (*Pantoea* sp., *Bacillus endophyticus*, *Bacillus* sp., *Bacillus pumilus* and *Microbacterium arborescens*) was designed for the decolorization of textile effluent containing dye. In the presence of vegetation with bacteria, color and TDS reduction was improved to 84 and 32%, respectively, after 48 h of treatment [41].

Augmentation of bacterial species to floating treatment wetland (FTWs) was also found capable for degrading and decolorizing dye-rich textile effluent. Both *Phragmites australis* and *Typha domingensis* wetland plants developed FTWs were introduced with the bacterial consortia of *Pseudomonas indoloxydans*, *Rhodococcus* sp. and *Acinetobacter junii*. The synergistic approach of *Phragmites australis* and the bacterial consortium showed 97.6% decolorization whereas the combination of *Typha domingensis* and the bacterial consortium decolorized 81.40% of dye containing textile effluent within 8 days. Moreover, the combinations were also found to remediate heavy metals as well [117].

Similarly, a bacterial augmented FTWs was also analyzed for the decolorization of Reactive Black 5. Plant *Phragmites australis* and bacterial consortium of *Acinetobacter junii*, *Pseudomonas indoloxydans*, and *Rhodococcus* sp. was found to decolorize 95.50% of dye within 15 days [116].

All the evident studies (Table 6) have shown that the synergistic approach of plant and microbe is the effective strategy for the decolorization of various dyestuffs from dye containing wastewater and soil.

Table 6 Plant–microbe synergism approach utilized for the decolorization and remediation dye and textile effluent

Synthetic dye/dye effluent	Plant species	Microorganism species	Decolorization potential (%)	References
Remazol Black B	*Zinnia angustifolia*	*Exiguobacterium aestuarii*	100	Khandare et al. [60]
Direct Red 5B	*Portulaca grandiflora*	*Pseudomonas putida*	100	Khandare et al. [58]
Navy Blue RX	*Petunia grandiflora*	*Bacillus pumilus*	96.86	Watharkar et al. [129]
Scarlet RR Textile effluent Dye Mixture	*Glandularia pulchella*	*Pseudomonas monteilii*	100 95 93	Kabra et al. [49]
Textile effluent	*Pogonatherum crinitum*	*Bacillus pumilus*	93	Watharkar et al. [130]
Navy Blue HE2R	*Ipomoea hederifolia*	*Cladosporium*	97	Patil et al. [90]
Textile effluent	*Brachiaria mutica*	Endophytic bacteria (*Microbacterium arborescens*, *Pantoea* sp., *Bacillus pumilus*, *Bacillus endophyticus* and *Bacillus* sp.)	84	Hussain et al. [41]
Textile effluent	*P. australis* *T. domingensis*	*Rhodococcus* sp., *Pseudomonas indoloxydans* and *Acinetobacter junii* consortia	97.6 81.40	Tara et al. [117]
Reactive Blue	*Eichhornia crassipes*	*Bacillus cereus*	57	Tusief et al. [121]
Reactive Black 5	*Phragmites australis*	*Rhodococcus* sp., *Pseudomonas indoloxydans* and *Acinetobacter junii* consortia	95.50	Tara et al. [116]

7 Plant–Microbe Synergism Approach: Selection of Microbe

Constructed wetlands (CWs) are the engineered wastewater treatment approach which has low maintenance cost and requires less energy. They are built based on the purification processes involved in natural occurring wetlands, such as the potential to reduce or eliminate contaminants from water flowing across them, thus enhancing the water quality [21]. CWs can be utilized as a one-of-kind treatment to remediate wastewater discharged by small communities because of their efficiency in reducing

BOD and COD as well as eliminating total suspended solids (TSS) and nutrients [125].

Several studies have recently indicated that by the introduction of microbe to the plant species present in constructed wetlands, could be a beneficial approach for enhancing the phytodepuration efficacy. Moreover, the microbe also promotes the plant growth and performance [93, 99]. The use of plant growth promoting bacteria (PGP) improves the efficiency of CWs. Previous researchers have isolated and selected the PGP bacteria from the rhizosphere and endosphere of commonly used plant species in CWs for the phytodepuration [103]. In some studies, the PGP bacteria had been obtained and isolated from the plant species which were grown in CWs whereas in few studies bacteria were isolated from different plants used in treating the same type of pollutant in CWs [111]. In contrast to these the PGP bacteria has also been isolated from the site polluted from the wastewater containing pollutants which has to be remediated [93, 99].

Moreover, the bacteria after isolation are analyzed for their decolorization potential of required dye or textile wastewater and their ability to promote plant growth. The strain exhibiting higher decolorization potential and plant growth is selected for the phytodepuration in constructed wetlands. Not only the single strain but the consortium of bacteria is also used in CWs. The bacterial culture is added into the wastewater for the bioaugmentation in CWs or the roots of plant species can also be dipped in bacterial suspension before growing in CWs [93, 103].

8   Analysis of the Metabolites and Degradation Products of Plant–Microbe Synergism

The previous studies showed that the microbe-assisted phytoremediation approach is significant to remediate the dye containing wastewater resulting in less toxic products or metabolites through the degradation and transformation of complex compounds. These metabolites were analyzed for their transformation, structure, toxicity through Fourier transform Infrared spectroscopy (FTIR), Ultraviolet–visible spectroscopy, Gas chromatography-mass spectrometry (GC–MS), High-performance thin layer chromatography (HPTLC) and High-performance liquid chromatography (HPLC).

If a dye is degraded into a less toxic metabolite, it can be visualized by the newer peaks formation or shift and fading of the existing wavelength bands in UV–Vis spectroscopy [112]. The HPLC analysis of the dye treated with the plant–microbe consortia also revealed the formation of metabolites by the parent dye transformation through the emergence of new peaks in the HPLC profile [58]. HPTLC can also provide information on the degradation or transformation of parent dye through the plant–microbe synergistic approach. Like, at 366 nm the HPLTC plate image and change in the retention factor (R_f) of dye showed that there is degradation of Navy Blue RX dye and the chromatic group in the dye had been eliminated. In

Petunia grandiflora-B. pumilus consortium treated sample exhibiting higher decolorization of dye, no major peaks were detected contrary to the individual treated sample (Watharkar et al. [129]. HPTLC analysis of Scarket RR decolorization by *Pseudomonas monteilii-G. pulchella* consortia also showed disappearance of peak with R_f 0.72 as well as the construction of new peaks supporting the total biotransformation of dye [49]. The HPTLC profile of *B. pumilus- P. crinitum* consortium also proved that the plant–microbe synergism approach is efficient for the degradation of dye effluent as the treated effluent showed the presence at different bands at the R_f 0.54, 0.64, 0.72 and 0.83 with an absorbance of 23.1, 95.30, 14.2, and 133.5 AU, respectively [130].

FTIR spectra also supports the biotransformation or degradation of parent dye. In the synergistic degradation of Direct red 5B dye, the product formed after the decolorization by the consortia of *P. putida* and *P. grandiflora* showed different peak elucidating C–C, N–O, S=O stretching showing that the products have sulfo groups and the reduction in their number shows more desulfonation of the dye and the disappearance of azo bond from the parent dye was also seen [58]. Likewise, after SSR dye and dye mixture decolorization the FTIR spectra of products formed also showed the biotransformation of the dye and dye mixture into different metabolites [49]. The appearance of different peak also showed the dye breakdown and formation of intermediates and less toxic metabolites [90].

The GC–MS has also analyzed the different metabolic pathways for the degradation of plant–microbe consortium treated dye wastewater based on the enzymatic inductions. After the treatment of Navy blue HE2R dye with the consortia of *I. hederifolia* and *C. cladosporioides*, the GC–MS analysis showed that the asymmetric cleavage of dye into 3,4 naphthalene-2-sulfonic acid through desulfonation is further cleaved by lignin peroxidase or veratryl alcohol oxidase leading to the formation of 3-naphthalene-2-sulfonic acid and intermediate. This was then oxidized by laccase and further reduced and underwent desulfonation to produce naphthalene-2-yl diazene [90].

9 Toxicity Evaluation of Metabolites and Dye Degraded Products Formed After Plant–Microbe Synergism

After the textile effluent containing dye treatment in VFCWs involving the plant-microbes consortia, the reduction in toxicity of wastewater was assessed by Fish toxicity analysis. Fish supplied with control wastewater treated in VFCWs which was not augmented with microbial and plant species died within 12 h whereas those exposed to the vegetated and inoculated VFCWs treated wastewater remained alive [41]. Similarly, Fish toxicity analysis was also done to evaluate the toxicity reduction in Reactive Black 5 dye by the synergetic approach of plant and microbe. In the untreated water all fish died whereas in the treated water none of the fish died even after 96 h [116].

Phytotoxicity analysis was also performed to evaluate the reduction in toxicity of the dye treated with the plant–microbe consortium. It was found that as compared to the textile effluent treated solely by bacteria or plant, the plumule and radicle lengths of *Triticum aestivum* were greater in seeds grown in the consortium treated textile effluent [117]. Similarly, the plant-bacteria consortia treated effluent showed 90% germination of *Sorghum vulgare* and 100% of *Phaseolus mungo* germination and increased the shoot and root length contrast to those supplied with untreated effluent [130]. Likewise, the non-toxicity of the decolorized solution of dye obtained by the treatment of plant–microbe consortium was also assessed. 90% germination of *S. vulgare* and 100% germination of *P. mungo* was showed by the metabolite solution, it also showed the increase in radical and plumule length (Watharkar et al. [129].

10   Conclusion

Although the bacteria and plants are potential to remediate environmental pollutants but they have certain limitations to their use. As it has been shown by various studies that the plant–microbe-based remediation approaches have emerged as an efficient method for the degradation and transformation of dyes into less toxic metabolites which are not causing any harm to environment and mankind. Moreover, they are promoting the plant growth and performance by decreasing the stress response. The use of naturally occurring synergism of plants and their roots associated microbiome can be used extensively as a consortium for the remediation of dyes from textile effluent as an eco-friendly and energy saving approach. This approach needs more research so that it can be used in large-scale remediation of dyes.

References

1. Abd Ellatif S, El-Sheekh MM, Senousy HH (2021) Role of microalgal ligninolytic enzymes in industrial dye decolorization. Int J Phytorem 23(1):41–52
2. Adams D (2019) Decolourisation of selected dyes by lettuce and mung bean seedlings: a potential phytoremediation strategy
3. Adnan LA, Hadibarata T, Sathishkumar P, Mohd Yusoff AR (2016) Biodegradation pathway of acid Red 27 by white-rot fungus Armillaria sp. F022 and phytotoxicity evaluation. CLEAN–Soil, Air, Water 44(3):239–246
4. Ahila KG, Ravindran B, Muthunarayanan V, Nguyen DD, Nguyen XC, Chang SW, … Thamaraiselvi C (2021) Phytoremediation potential of freshwater macrophytes for treating dye-containing wastewater. Sustainability 13(1):329
5. Ahmad M (2017) Applications of biosorbents for industrial wastewater treatment (Doctoral dissertation)
6. Akansha K, Chakraborty D, Sachan SG (2019) Decolorization and degradation of methyl orange by Bacillus stratosphericus SCA1007. Biocatal Agric Biotechnol 18:101044
7. Al-Baldawi IA, Abdullah SRS, Almansoory AF, Hasan HA, Anuar N (2020) Role of Salvinia molesta in biodecolorization of methyl orange dye from water. Sci Rep 10(1):1–9

8. Allen SJ, Mckay G, Porter JF (2004) Adsorption isotherm models for basic dye adsorption by peat in single and binary component systems. J Colloid Interface Sci 280(2):322–333

9. Anjaneya O, Shrishailnath SS, Guruprasad K, Nayak AS, Mashetty SB, Karegoudar TB (2013) Decolourization of Amaranth dye by bacterial biofilm in batch and continuous packed bed bioreactor. Int Biodeterior Biodegradation 79:64–72

10. Arteaga LC, Zavaleta MP, Eustaquio WM, Bobadilla JM (2018) Removal of aniline blue dye using live microalgae Chlorella vulgaris. J Energy Environ Sci 2(1):6–12

11. Arulazhagan P (2016) A study on microbial decolourization of reactive red M8B by Bacillus subtilis isolated from dye contaminated soil samples. Int J Curr Res Biol Med 1(1):1–13

12. Ayele A, Getachew D, Kamaraj M, Suresh A (2021) Phycoremediation of synthetic dyes: an effective and eco-friendly algal technology for the dye abatement. J Chem 2021

13. Balseiro-Romero M, Gkorezis P, Kidd PS, Van Hamme J, Weyens N, Monterroso C, Vangronsveld J (2017) Use of plant growth promoting bacterial strains to improve Cytisus striatus and Lupinus luteus development for potential application in phytoremediation. Sci Total Environ 581:676–688

14. Bankole PO, Adekunle AA, Obidi OF, Chandanshive VV, Govindwar SP (2018) Biodegradation and detoxification of Scarlet RR dye by a newly isolated filamentous fungus, Peyronellaea Prosopidis. Sustain Environ Res 28(5):214–222

15. Bansal P, Sud D (2012) Photodegradation of commercial dye, CI Reactive Blue 160 using ZnO nanopowder: Degradation pathway and identification of intermediates by GC/MS. Sep Purif Technol 85:112–119

16. Bera SP, Tank SK (2021) Bioremedial approach of Pseudomonas stutzeri SPM-1 for textile azo dye degradation. Arch Microbiol 203(5):2669–2680

17. Berradi M, Hsissou R, Khudhair M, Assouag M, Cherkaoui O, El Bachiri A, El Harfi A (2019) Textile finishing dyes and their impact on aquatic environs. Heliyon 5(11):e02711

18. Bharagava RN, Purchase D, Saxena G, Mulla SI (2019) Applications of metagenomics in microbial bioremediation of pollutants: from genomics to environmental cleanup. In: Microbial diversity in the genomic era. Academic Press, pp 459–477

19. Can-Terzi B, Goren AY, Okten HE, Sofuoglu SC (2021) Biosorption of methylene blue from water by live Lemna minor. Environ Technol Innov 22:101432

20. Carolin CF, Kumar PS, Joshiba GJ (2021) Sustainable approach to decolourize methyl orange dye from aqueous solution using novel bacterial strain and its metabolites characterization. Clean Technol Environ Policy 23(1):173–181

21. Carvalho PN, Arias CA, Brix H (2017) Constructed wetlands for water treatment: new developments

22. Chakraborty S, Basak B, Dutta S, Bhunia B, Dey A (2013) Decolorization and biodegradation of congo red dye by a novel white rot fungus Alternaria alternata CMERI F6. Biores Technol 147:662–666

23. Chandanshive VV, Rane NR, Gholave AR, Patil SM, Jeon BH, Govindwar SP (2016) Efficient decolorization and detoxification of textile industry effluent by Salvinia molesta in lagoon treatment. Environ Res 150:88–96

24. Chatterjee S, Dey S, Sarma M, Chaudhuri P, Das S (2020) Biodegradation of Congo Red by Manglicolous Filamentous Fungus Aspergillus flavus JKSC-7 Isolated from Indian Sundabaran Mangrove Ecosystem. Appl Biochem Microbiol 56(6):708–717

25. Chen SH, Cheow YL, Ng SL, Ting ASY (2019) Biodegradation of triphenylmethane dyes by non-white rot fungus Penicillium simplicissimum: enzymatic and toxicity studies. Int J Environ Res 13(2):273–282

26. Chin JY, Chng LM, Leong SS, Yeap SP, Yasin NHM, Toh PY (2020) Removal of synthetic Dye by chlorella vulgaris microalgae as natural adsorbent. Arab J Sci Eng 45(9):7385–7395

27. Chowdhury MF, Khandaker S, Sarker F, Islam A, Rahman MT, Awual MR (2020) Current treatment technologies and mechanisms for removal of indigo carmine dyes from wastewater: a review. J Mol Liq 114061

28. Chu WL, See YC, Phang SM (2009) Use of immobilised Chlorella vulgaris for the removal of colour from textile dyes. J Appl Phycol 21(6):641–648

29. Dellamatrice PM, Silva-Stenico ME, Moraes LABD, Fiore MF, Monteiro RTR (2017) Degradation of textile dyes by cyanobacteria. Brazilian J Microbiol 48(1):25–31
30. Ejder-Korucu M, Gürses A, Dogar C, Sharma S, Acikyildiz M (2015) Removal of organic dyes from industrial effluents: an overview of physical and biotechnological applications
31. Ekanayake MS, Udayanga D, Wijesekara I, Manage P (2021) Phytoremediation of synthetic textile dyes: biosorption and enzymatic degradation involved in efficient dye decolorization by Eichhornia crassipes (Mart.) Solms and Pistia stratiotes L. Environ Sci Pollut Res 28(16):20476–20486
32. El-Sheekh MM, Abou-El-Souod GW, El Asrag HA (2018) Biodegradation of some dyes by the green alga Chlorella vulgaris and The Cyanobacterium Aphanocapsa elachista. Egypt J Bot 58(3):311–320
33. Eskandari F, Shahnavaz B, Mashreghi M (2019) Optimization of complete RB-5 azo dye decolorization using novel cold-adapted and mesophilic bacterial consortia. J Environ Manage 241:91–98
34. Ewadh HM (2020) Removal of methylene blue by coontail (Ceratophyllum demersum) using phytoremediation concept. Plant Arch 20(1):2677–2681
35. Favero BM, Favero AC, Taffarel SR, Souza FS (2018) Evaluation of the efficiency of coagulation/flocculation and Fenton process in reduction of colour, turbidity and COD of a textile effluent. Environ Technol
36. Franciscon E, Mendonca D, Seber S, Morales DA, Zocolo GJ, Zanoni MB, … Umbuzeiro GA (2015) Potential of a bacterial consortium to degrade azo dye Disperse Red 1 in a pilot scale anaerobic–aerobic reactor. Process Biochem 50(5):816–825
37. Gaviria-Arroyave MI, Osorio-Echavarría J, Gómez-Vanegas NA (2018) Evaluating the scale-up of a reactor for the treatment of textile effluents using Bjerkandera sp. Revista Facultad De Ingeniería Universidad De Antioquia 88:80–90
38. Guo G, Tian F, Zhang C, Liu T, Yang F, Hu Z, … Ding K (2019) Performance of a newly enriched bacterial consortium for degrading and detoxifying azo dyes. Water Sci Technol 79(11):2036–2045
39. Hanis KKA, Nasri AM, Farahiyah WW, Rabani MM (2020, April) Bacterial Degradation of Azo Dye Congo Red by Bacillus sp. In: Journal of Physics: Conference Series, vol 1529, no 2. IOP Publishing, p 022048
40. He XL, Song C, Li YY, Wang N, Xu L, Han X, Wei DS (2018) Efficient degradation of azo dyes by a newly isolated fungus Trichoderma tomentosum under non-sterile conditions. Ecotoxicol Environ Saf 150:232–239
41. Hussain Z, Arslan M, Malik MH, Mohsin M, Iqbal S, Afzal M (2018) Treatment of the textile industry effluent in a pilot-scale vertical flow constructed wetland system augmented with bacterial endophytes. Sci Total Environ 645:966–973
42. Ilyas S, Rehman A (2013) Decolorization and detoxification of Synozol red HF-6BN azo dye, by Aspergillus niger and Nigrospora sp. Iran J Environ Health Sci Eng 10(1):1–9
43. Imron MF, Kurniawan SB, Soegianto A, Wahyudianto FE (2019) Phytoremediation of methylene blue using duckweed (Lemna minor). Heliyon 5(8):e02206
44. Indrani J (2020) Microbial enzymatic system in the degradation of textile dyes
45. Ismail M, Akhtar K, Khan MI, Kamal T, Khan MA, Asiri AM, … Khan SB (2019) Pollution, toxicity and carcinogenicity of organic dyes and their catalytic bio-remediation. Curr Pharm Des 25(34):3645–3663
46. Ito T, Shimada Y, Suto T (2018) Potential use of bacteria collected from human hands for textile dye decolorization. Water Res Ind 20:46–53
47. Jadhav JP, Kalyani DC, Telke AA, Phugare SS, Govindwar SP (2010) Evaluation of the efficacy of a bacterial consortium for the removal of color, reduction of heavy metals, and toxicity from textile dye effluent. Biores Technol 101(1):165–173
48. Javaid R, Qazi UY (2019) Catalytic oxidation process for the degradation of synthetic dyes: an overview. Int J Environ Res Public Health 16(11):2066
49. Kabra AN, Khandare RV, Govindwar SP (2013) Development of a bioreactor for remediation of textile effluent and dye mixture: a plant–bacterial synergistic strategy. Water Res 47(3):1035–1048

50. Kabra AN, Khandare RV, Waghmode TR, Govindwar SP (2012) Phytoremediation of textile effluent and mixture of structurally different dyes by Glandularia pulchella (Sweet) Tronc. Chemosphere 87(3):265–272

51. Kagalkar AN, Jagtap UB, Jadhav JP, Bapat VA, Govindwar SP (2009) Biotechnological strategies for phytoremediation of the sulfonated azo dye Direct Red 5B using Blumea malcolmii Hook. Biores Technol 100(18):4104–4110

52. Katheresan V, Kansedo J, Lau SY (2018) Efficiency of various recent wastewater dye removal methods: a review. J Environ Chem Eng 6(4):4676–4697

53. Kaushal J, Mahajan P (2021) Kinetic evaluation for removal of an anionic diazo Direct Red 28 by using phytoremediation potential of Salvinia molesta Mitchell. Bull Environ Contam Toxicol 1–6

54. Kaushal J, Mahajan P (2015) Exploring the Phytoremediation Potential of Salvinia molesta for the degradation of Malachite green dye. Res J Chem Environ 19

55. Khan R, Fulekar MH (2017) Mineralization of a sulfonated textile dye Reactive Red 31 from simulated wastewater using pellets of Aspergillus bombycis. Biores Bioprocess 4(1):1–11

56. Khan SA, Mehmood S, Iqbal A, Hamayun M (2020) Industrial polluted soil borne fungi decolorize the recalcitrant azo dyes Synozol red HF–6BN and Synozol black B. Ecotoxicol Environ Saf 206:111381

57. Khan S, Malik A (2016) Degradation of reactive black 5 dye by a newly isolated bacterium Pseudomonas entomophila BS1. Can J Microbiol 62(3):220–232

58. Khandare RV, Kabra AN, Awate AV, Govindwar SP (2013) Synergistic degradation of diazo dye Direct Red 5B by Portulaca grandiflora and Pseudomonas putida. Int J Environ Sci Technol 10(5):1039–1050

59. Khandare RV, Kabra AN, Tamboli DP, Govindwar SP (2011) The role of Aster amellus Linn. in the degradation of a sulfonated azo dye Remazol Red: a phytoremediation strategy. Chemosphere 82(8):1147–1154

60. Khandare RV, Rane NR, Waghmode TR, Govindwar SP (2012) Bacterial assisted phytoremediation for enhanced degradation of highly sulfonated diazo reactive dye. Environ Sci Pollut Res 19(5):1709–1718

61. Khataee AR, Dehghan G, Ebadi A, Zarei M, Pourhassan M (2010) Biological treatment of a dye solution by Macroalgae Chara sp.: effect of operational parameters, intermediates identification and artificial neural network modeling. Biores Technol 101(7):2252–2258

62. Khataee AR, Movafeghi A, Torbati S, Lisar SS, Zarei M (2012) Phytoremediation potential of duckweed (Lemna minor L.) in degradation of CI Acid Blue 92: Artificial neural network modeling. Ecotoxicol Environ Saf 80:291–298

63. Khataee AR, Movafeghi A, Vafaei F, Salehi Lisar SY, Zarei M (2013) Potential of the aquatic fern Azolla filiculoides in biodegradation of an azo dye: modeling of experimental results by artificial neural networks. Int J Phytorem 15(8):729–742

64. Khosravi R, Hossini H, Heidari M, Fazlzadeh M, Biglari H, Taghizadeh A, Barikbin B (2017) Electrochemical decolorization of reactive dye from synthetic wastewater by mono-polar aluminum electrodes system. Int J Electrochem Sci 12:4745–4755

65. Kishor R, Purchase D, Saratale GD, Ferreira LFR, Bilal M, Iqbal HM, Bharagava RN (2021) Environment friendly degradation and detoxification of Congo red dye and textile industry wastewater by a newly isolated Bacillus cohnni (RKS9). Environ Technol Innov 22:101425

66. Kour D, Kaur T, Devi R, Yadav A, Singh M, Joshi D, … Saxena AK (2021) Beneficial microbiomes for bioremediation of diverse contaminated environments for environmental sustainability: present status and future challenges. Environ Sci Pollut Res 1–23

67. Krishnamoorthy R, Jose PA, Ranjith M, Anandham R, Suganya K, Prabhakaran J, … Kumutha K (2018) Decolourisation and degradation of azo dyes by mixed fungal culture consisted of Dichotomomyces cejpii MRCH 1-2 and Phoma tropica MRCH 1-3. J Environ Chem Eng 6(1):588–595

68. Kumar AS, Arunagirinathan N, Vijayanand S, Hemapriya J, Indra V (2017) Bioremediation and detoxification of a textile azo dye-evans blue by bacterial strain akip-2. J Curr Microbiol App Sci 6:2687–2694

69. Kurade MB, Ha YH, Xiong JQ, Govindwar SP, Jang M, Jeon BH (2021) Phytoremediation as a green biotechnology tool for emerging environmental pollution: a step forward towards sustainable rehabilitation of the environment. Chem Eng J 129040
70. Kurade MB, Waghmode TR, Jadhav MU, Jeon BH, Govindwar SP (2015) Bacterial–yeast consortium as an effective biocatalyst for biodegradation of sulphonated azo dye Reactive Red 198. RSC Adv 5(29):23046–23056
71. Lade H, Kadam A, Paul D, Govindwar S (2015) Biodegradation and detoxification of textile azo dyes by bacterial consortium under sequential microaerophilic/aerobic processes. EXCLI J 14:158
72. Lellis B, Fávaro-Polonio CZ, Pamphile JA, Polonio JC (2019) Effects of textile dyes on health and the environment and bioremediation potential of living organisms. Biotechnol Res Innov 3(2):275–290
73. Lokhande VH, Kudale S, Nikalje G, Desai N, Suprasanna P (2015) Hairy root induction and phytoremediation of textile dye, Reactive green 19A-HE4BD, in a halophyte, Sesuvium portulacastrum (L.) L. Biotechnol Rep 8:56–63
74. Lu T, Zhang Q, Yao S (2017) Efficient decolorization of dye-containing wastewater using mycelial pellets formed of marine-derived Aspergillus niger. Chin J Chem Eng 25(3):330–337
75. Ma Y, Prasad MNV, Rajkumar M, Freitas H (2011) Plant growth promoting rhizobacteria and endophytes accelerate phytoremediation of metalliferous soils. Biotechnol Adv 29(2):248–258
76. Madhav S, Ahamad A, Singh P, Mishra PK (2018) A review of textile industry: wet processing, environmental impacts, and effluent treatment methods. Environ Qual Manage 27(3):31–41
77. Mahajan P, Kaushal J (2020) Phytoremediation of azo dye methyl red by macroalgae Chara vulgaris L.: kinetic and equilibrium studies. Environ Sci Pollut Res 27(21):26406–26418
78. Mahajan P, Kaushal J, Upmanyu A, Bhatti J (2019) Assessment of phytoremediation potential of Chara vulgaris to treat toxic pollutants of textile effluent. J Toxicol
79. Mahmood R, Sharif FAIZA, Ali S, Hayyat MU, Cheema TA (2012) Isolation of indigenous bacteria and consortia development for decolorization of textile dyes. Biol Pakistan 58:53–60
80. Mishra S, Maiti A (2018) The efficacy of bacterial species to decolourise reactive azo, anthroquinone and triphenylmethane dyes from wastewater: a review. Environ Sci Pollut Res 25(9):8286–8314
81. Modi HA, Rajput G, Ambasana C (2010) Decolorization of water soluble azo dyes by bacterial cultures, isolated from dye house effluent. Biores Technol 101(16):6580–6583
82. Mohamed WSED (2016) Isolation and screening of reactive dye decolorizing bacterial isolates from textile industry effluent. Int J Microbiol Res 7(1):01–08
83. Mokif LA (2019) Removal methods of synthetic dyes from industrial wastewater: a review. Mesopotamia Environ J 5(1)
84. Movafeghi A, Khataee AR, Torbati S, Zarei M, Lisar SS (2013) Bioremoval of CI Basic Red 46 as an azo dye from contaminated water by Lemna minor L.: modeling of key factor by neural network. Environ Prog Sustain Energy 32(4):1082–1089
85. Munck C, Thierry E, Gräßle S, Chen SH, Ting ASY (2018) Biofilm formation of filamentous fungi Coriolopsis sp. on simple muslin cloth to enhance removal of triphenylmethane dyes. J Environ Manage 214:261–266
86. Muthusamy S, Govindaraj D, Rajendran K (2018) Phytoremediation of textile dye effluents. In: Bioremediation: applications for environmental protection and management. Springer, Singapore, pp 359–373
87. Pandey AK, Sarada DV, Kumar A (2016) Microbial decolorization and degradation of reactive red 198 azo dye by a newly isolated Alkaligenes species. Proc Nat Acad Sci India Sect B Biol Sci 86(4):805–815
88. Pathak VV, Kothari R, Chopra AK, Singh DP (2015) Experimental and kinetic studies for phycoremediation and dye removal by Chlorella pyrenoidosa from textile wastewater. J Environ Manage 163:270–277
89. Patil AV, Jadhav JP (2013) Evaluation of phytoremediation potential of Tagetes patula L. for the degradation of textile dye Reactive Blue 160 and assessment of the toxicity of degraded metabolites by cytogenotoxicity. Chemosphere 92(2):225–232

90. Patil SM, Chandanshive VV, Rane NR, Khandare RV, Watharkar AD, Govindwar SP (2016) Bioreactor with Ipomoea hederifolia adventitious roots and its endophyte Cladosporium cladosporioides for textile dye degradation. Environ Res 146:340–349
91. Pereira L, Alves M (2012) Dyes—environmental impact and remediation. In: Environmental protection strategies for sustainable development. Springer, Dordrecht, pp 111–162
92. Pokharia A, Ahluwalia SS (2016) Decolorization of xenobiotic azo dye-Black WNN by immobilized Paenibacillus alvei MTCC 10625. Int J Environ Bioremed Biodegrad 4:35–46
93. Prum C, Dolphen R, Thiravetyan P (2018) Enhancing arsenic removal from arsenic-contaminated water by Echinodorus cordifolius—endophytic Arthrobacter creatinolyticus interactions. J Environ Manage 213:11–19
94. Qu W, Hong G, Zhao J (2018) Degradation of malachite green dye by Tenacibaculum sp. HMG1 isolated from Pacific deep-sea sediments. Acta Oceanologica Sinica 37(6):104–111
95. Rane NR, Chandanshive VV, Khandare RV, Gholave AR, Yadav SR, Govindwar SP (2014) Green remediation of textile dyes containing wastewater by Ipomoea hederifolia L. RSC Adv 4(69):36623–36632
96. Rane NR, Chandanshive VV, Watharkar AD, Khandare RV, Patil TS, Pawar PK, Govindwar SP (2015) Phytoremediation of sulfonated Remazol Red dye and textile effluents by Alternanthera philoxeroides: an anatomical, enzymatic and pilot scale study. Water Res 83:271–281
97. Rane NR, Khandare RV, Watharkar AD, Govindwar SP (2017) Phytoremediation as a green and clean tool for textile dye pollution abatement. In: Phytoremediation of environmental pollutants. CRC Press, pp 327–360
98. Rápó E, Posta K, Csavdári A, Vincze BÉ, Mara G, Kovács G, … Tonk S (2020) Performance comparison of Eichhornia crassipes and Salvinia natans on azo-dye (Eriochrome Black T) phytoremediation. Crystals 10(7):565
99. Rehman K, Imran A, Amin I, Afzal M (2018) Inoculation with bacteria in floating treatment wetlands positively modulates the phytoremediation of oil field wastewater. J Hazard Mater 349:242–325
100. Saini RD (2017) Textile organic dyes: polluting effects and elimination methods from textile waste water. Int J Chem Eng Res 9(1):121–136
101. Salem OM, Abdelsalam A, Boroujerdi A (2021) Bioremediation potential of Chlorella vulgaris and Nostoc paludosum on azo dyes with analysis of metabolite changes. Baghdad Sci J 18(3):0445
102. Salem SS, Mohamed A, El-Gamal M, Talat M, Fouda A (2019) Biological decolorization and degradation of azo dyes from textile wastewater effluent by Aspergillus niger. Egypt J Chem 62(10):1799–1813
103. Salgado I, Cárcamo H, Carballo ME, Cruz M, del Carmen Durán M (2018) Domestic wastewater treatment by constructed wetlands enhanced with bioremediating rhizobacteria. Environ Sci Pollut Res 25(21):20391–20398
104. Sarwa P, Vijayakumar R, Verma SK (2014) Adsorption of acid red 66 dye from aqueous solution by green microalgae Acutodesmus obliquus strain PSV2 isolated from an industrial polluted site. Open Access Library J 1(3):1–8
105. Shanmugam L, Ahire M, Nikam T (2020) Bacopa monnieri (L.) Pennell, a potential plant species for degradation of textile azo dyes. Environ Sci Pollut Res 1–15
106. Sharma J, Sharma S, Soni V (2021a) Classification and impact of synthetic textile dyes on Aquatic Flora: a review. Reg Stud Marine Sci 101802
107. Sharma R, Saini H, Paul DR, Chaudhary S, Nehra SP (2021) Removal of organic dyes from wastewater using Eichhornia crassipes: a potential phytoremediation option. Environ Sci Pollut Res 28(6):7116–7122
108. Sharma S, Kaur A (2018) Various methods for removal of dyes from industrial effluents-a review. Indian J Sci Technol 11:1–21
109. Singh G, Dwivedi SK (2020) Decolorization and degradation of Direct Blue-1 (Azo dye) by newly isolated fungus Aspergillus terreus GS28, from sludge of carpet industry. Environ Technol Innov 18:100751

110. Singh RL, Singh PK, Singh RP (2015) Enzymatic decolorization and degradation of azo dyes–a review. Int Biodeterior Biodegradation 104:21–31
111. Syranidou E, Christofilopoulos S, Gkavrou G, Thijs S, Weyens N, Vangronsveld J, Kalogerakis N (2016) Exploitation of endophytic bacteria to enhance the phytoremediation potential of the wetland helophyte Juncus acutus. Front Microbiol 7:1016
112. Tahir U, Sohail S, Khan UH (2017) Concurrent uptake and metabolism of dyestuffs through bio-assisted phytoremediation: a symbiotic approach. Environ Sci Pollut Res 24(29):22914–22931
113. Talha MA, Goswami M, Giri BS, Sharma A, Rai BN, Singh RS (2018) Bioremediation of Congo red dye in immobilized batch and continuous packed bed bioreactor by Brevibacillus parabrevis using coconut shell bio-char. Biores Technol 252:37–43
114. Tan KA, Morad N, Ooi JQ (2016) Phytoremediation of methylene blue and methyl orange using Eichhornia crassipes. Int J Environ Sci Dev 7(10):724
115. Tang W, Xu X, Ye BC, Cao P, Ali A (2019) Decolorization and degradation analysis of Disperse Red 3B by a consortium of the fungus Aspergillus sp. XJ-2 and the microalgae Chlorella sorokiniana XJK. RSC Adv 9(25):14558–14566
116. Tara N, Iqbal M, Habib F, Khan QM, Iqbal S, Afzal M, Brix H (2021) Investigating degradation metabolites and underlying pathway of azo dye "Reactive Black 5" in bioaugmented floating treatment wetlands
117. Tara N, Iqbal M, Mahmood Khan Q, Afzal M (2019) Bioaugmentation of floating treatment wetlands for the remediation of textile effluent. Water Environ J 33(1):124–134
118. Tkaczyk A, Mitrowska K, Posyniak A (2020) Synthetic organic dyes as contaminants of the aquatic environment and their implications for ecosystems: a review. Sci Total Environ 717:137222
119. Torbati S (2017) Feasibility study on phytoremediation of malachite green dye from contaminated aqueous solutions using watercress (Nasturtium officinale). Iran J Health Environ 9(4)
120. Torbati S (2019) Toxicological risks of Acid Bordeaux B on duckweed and the plant potential for effective remediation of dye-polluted waters. Environ Sci Pollut Res 26(27):27699–27711
121. Tusief MQ, Malik MH, Mohsin M, Asghar HN, Iqbal M, Mahmood N (2020) Eco-friendly degradation of reactive blue dye enriched textile water by floating treatment wetlands (FTWs) system applying the strategy of plant-bacteria partnership (Part-B). Pakistan J Sci Ind Res Ser A Phys Sci 63(1):40–47
122. Uday USP, Bandyopadhyay TK, Bhunia B (2016) Bioremediation and detoxification technology for treatment of dye (s) from textile effluent. Text Wastewater Treat 75–92
123. Velayutham K, Madhava AK, Pushparaj M, Thanarasu A, Devaraj T, Periyasamy K, Subramanian S (2018) Biodegradation of Remazol Brilliant Blue R using isolated bacterial culture (Staphylococcus sp. K2204). Environ Technol 39(22):2900–2907
124. Waghmode TR, Kurade MB, Kabra AN, Govindwar SP (2012) Degradation of Remazol Red dye by Galactomyces geotrichum MTCC 1360 leading to increased iron uptake in Sorghum vulgare and Phaseolus mungo from soil. Biotechnol Bioprocess Eng 17(1):117–126
125. Wang J, Tai Y, Man Y, Wang R, Feng X, Yang Y, … Cai N (2018) Capacity of various single-stage constructed wetlands to treat domestic sewage under optimal temperature in Guangzhou City, South China. Ecol Eng 115:35–44
126. Wang N, Chu Y, Zhao Z, Xu X (2017) Decolorization and degradation of Congo red by a newly isolated white rot fungus, Ceriporia lacerata, from decayed mulberry branches. Int Biodeterior Biodegradation 117:236–244
127. Waqas R, Arshad M, Asghar HN, Asghar M (2015) Optimization of factors for enhanced phycoremediation of reactive blue azo dye. Int J Agric Biol 17(4)
128. Watharkar AD, Khandare RV, Kamble AA, Mulla AY, Govindwar SP, Jadhav JP (2013) Phytoremediation potential of Petunia grandiflora Juss., an ornamental plant to degrade a disperse, disulfonated triphenylmethane textile dye Brilliant Blue G. Environ Sci Pollut Res 20(2):939–949

129. Watharkar AD, Rane NR, Patil SM, Khandare RV, Jadhav JP (2013) Enhanced phytotransfor-
     mation of Navy Blue RX dye by Petunia grandiflora Juss. with augmentation of rhizospheric
     Bacillus pumilus strain PgJ and subsequent toxicity analysis. Biores Technol 142:246–254
130. Watharkar AD, Khandare RV, Waghmare PR, Jagadale AD, Govindwar SP, Jadhav JP
     (2015) Treatment of textile effluent in a developed phytoreactor with immobilized bacte-
     rial augmentation and subsequent toxicity studies on Etheostoma olmstedi fish. J Hazard
     Mater 283:698–704
131. Yan A, Wang Y, Tan SN, Yusof MLM, Ghosh S, Chen Z (2020) Phytoremediation: a promising
     approach for revegetation of heavy metal-polluted land. Front Plant Sci 11
132. Yusuf M (2019) Synthetic dyes: a threat to the environment and water ecosystem. Text Clothing
     11–26
133. Zhou Y, Lu J, Zhou Y, Liu Y (2019) Recent advances for dyes removal using novel adsorbents:
     a review. Environ Pollut 252:352–365

Role of Fungi in the Removal of Synthetic Dyes from Textile Industry Effluents

P. Santhilatha, B. Haritha, and L. Suseela

Abstract The present review focuses on the role of fungi in the biological degradation/decolourization of textile wastes and effluents. The major decomposers of xenobiotics are fungi and bacteria. Among both of them, fungi occupy a key role in dye degradation/decolourization due to their efficiency in the production of dye decolourizing enzymes as absorbents. The present study highlighted the advancement of the fungal degradation of textile effluents by the usage of effective fungal strains. The biological degradation by using fungi or mycoremediation on textile effluents or dyes using effective fungi is a comparatively economic and eco-friendly process, to mineralize the degrading compounds of textile dyes. Fungi play a significant role in the biodegradation/decolourization of textile dyes through the enzymatic activity that depends on absorption, adsorption, and accumulation of recalcitrant compounds from the effluents.

Keywords Fungi · Mycoremediation · Dye degradation · Textile effluents · Fungal enzymes

1 Introduction

Environmental pollution caused by the release of various effluents as a consequence of progress in industrialization is now becoming a global problem. As a result of rapid industrialization and urbanization, several chemicals like dyes, pigments, and aromatic compounds, etc., used in various industries such as textiles, printing, pharmaceuticals, food, toys, paper, plastic, and cosmetics are being produced and released

B. Haritha · L. Suseela (✉)
Department of Bio-Sciences and Biotechnology, Krishna University, Machilipatnam, Krishna District, Andhra Pradesh, India

P. Santhilatha
Onco-Stem Cell Research Laboratory, Department of Biochemistry and Bioinformatics, Institute of Science, GITAM Deemed to be University, Visakhapatnam, India

© The Author(s), under exclusive license to Springer Nature Singapore Pte Ltd. 2022
A. Khadir and S. S. Muthu (eds.), *Biological Approaches in Dye-Containing Wastewater*,
Sustainable Textiles: Production, Processing, Manufacturing & Chemistry,
https://doi.org/10.1007/978-981-19-0526-1_7

into the environment without any proper treatment that ultimately affect our day-to-day life [28]. These pollutants by accumulating in the earth at toxic levels spoil the natural ecosystem.

Dyes in the form of environmental pollutants are too hard to degrade by using chemical and physical methods. Dyes owing to their complex chemical structures remain stable/recalcitrant to decolourization in water and soil [36]. As the available physico-chemical treatment methods are more labour-intensive, highly uneconomical with methodological challenges [5], there is an urgent need to reduce the negative effects of dye pollutants by developing eco-friendly biological treatment techniques. The biological treatment methods serve as the best alternatives to the existing expensive, and commercially or environmentally unattractive physico-chemical technologies, and can be applied to treat majority of the dyes that are released as industrial effluents [19]. Microbes show greater efficacy in dye degradation by their effective enzyme systems [30]. Recent research has shown the importance of fungi in dye degradation because of their high biomass yield compared to bacteria [21]. Dyes belonging to different chemical groups have been found to react to fungal oxidation because of the ability of fungi to secrete highly oxidative, and nonspecific ligninolytic enzymes [8, 11, 39]—lignin peroxidase (LiP), manganese peroxidase (MnP), and laccase [45].

The use of fungi in the process of degradation or decolorization of dyes by bio-adsorption has been reported by Fu and Viraraghavan [16]. Wong and Yu [49] reported the role of fungi in dye removal by mineralization. Biosorption is considered as the most advantageous method compared to other types for the treatment of effluent waters [22]. Fungi because of their ability to produce large fungal hyphal systems have been used as the most effective dye degradation microbes by bio-adsorption [15].

2   Textile Industries-Effluent Waste Waters-Synthetic Dyes-Environmental and Human Health Issues

Release of some unwanted toxic elements into the atmosphere is one of the bad consequences of rapid industrialization. These pollutants once accumulate and reach up to the toxic level in the biosphere, pose different environmental problems that ultimately disturb the natural ecosystems. The worldwide usage of synthetic dyes in the textile industry has been increased tremendously in the recent years [31]. This in turn increases the environmental pollution in the form of wastewater disposal directly into the surrounding water bodies [12]. These textile effluents change the quality of the water that can be evident by means of total organic carbon (TOC), biological oxygen demand (BOD), chemical oxygen demand (COD), colour, pH, and by the existence of hazardous synthetic compounds such as azodyes and heavy metals [2, 5, 29]. Depending upon their chemical structure, the dyes used in textile industries are categorized as azo, diazo, cationic, basic, anthraquiones [35]. Annually more than

100,000 commercial dyes with over 7×10^5 tonnes of dyestuff are produced [38, 51]. More than 8000 different synthetic chemical products are listed in the Colour Index that are being used in the dyeing processes [44]. Indiscriminate release of synthetic textile dyes as effluents became a menace as they badly pollute our environment. Lack of proper effluent wastewater treatment methods results in both soil and water pollution. Pollutants that are disposed into the environment pose a great threat to both flora, and fauna of our ecosystem [12]. In developing nations, these textile dyes make a prominent portion of textile disposal and municipal waste [24]. Dye pollutants released into the water bodies cause a reduction in light penetration capacity that in turn affects the photosynthetic rate of aquatic flora [14]. Most of the synthetic dyes used by the textile industries are toxic, and even carcinogenic to human beings because they are mainly made from benzidine, and various aromatic compounds [5]. Azo dyes which comprises of 70% of all textile dyes results in the production of amines by their reductive cleavage that are found to be mutagenic to humans [4, 10, 48]. Accidental ingestion of Azo dyes results in their retention in the lower intestine by the intestinal microflora. Besides, these effluents also contain heavy metals in a huge quantity that are highly carcinogenic and mutagenic. These include Cd, Cr, Co, Cu, Hg, Ni, Mg, Fe, and Mn [11, 43]. Considering all these, textile effluents should be properly treated for toxic dye removal for their safer discharge into the environment.

3 Available Remediation Methods to Treat Textile Effluents

There are a variety of different treatment methods that can be applied to treat textile effluents. These include physico-chemical methods, chemical methods, and biological methods. The following figure shows the available methods for textile wastewater treatment (Fig. 1). Of all these methods, biological methods were found to be more effective and environment friendly.

4 Role of Microorganisms in the Bioremediation of Textile Effluents

4.1 Degradation of Dyes by Bacteria

Bacterial oxido-reductive enzymes can effectively bioremediate synthetic dyes. This will allow the usage of bacterial strains in the degradation of dyestuff which will degrade more complex toxic metabolites to less complex metabolites [27, 41]. Bacteria isolated from textile wastewater disposal sites can effectively bioremediate the dyes because of the presence of already activated enzyme systems. *Bacillus sp*

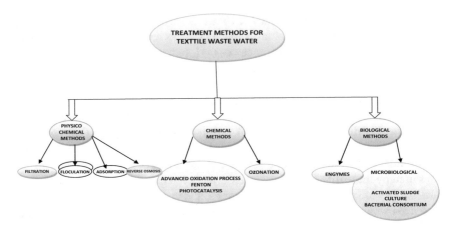

Fig. 1 Methods for textile wastewater treatment

and *Aeromonas hydrophila* were found to be potential microbes as early as in 1970s in the bioremediation of various industrial effluents [50].

4.2 Degradation of Dyes by Algae

Recent investigations suggest that photosynthetic autotrophic organisms like algae and cyanobacteria can able to decolourize azo dyes by an induced form of azoreductase [47]. Algae are also used in biological adsorption and some species of algae like *Chlorella* and *Oscillatoria* can degrade azo dyes and break them into their aromatic amines, and further degrade into simpler organic compounds or CO_2 [40].

4.3 Degradation of Dyes by Yeasts

Yeast also produces azoreductases that help in the reductive breakdown of azo groups (–N=N–). Some of the yeasts able to degrade dyes include—*Candida zeylanoides* and *C.oleophila*. The principle behind this is the formation of respective amines by degrading azo bond by reduction. Biodegradation by *Saccharomyces cerevisiae* MTCC 463 on Methyl Red showed different enzymatic activities viz., activities of laccase, aminopyrine N-demethylase, lignin peroxidase, NADH–DCIP reductase, and azoreductase. Oxidative enzymes like lignin peroxidase and laccase further degrade the products into their corresponding aliphatic amines [20]. *S. cerevisiae*, while its growth in molasses medium containing dyes showed the accumulation of reactive dye (Remazol Black B and Remazol Red RB). Some Ascomycete fungal species like yeast viz., *Candida tropicalis, and Debaryomyces polymorphus* have been observed to degrade azo dyes. The highly reactive textile effluents that are easily degraded by *Galactomyces geotrichum* MTCC 1360 can also degrade azo dyes [40].

4.4   Degradation of Dyes by Fungi

For the mineralization of synthetic dyes, many studies have reports on white-rot fungi [27, 47]. They involve in the production of varied extracellular oxido-reductases that break lignins and corresponding aromatic compounds. Their structurally nonstereoselective and nonspecific enzymes include—manganese peroxidase (MnP), lignin peroxidase (LiP), and laccase. *Phanerochaete chrysosporium* and *Neurospora crassa* produces laccase which has been extensively studied for the degradation of pigments and phenols from effluents [42]. In addition to the above, the other strains like *Trametes versicolor, Aspergillus ochraceus, Bjerkandera adusta, Pleurotus sp* and *Phlebia*, etc., have also be shown great role in the degradation process [40]. For the wastewater treatment, the usage of gel entrapment and adherence to a matrix, enhanced the results in decolourization process [42]. But the fact that white-rot fungi are not actually observed in effluents makes the enzymes unreliable. Besides this, other disadvantages with white-rot fungi are their huge growth period, the dependency on growth nutrients, and the long hydraulic retention time that limits the complete decolorization process [23]. Table 1 shows the fungal species

Table 1 Fungal species and the corresponding synthetic dyes degraded or removed

S. No	Name of the Fungal species	Synthetic dye removed	References
1	*Aspergillus flavus, A.wentii*	Dyes Acid Blue and Yellow MGR	Agnes et al. [1]
2	*Alternaria solani*	Malachite Green	Ali et al. [3]
3	*Pencillium chrysogenum, Aspergillus niger and Cladosporium sp.*	Azo Dye Red 3BN	Kumar Praveen and Bhat [25]
4	*Peyronellaea prosopidis*	Scarlet RR Dye	Bankole et al. [7]
5	*Sachharomyces cerevicea*	Remazol Blue, Remazol black B, Remazol Red RB	Saratale et al. [40]
6	*Trichosporon beigii*	Navy Blue HER	Saratale et al. [40]
7	*Candida oleophila, C. tropicalis, C. Zeylanoides, Galactomyces geotrichum*	Azo Dye	Saratale et al. [40]
8	*Phanerochaete chryosporium, Neurospora crassa*	Phenols	Saratale et al. [40]
9	*Phanerochaete chrysosporium*	Dioxins and Polychlorinated Biphenyls and other chloro organics	Reddy [37] Eaton [13]

and the corresponding synthetic dyes degraded or removed.

5   Role of Fungi in the Removal of Synthetic Dyes from Textile Industry Effluents

As the usage of synthetic dyes is enhancing day by day, there is an increased alarming effect on environment because the discharge of these textile wastes may cause substantial ecological damage. Use of microbes as biotic factors in decolorization of dyes is an eco-friendly and low cost alternative to chemical methods. The most of the textile mill effluents contained azo dye as their major constituent. In solid state fermentation, the fungal cultures are inoculated to produce laccase and the obtained enzyme was used in decolorization of synthetic dyes in both solid medium and in liquid broth medium. The result shows the effective decolorization of synthetic dyes by the fungal enzyme laccase. *Pleurotus ostreatus*, a common fungus, produces laccase enzyme which works effectively in decolorization of blue HFRL dye [26].

Application of microbes for the destruction of xenobiotics shows a green solution to the problem of industrial effluent pollution. Microorganisms have been blessed by nature, and hence have the ability of degrading any type of pollutants. Various fungi have the capacity to dissociate recalcitrant organic compounds into simple substances, and there by achieve complete mineralization. Degradation of triphenyl-methane dye and malachite green was observed with the use of two fungi, *Aspergillus flavus* and *Alternaria solani* and both were able to decolorize malachite green of various concentrations within 6 days [3].

Textile effluent and aqueous solution of dyes, Acid blue and yellow MGR were effectively decolourized by the fungi *Aspergillus flavus* and *Aspergillus wentii* when grown on Yeast-Glucose Peptone (YGP) and Potato Dextrose medium (PD broth) at room temperature with incubation of 24 h in rotary shaker at 170 rpm. The percentage of decolorization was calculated as 98.47% in effluent, by immobilized strain of *A. flavus* around 94.41% for Acid blue dye, and 95.88% by immobilized strain of *A. wentii* for Yellow MGR dye. In PD medium, the free cells of *A. wemntii* had a capacity to decolorize effluent (88.58%), with Yellow MGR dye (90.48%) and with Acid blue (93.05%) [1].

The fungal species *Penicillium chrysogenum, Aspergillus niger* and *Cladosporium* sp. found to effectively decolorize Azo Dye Red 3 BN in potato dextrose (PD) medium having 0.01% of Red 3 BN. Ideal conditions for *P. chrysogenum* to decolorize the dye were found to be 1% yeast extract, 1% maltose, 27° C, pH 8, and 2% inoculums and for *Cladosporium* sp., 1% peptone, 1% maltose, 37° C, pH 6, and 10% inoculums. Optimal conditions for *Aspergillus niger* were found to be 1% yeast extract, 1% maltose, pH 8, 27 °C, and 10% inoculum. The maximum decolorization recorded by *Cladosporium* sp. was 98.18%, by *P. chrysogenum* under optimum conditions was 99.56%, and that by *A. niger* was 98.64%. The above observations opened scope for using these fungi in the treatment of textile effluent for the degradation of Azo Dye Red 3 BN [25].

The biosorption potential of selected biomass of *Cunninghamella elegans* has shown a very effective degradation of textile effluents with that of the three fungal members *Acremonium strictum, Penicillium sp* and *Acremonium sp.* in dye biosorption [33]. *C. elegans* was the most efficient (degradation of dye up to 97% within 30 min) for all simulated dye baths. *A. strictum* was less active and able to produce a quick decolorization of only three of the simulated dye baths (up to 90% within 30 min). The biomasses of *C. elegans* and *A. strictum* were analyzed for the treatment of nine dye baths. *A. strictum* was found to be effective at neutral or acidic pH and *C. elegans* showed its effectiveness towards exhausted dye baths categorized by various classes of dyes (disperse, acid, vat, reactive) that show changes in ionic strength and pH. In an industrial setting, the effect of pH in the process of biosorption was analyzed and that was shown a realistic analysis on the validity of the lab results. *C. elegans* showed effective function from pH 3 to pH 11 and hence can be considered as an excellent and versatile biosorbent material.

A recalcitrant compound like Direct Red dye is one of the important variety of synthetic dye used in textile industries. *Aspergillus* sp are used for degradation of this dye containing effluents. Under static and shaking conditions, batch experiments were conducted for the degradation of Direct Red dye using *Aspergillus flavus* and *Aspergillus niger* [18]. At static and shaking conditions, with a batch of 50 mg/L, 97% degradation was achieved with *A.niger* in 48 h but with that of *A. flavus,* percentage of degradation was observed to be 78%. The decolorization followed Freundlich model during the adsorption isotherm studies, for the organisms with a regression coefficient of 0.833, and this brings out the capability of *Aspergillus sp.* to decolourize reactive dyes and the potential of these fungi for the effective degradation of textile waste liquid [38].

For the degradation of the dye, Acid orange 10, the fungal strain, *Phanerochaete chrysosporium* MTC 787 was used [34]. The starting pH and oxidation–reduction potential (ORP) for effective degradation was observed at 4.5 and at −250 mV, respectively, and the extent of dye degradation enhanced with the increase of dye concentration to 100 g L − 1 and then reduced with more dye concentrations. With enhancing the initial biomass concentration, the rate of dye decolorization and enzyme utilization also enhanced linearly. To describe the rate of substrate inhibition and enzyme utilization as a function of the starting substrate and flow rate, a kinetic model was adopted. From the experimental results, the kinetic constants were determined and they were—a starting biomass concentration should be of 3.2 × 10^5 cell/mL to obtain high rates of degradation percentage and to avoid inhibition of substrate [8].

Water pollution through industrial wastes, mainly in the form of effluent, is one of the biggest problems. These effluents have high concentrations of chemical oxygen demand (COD), phenol and its derivatives, metals, organic compounds, inorganic nutrients, polypeptides, cyanides, chlorinated lignin, and dyes. Bioremediation of toxic textile effluents by microbes provides an effective procedure to substitute the conventional removal processes. Fungal strains have a large capability of treating wastes discharged from various textiles. The enzymes produced by fungi made them effective degraders, and by biosorption they can remove toxic metals rendering the effluents more environment friendly [32].

6 Conclusion

Decolourisation of dye is a complex process that might be done in lab scale only. The use of microorganisms has many advantages as running costs are very much low and the end products are completely mineralized with no toxicity. Hence, there is a need to search for microbes that can effectively convert the textile effluent into useful liquid waste. Microbes are gaining much attention in the synthetic dye degradation owing to their capacity to decolourize the dyes with much efficacy and at low cost. Compared to other microbes, fungi are the best choice because of their ability to produce greater biomass which facilitates increased adsorption of dye molecules besides their low cost. Application of genetic engineering tools to isolate potential enzyme candidates from novel fungal strains could increase the specific activity of the enzymes towards varied dye molecules.

7 Future Directions

From the above studies we can clearly observe the importance of fungi in the bio-degradation of dangerous dyes and effluents among microbial world. But with studies on the growth in a normal medium, fungi showed less efficient mycoremediation with regards to decolourizing dye molecules, and this was noticed with the usage of same potential fungal strains. In dye degradation studies, a reduction in fungal growth with the usage of few specific dye compounds clearly showed the inability of the used fungal strains, and this could be due to the toxicity of the dye that inhibits mycelial growth, and as well the production of enzymes. Usage of molecular tools and techniques to recognize specific genes that code for potential enzymes from potential fungal strains could offer the best solution to treat textile dye effluents.

References

1. Agnes MD, Rajeshwari S, Venckatesh R (2012) Decolorization efficiencies of dyes and effluent by free and immobilized fungal isolates. Int J Environ Sci Res 1(4):109–113
2. Akan JC, Abdulrahman FI, Ayodele JT (2009) Impact of tannery and textile effluent on the chemical characteristics of Challawa River, Kano State, Nigeria. Aust J Basic Appl Sci 3(3):1933–1947
3. Ali H, Ahmad W, Taqweemul H (2009) Decolorization and degradation of malachite green by *Aspergillus flavus* and *Alternaria solani*. Afr J Biotechnol 8(8):1574–1576

4. Asgher M, Shah SAH, Ali M, Legge RL (2006) Decolorization of some reactive textile dyes by white rot fungi isolated in Pakistan. World J Microbiol Biotech 22(1):89–93
5. Asgher M, Yasmeen Q, Iqbal HMN (2013) Enhanced decolorization of solar brilliant red 80 textile dye by an indigenous white rot fungus *Schizophyllum commune* IBL-06. Saudi J Biol Sci 20:347e52
6. Banat IM, Nigam P, Singh D (1996) Microbial decolorization of textile-dye-containing effluents: a review. Biores Technol 58:217–227
7. Bankole PO, Obidi OF Chandanshive V, Govindwar SP, Adedotun AA (2018) Biodegradation and detoxification of Scarlet RR dye by a newly isolated filamentous fungus, *Peyronellaea prosopidis*. Sustain Environ Res 28(5):214–222
8. Borchert M, Libra JA (2001) Decolorization of reactive dyes by the white rot fungus *TrametesVersicolor* in sequencing batch reactors. Biotechnol Bioeng 75:313–321
9. Chander M, Arora DS (2007) Evaluation of some white-rot fungi for their potential to decolourise industrial dyes. Dyes Pigments 72:192e8
10. Chung KT, Stevens SE, Cerniglia CE (1992) The reduction of azo dyes by the intestinal microflora. Crit Rev Microbiol 18(3):175–190
11. Delclos KB, Tarpley WG, Miller EC (1984) 4-Aminoazobenzene and N, N-dimethyl-4-aminoazobenzene as equipotent hepatic carcinogens in male C57BL/6XC3H/He F1 mice and characterization of N-(deoxyguanosine-8-yl)-4-aminoazobenzene as the major persistent hepatic DNA-bound dye in these mice. Cancer Res 44(6):2540–2550
12. Dos Santos AZ, Neto JMC, Tavares CRG, Da Costa SMG (2004) Screening of filamentous fungi for the decolorization of a commercial reactive dye. J Basic Microb 44:288e95
13. Eaton DC (1985) Mineralization of polychlorinated biphenyls by *Phanerochaete chrysosporium*: a Ligninolytic Fungus. Enzyme Microb Technol 7:194–196
14. Eichlerov_a I, Homolka L, Nerud F (2006) Synthetic dye decolorization capacity of white rot fungus *Dichomitus squalens*. Bioresour Tech 97:2153e9
15. Elbanna K, Hassan G, Khider M, Mandour R (2010) Safe biodegradation of textile azo dyes by newly isolated lactic acid bacteria and detection of plasmids associated with degradation. J Bioremed Biodegrad 1:1000112
16. Fu Y, Viraraghavan T (2002) Dye biosorption sites in *Aspergillus niger*. Bioresour Tech 82:139e45
17. Fu YZ, Viraraghavan T (2000) Removal of a dye from an aqueous solution by the fungus *Aspergillus niger*. Water Qual Res J Can 35:95e111
18. Ganappriya M, Logambal K, Ravikuma R (2012) Investigation of Direct Red Dye using *Aspergillus niger* and *Aspergillus flavus* under static and shacking conditions with modeling. Int J Sci Environ Tech 1(3):144–215
19. Iqbal HMN, Asgher M (2013) Characterization and decolorization applicability of xerogel matrix immobilized manganese peroxidase produced from *Trametes versicolor* IBL-04. Protein Pept Lett 20:591e600
20. Jafaria N, Soudib MR, Kasra-Kermanshahi R (2014) Biodegradation perspectives of azo dyes by yeasts. Microbiology 484–497
21. Kabbout R, Taha S (2014) Biodecolorization of textile dye effluent by biosorption on fungal biomass materials. Phys Proc 55:437e44
22. Kaushik P, Malik A (2009) Fungal dye decolourization: recent advances and future potential. Environ Int 35:127e41
23. Khan R, Bhawana P, Fulekar MH (2013) Microbial decolorization and degradation of synthetic dyes: a review. Rev Environ Sci BioTech 12:75–97
24. Khandare RV, Kabra AN, Kadam AA, Govindwar SP (2013) Treatment of dye containing wastewaters by a developed lab scale phytoreactor and enhancement of its efficacy by bacterial augmentation. Int Biodeter Biodegr 78:89e97
25. Kumar Praveen GN, Bhat SK (2012) Fungal degradation of azo dye-red 3BN and optimization of physico-chemical parameters. ISCA J Biol Sci 1(2):17–24
26. Malini Devi V, Inbathamiz L, Mekalai Ponnu T, Premalatha S, Divya M (2012) Dye decolorization using fungal laccase. Bull Env Pharmacol Life Sci 1(3):67–71

27. McMullan G, Meehan C, Conneely A, Kirby N, Robinson T, Nigam P, Banat IM, Marchantand R, Smyth WF (2001) Microbial decolourisation of degradation of textile dyes. Appl Micro Biotech 56:81–87
28. Mohana S, Shrivastava S, Divehi J, Medawar D (2008) Bioresour Tech 99:562–569
29. O'Neil C, Hawkes FR, Hawkes DL (1999) Colour in textile effluents-sources, measurement, discharge consents and simulation: a review. J Chem Technol Biotech 74(11):1009–1018
30. Oves M, Khan MS, Zaidi (2013) Biosorption of heavy metals by Bacillus thuringiensis strain OSM29 originating from industrial effluent contaminated north Indian soil. Saudi J Biol Sci 20:121e9
31. Pandey A, Singh P, Iyengar L (2007) Bacterial decolorization and degradation of azo dyes. Int Biodeter Biodegr 59:73e84
32. Ponraj M, Gokila K, Zambare V (2011) Bacterial decolorization of textile dye-orange 3R'. Int J Adv Biotech Res 2(1):168–177
33. Prigione V, Grosso I, Tigini V, Anastasi A, Varese GC (2012) Fungal waste-biomasses as potential low-cost biosorbents for decolorization of textile wastewaters. Water 4:770–780
34. Radha K, Balu V (2014) Kinetic study on decolorization of the dye acid orange the fungus *phanerochate chrysosporium*. J Modern Appl Sci 3(7):38–47
35. Rajendran P, Gunasekaran P (2006) Potential process implicated in bioremediation of textile effluents. MJP Publishers, pp 138–159
36. Rane NR, Chandanshive VV, Khandare RV, Gholave AR, Yadav SR, Govindwar SP (2014) Green remediation of textile dyes containing wastewater by *Ipomoea hederifolia* L. RSC Adv 4:36623e32
37. Reddy CA (1995) The potential for white rot fungi in the treatment of pollutants. Curr Opin Biotechnol 6:320–328
38. Robinson T, McMullan G, Marchant R (2001) Remediation of dyes in textile effluent: a critical review on current treatment technologies with a proposed alternative. Bioresour Tech 77(3):274–255
39. Sabrien AO (2016) Decolorization of different textile dyes by isolated *Aspergillus niger*. J Environ Sci Technol 9:149e56
40. Saratale RG, Saratale GD, Chang JS, Govindwar SP (2011) Bacterial decolorization and degradation of azo dyes: A review. J Taiwan Inst Chem Eng 42(1):138–157
41. Shertate RS, Thora PR (2013) Bio decolorizarion and degradation of textile diazo dye reactive blue 171 by *Marinobactor* Sp. N.B-6 -A bioremedial aspect. Int J Pharm Biol Sci 3(1):330–442
42. Singh L, Singh VP (2015) Textile dyes degradation: a microbial approach for biodegradation of pollutants. In: Microbial degradation of synthetic dyes in waste waters, environmental science and engineering. Springer International Publishing Switzerland
43. Singh L, Singh VP (2017) Decolourization of azo (acid red) and anthraquinonic (basic blue) dyes by the fungus *Aspergillus flavus*. Int J Biomed Engine Cli Sci 3(1):1–5
44. Society of Dyers and Colourists (1976) Colour index, 3rd edn. Society of Dyers and Colourists, Yorkshire, UK
45. Toh YC, Yen JJL, Obbard JP, Ting YP (2003) Decolourisation of azo dyes by white-rot fungi (WRF) isolated in Singapore. Enzyme Microb Tech 33:569e75
46. Tripathi ASK, Harsh NSK, Gupta N (2007) Fungal treatment of industrial effluents: a mini review. Life Sci J 4(2):78–81
47. Vijayaraghavan K, Yun Y (2014) A study on biosorption potential of *Aspergillus* sp. of tannery effluent. Adv Biosci Biotech 5(10):266–291
48. Weisburger JH (2002) Comments on the history and importance of aromatic and heterocyclic amines in public health. Mutat Res 506–507:9–20
49. Wong Y, Yu J (1999) Laccase-catalyzed decolorization of synthetic dyes. Water Res 33:3512e20
50. Wuhrmann K, Mechsner K, Kappeler T (1980) Investigations on rate determining factors in the microbial reduction of azo dyes. Eur J Appl Microbiol Biotechnol 9:325–338
51. Zollinger H (1987) Colour chemistry: synthesis. VCH Publishers, New York, Properties of Organic Dyes and Pigments, pp 92–100

The Applicability of the Microalgae-Based Systems in Textile Dye Industrial Wastewater

Rafaela Basso Sartori, Paola Lasta, Patrícia Arrojo da Silva, Álisson Santos de Oliveira, Leila Queiroz Zepka, and Eduardo Jacob-Lopes

Abstract The textile industries are responsible for one of the biggest environmental pollution problems in the world, as they release many undesirable compounds. Textile wastewater contains synthetic dyes mixed with various other contaminants that can lead to a serious threat to public health and the environment when released into bodies of water without specific treatment. During the past few decades, many treatment techniques have been implemented, but few have become viable due to their low efficiency and high cost. In this sense, wastewater treatment using microalgae-based technology can be a global solution for the recovery of wastewater resources and simultaneously supply accessible raw material for different types of bioproducts. This chapter initially presents the main compositions and characteristics of textile wastewater as well as the main methods of conventional treatment. Finally, it discusses the potential of microalgae in the bioremediation of effluents and the applicability of their biomass after wastewater treatment.

Keywords Microalgae · Effluent · Treatment · Bioremediation · Biomass · Environment

1 Introduction

The textile industry represents one of the oldest, most important, and high growth sectors in the world [54]. The production of textiles is usually carried out through the processes of desizing, bleaching, dyeing, printing, and finishing techniques [38, 63]. Considering the high consumption of water in these processes, the textile industry is considered to be a high polluter when compared to other industrial sectors, producing almost 20% of all industrial wastewater [14, 61]. Furthermore, a report issued by the United States Environmental Protection Agency (USEPA) suggests that approximately more than 70 kg of water is used to generate less than 0.5 kg of a textile product [54, 110].

R. B. Sartori · P. Lasta · P. A. da Silva · Á. S. de Oliveira · L. Q. Zepka · E. Jacob-Lopes (✉)
Bioprocess Intensification Group, Federal University of Santa Maria (UFSM), Santa Maria, RS 97105-900, Brazil

Textile effluents contain a total of more than 70 toxic substances, various chemicals, and high concentrations of dyes. The use of synthetic dyes has increased dramatically in the textile industries due to their cost-effective relationship, better stability to light, temperature, microbial attack, and a greater variety of colors over natural dyes [33]. Most of these components have low biodegradability and are more recalcitrant. Therefore, when released into bodies of water without treatment, they pose a serious threat to health and the environment [11].

Considering that traditional municipal sanitary treatment processes are not able to degrade these compounds, industrial wastewater, therefore, needs more specific treatments before being discharged into water bodies. Consequently, stricter environmental regulation and standards bodies have also been promoted in recent decades to oversee the discharge of these effluents. Despite the fact that each region has specific legislation, some control parameters (such as chemical oxygen demand, suspended solids, salinity, color, and detergents or oils) can be observed in general [46, 82, 93].

The effluent treatment of the textile industries mainly involves physical and chemical methods, in which they can present several disadvantages, such as expensive, less efficient, limited application, and can often give rise to secondary pollution. On the other hand, biological methods are economical, safe, and environmentally friendly [35, 66]. Bioremediation is one of the most current approaches and was introduced to describe the process that uses biological agents to remove environmental pollutants. The bioremediation process uses several biological agents, such as bacteria, fungi, algae, and cyanobacteria as the main tools in the treatment of effluents [38].

Recently, microalgae have received considerable attention in the bioremediation of wastewater by means of its carbon dioxide fixation potential through photosynthesis [35]. Microalgae are capable of removing a wide range of toxic substances, such as nitrogen, phosphorus, and heavy metals from different types of wastewater. In textile effluents, many studies have revealed that microalgae were able to biodegrade and discolor more than 30 synthetic dye compounds into simpler compounds [14, 112]. Furthermore, the cultivation of microalgae in wastewater has gained visibility due to its efficiency in the production of high-value biomass with reduced costs and less environmental impacts, since nutrients and freshwater are not requirements [39].

In this sense, the applicability of the microalgae-based systems in the bioremediation of textile dye industrial wastewater has been privileged and proposed as an alternative approach without any negative or adverse effects on the environment, with the capacity to produce in parallel, a multitude of bioproducts.

2   Textiles Wastewater: Origin and Composition

The composition of textile wastewater varies according to the nature of the product manufactured by these industries [54]. They carry a large number of pollutants, usually detergents, surfactants, heavy metals, inorganic salts, solvents, oils, but mainly by dyes [6]. The dyes are used to give color to the fabric due to the presence of chromophoric groups in its structure and can be divided into natural or synthetic paints. synthetic dyes are produced more easily, which makes them more widely used in textile industries [11].

Synthetic dyes are described into different groups according to their chemical structure (for example, azo (representing almost 60% of all dyes), anthraquinone, sulfur, phthalocyanine, and triarylmethane) and according to their mode of application (for example, reactive, direct, dispersed, basic, and tank dyeing) [86, 111]. Globally, more than 10,000 different dyes are used in the textile industry and more than 7×10^5 tonnes of synthetic dyes are produced annually [11, 101]. These inks are discharged in wastewater to a varying extent (10–60%) and may cause waste almost 300,000 tonnes of dyes per year [35, 67, 89].

Textile wastewater is rich in strong color, pH, salinity, temperature, high concentration of suspended solids (TSS), total dissolved solids (TDS), nitrogen (N), phosphorus (P), sulfates (S), biological oxygen demand (BOD), chemical oxygen demand (COD), and non-biodegradable organic compounds. Also contain traces of heavy metals like potassium (K), sodium (Na), lead (Pb), chromium (Cr), iron (Fe), copper (Cu), and zinc (Zn) [80]. The release of these nutrients not only leads to aesthetic pollution, but can also lead to eutrophication in water bodies and, consequently, several risks to the ecosystem. [96]. In addition, these dyes are extremely dangerous to human health, as their degradation products were considered carcinogenic and mutagenic [62, 105]. Table 1 presents the main characteristics found in textile industrial wastewater.

3 Conventional Treatment Methodologies and Their Limitations

It's known, that wastewater contains synthetic dyes that can harm the environment. Notably, due to environmental and health concerns associated with wastewater effluents, different separation techniques have been utilized to remove dyestuff from watery solutions. These techniques are divided into three categories: physical, physical–chemical, and biological methods [31, 47]. The treatment process often involves combinations of these methods [54]. The different conventional methods used are discussed below:

3.1 Physical Methods

Physical treatment refers to the elimination of contaminants without changing their biochemical characteristics. The physical method includes processes such as filtration, screening, aeration, sedimentation, ion exchange, irradiation, etc. These methods are reasoned on the mass transfer strategy. Its main advantage is to have

Table 1 General
characteristics of textile
wastewater

Characteristics	Value	References
pH	6–10	[80]
Temperature (°C)	35–45	[37]
Color (Pt–Co)	50–2,500	[37]
Oil and grease (mg/L)	10–30	[37]
Dye (mg/L)	70–700	[80]
TDS (mg/L)	2,900–3,100	[57]
TSS (mg/L)	15–8,000	[57]
BOD (mg/L)	80–6,000	[111]
COD (mg/L)	150–10,000	[111]
NH_4^{+1} (mg/L)	50	[35]
NO_2^{-1} (mg/L)	350	[35]
PO_4^{-1} (mg/L)	4–17	[35]
SO_4^{-2} (mg/L)	50–900	[35]
K (mg/L)	30–50	[50]
Na (mg/L)	400–2,000	[50]
Cr (mg/L)	<10	[38]
Pb (mg/L)	<10	[38]
Fe (mg/L)	<10	[38]
Ar (mg/L)	<10	[38]
Zn (mg/L)	<10	[38]

simple and flexible processes, requiring simple equipment and being able to be adapted to different treatment formats [4, 41].

The filtration method is a quick and effective process for removing all types of dyes. Filtration technologies to treat textile effluents include microfiltration, ultra-filtration, nanofiltration, and reverse osmosis. All of these technologies are effective in removing color, organic salts, and suspended impurities from wastewater in the textile industry [47, 59]. Microfiltration is the most competitive method owing to the larger pore size for its building modules and can act as a pre-treatment option before additional treatment using reverse osmosis or nanofiltration to satisfy the reuse criteria for the textile industry. However, the filtration method has some disadvantages, such as a practical limitation to the lower flow rate of wastewater, high cost of the membrane system, high working pressure to push the flow of wastewater through the membrane, and inability to reduce the content of solids dissolved [81].

With regard to screening, it is one of the oldest physical methods, it has the ability to remove large floating solid materials that can be a barrier to the next treatment process. Aeration, in turn, is a treatment based on a gaseous substance absorbed in many of the hybrid treatment processes. In this treatment, the air is inserted into the liquid medium to capture dissolved contaminants, gaseous contaminants, and volatile organic contaminants. However, this is the first process adopted at much large-scale wastewater treatment [4].

Sedimentation is a treatment in which the materials containing dye are separated from the mixture separate suspended solids, and coagulated mass from the remaining effluent [54]. In the water treatment process, due to gravity, uniform segregation of these suspended residues occurs. Particularly, these dye-related particles are entertained by strong solvent turbulence and then left to stand for a period until the entire sediment settles against a barrier. It is worth adding that this is a totally physical process where the loss of dye or residual water can be avoided. However, this process is quite tedious and seems impossible when the scale of these residues increases [8].

Already, ion exchange has great limitations for the elimination of dyes in textile effluents, being of a very specific nature for dyes and impurities present in effluents [41]. Ion exchange can only be worn to remove unwanted anions and cations from wastewater. The main disadvantage of this method in the treatment is its inefficiency in the removal of several types of dyes. Already, the advantages include the recovery of the adsorbent after successful regeneration, the efficient removal of soluble dyes, and the recovery of the solvent after use [81].

The irradiation process is more suitable for low volume discoloration, sufficient amounts of dissolved oxygen are necessary for organic substances to be degraded of dyes [41]. The irradiation technologies offer several advantages in terms of energy savings, ease of use, low environmental impact, high treatment speed, and savings. The modification of fiber surfaces using irradiation methods is gaining interest for use in coloring and finishing textiles as new treatments without affecting other volume properties [74].

3.2 Physico-Chemical Methods

The main methods of physical–chemical treatment used are coagulation, flocculation, adsorption, advanced oxidative processes, and electrochemical techniques. These treatments are worn to remove color and reduce BOD/COD, in the primary and secondary treatment stages, respectively. Generally, chemical methods result in the degradation of dyes, dissolved or colloidal organic contaminants by oxidation [54].

Coagulation/flocculation is utilized to remove organic materials by removing BOD, COD, total dissolved solids (TDS), and color from the effluent. Basically, this method depends on the law of adding coagulants that are related to pollutants forming clots/flakes and, subsequently, the precipitates that are removed by flotation, sedimentation, filtration to form a sludge that is treated to reduce its toxicity. The high cost of sludge treatment and the restrictions on disposal in the environment is the main disadvantage of this process [65].

Adsorption is one effective treatment method for removing organic and inorganic contaminants; is simple and suitable for removing several pollutants, is non-destructive and pollutants are transferred from one phase to another. Additionally, the regeneration and elimination of adsorbents increase operating costs. The performance of the process depends on the type of the adsorbent material, often being quite expensive. Therefore, increased the interest in the generation and application of low-cost adsorbents [16]. Activated carbon, kaolin, and silicon polymers are used as adsorbents with selective adsorption ability for dyes, but activated carbon is the most used in relation to other popular adsorbents [104].

The advanced oxidation process includes many different techniques that have a common aspect, can lead to the generation of hydroxyl radicals. In general, the hydroxyl radical is a powerful, non-selective oxidant, capable of reacting through three mechanisms: hydrogen abstraction, electron transfer, or radical addition. These mechanisms are able to degrade complex chemical structures, such as biorefractory compounds, mainly anthropogenic pollutants that are difficult to be attacked by microorganisms. It is worth adding that advanced oxidation processes are considered expensive, requiring amounts of electrical energy or chemical reagents [83].

Electrochemical processes, such as electrocoagulation, anodic oxidation, indirect electrochemical oxidation, sono-electro-Fenton, electro-Fenton, photo-electro-Fenton, bio electro-Fenton, peroxicoagulation, and electroperoxone processes have received much attention own to their efficiency in the treatment of high-resistance waste [77]. Notably, there is a greater interest in electrocoagulation and electrochemical oxidation techniques. Electrochemical oxidation favors the degradation of pollutants, such as metals and dyes of an electrochemical particle. Already, electrocoagulation involves the self-generation of coagulants during the electrochemical process, so that complementary chemicals do not need to be added to the treatment [102].

3.3 Biological Methods

Biological approaches are treatment methods green, inexpensive, and ecological, which can be used effectively in the treatment of industrial wastewater. Microorganisms such as microalgae, fungi, bacteria, and yeasts can detoxify, discolor, mineralize, and degrade a number of wastewater pollutants, using their different metabolic pathways and biosorption processes [59].

Biological treatment is considered the most effective way of removing wastewater rich in organic components. Microorganisms play a crucial paper in the mineralization of xenobiotic compounds and complex organic molecules. An advantage of biological treatment over physical–chemical methods is that more than 70% of organic materials, determined by COD analysis, can be converted into biosolids. However, the biological removal of dyes from industrial wastewater in the manufacture of textiles and dyes can be classified into three categories: treatment aerobic, anaerobic, and anaerobic–aerobic combination. Aerobic treatment involves the use of free oxygen dissolved in wastewater to degrade organic components in the presence of microorganisms. Already, anaerobic treatment occurs in the absence of free oxygen and converts organic compounds into methane and carbon dioxide [102].

As for microorganisms, fungus culture has the capacity to acclimatize its metabolism to changes in environmental conditions, being vital for its existence [47]. Despite different strains of fungi being successfully used in the decolorization and degradation of azo dyes, in parallel, there are several problems for the removal of dyes from textile wastewater, such as large production volume, biomass control, and the nature of dyes synthetics. Although those problems, fungal-mediated decolorization is a possibility to replace present treatment processes [103]. The main microorganisms responsible for the biodegradation of organic compounds by fungi white-rot are *Hirschioporus larincinus, Inonotus hispidus, Phanerochaete chrysosporium, Phlebia tremellosa, Coriolus versicolor,* etc. [96].

Bacteria are suited to fine for mineralization and decolorization complete of azo dyes as they are easy to cultivate and grow rapidly. The process of decolorization by a bacterial system may be anaerobic or aerobic or involve a combination of both [103]. Bacteria in anaerobic conditions are able to discolor azo dyes, i.e., azo bonds are broken by the enzyme azoredutase in the presence of redox mediators. However, there are also some species of bacteria capable of discoloring wastewater under aerobic conditions [83]. Some examples of the main bacteria used in the biological treatment are *Aeromonas Hidrofilia, Acetobacter liquefaciens, Bacillus cetreus, Bacillus subtilis, Klebsiella pneunomoniae*, species of *Pseudomonas, Sphingomonas,* etc. [96].

Typically, there are some reports of biodegradation by yeast strains, such as *Candida zeylanoides, Candida zeylanoides,* and *Issatchenkia occidentalis.* Yeasts have advantages in comparison to bacteria and fungi, they grow quickly as bacteria and have the capacity to resist unfavorable environments, such as fungi. Furthermore, in yeast, the ferric reductase system participates in the extracellular reduction of dyes. In general, yeasts show better decolorization and biodegradation activities at acidic or neutral pH [84]. It is worth adding that yeast-mediated biodecoloration occurs by two processes: biosorption or biodegradation. Biosorption involves the union of solute with cell biomass in a process that involves energy consumption. Already, in the process of biodegradation, several changes occur in the molecular composition of the dye until the total mineralization of the dye molecule in simpler substances such as CO_2, H_2O, or CH_4 [68].

Finally, the microalgae can be found in either fresh or saltwater, and are currently being studied as great bioabsorbents. These microorganisms have the greatest biosorption potential and electrostatic attraction force for contaminants in the growth of wastewater, through bioremediation [11]. It is worth mentioning that microalgae do not need preservation, as their growth depends heavily on sunlight and carbon dioxide, which do not generate secondary waste [97].

4   Bioremediation of Textile Wastewater by Microalgae

The vast majority of textile-producing industries have effluent treatment stations (ETPs). However, often, despite being treated, the wastewater released generally does not meet the criteria established by environmental protection authorities in relation to pollution [88]. The process itself needs optimization seeking greater quality, process applicability, and cost reduction [48, 88].

An alternative treatment option that can make bioremediation an efficient process, with environmental awareness and cost reduction, is the application of microalgae and other microorganisms. These cultures can act photoautotrophically, heterotrophically, and or mixotrophically, and have aroused interest in their exploration, due to their easy and quick capacity for adaptation and assimilation of nutrients, such as Co_2 [10]. According to Aragaw and Asmare [5], the action of photoautotrophic microalgae is more efficient, non-pathogenic, economically viable, and sustainable.

Microalgae and cyanobacteria, through the action of inducible enzymes, are able to convert azo dyes (NN double bond) to simpler compounds, reducing this double bond and converting to aromatic amines (NH_2 single bond) [9]. The composition of the microalgae cell wall is composed of different functional groups, which end up acting as binding sites and thus helps in the ease of contaminant biosorption [105].

In addition, besides acting in the treatment of effluents, it produces biomass rich in fatty acids (mainly C16, C18), which can later be used in the production of biofuels such as bioethanol, biodiesel, and biohydrogen [9]. It can also be applied in the manufacture of bioelectrodes, acting as a conductive material for microbial fuel cells (MFC), this type of application makes them capable of converting solar energy into bioelectricity through the application of biomass [36]. The process of inserting microalgae in the treatment of this waste is called phycoremediation.

4.1   Phycoremediation of Textile Wastewater Using Microalgae

Phycoremediation is a biological method using algae, responsible for removing and degrading pollutants present in wastewater (nutrients, heavy metals, color, and xenobiotics) [78]. The term was recently coined [53], but the investigation above the use of algae in the remediation of wastewater started decades ago [79].

The biological methodology applying microorganisms with the power to absorb pollutants in the effluent or reduce them to smaller molecules easily managed is an economically viable and sustainable method. Among the processes are enzymatic degradation of dyes, removal of color and many heavy metals by the biosorption process in the primary treatment stage of textile wastewater, degradation of organic and inorganic chemicals, and a wide spectrum of pollutants [54].

During bioremediation, simultaneously bioconversion and biosorption occur, bioconversion is a process in which microalgae use dyes as a carbon source and convert them into metabolites. After this process, the microalgae act as biosorbents, where the dyes are adsorbed on their surface [28]. An advantage is that both live and dead algae can participate in the processes, however dead algae only act on the adsorption of dyes [35].

According to Cheriaa et al. [27], the application of *Chlorella* in the bioremediation of different textile dyes was responsible for discoloring different dyes, ranging from 72 to 89.3% of effectiveness. El-Kassas and Mohammad (2014), using the same family (*Chlorella*) in textile wastewater, obtained a 70% reduction in COD. Acuner and Dilek [3], concluded that the microalgae *Chlorella Vulgaris* is capable of degrading 63–69% of the mono-azo dyes into converting them into simple aromatic substances.

When applied as textile wastewater binders, *Spirulina platensis* was able to remove 97% of the reactive red 120 (RR-120) reaching maximum biosorption capacity [18]. *Scenedesmus quadricauda* was applied to remove bright blue from remazol R (RBBR) [34]. *Cosmarium sp.* applied to the removal of malachite green (MG) demonstrated the efficiency of discoloration, reuse, and stability [29]. Some steps must be followed for the application of these microorganisms in the treatment of textile wastewater, from the correct choice of the strain, the type of bioreactor, and the secondary metabolites of interest.

4.1.1   Steps for the Phycoremediation of Textile Effluents Using Microalgae

Textile wastewater benefits the growth of microalgae, as it contains dyes that serve as a carbon source, nitrates that are a source of nitrogen, phosphates as a source of phosphorus, and various metals that serve like micronutrients, thus serving as a complete culture medium, promising and for low cost for the application. As they are photosynthetic microorganisms, they act by sequestering CO_2 from the medium and removing other nutrients for their growth, removing approximately 30 to 91% of these. Different phases of this process can be seen in Fig. 1 [17, 28, 35].

The first thing to be done is to analyze and select which is the best and most robust microalgae culture to perform this phycoremediation. Bearing in mind that it must have the potential to pre-treat textile wastewater, as well as being the best candidate for the accumulation of secondary metabolites for the use of biomass. They must be resistant to different environmental conditions; be resistant to shear force; high cell productivity in the presence of different micro and macronutrients and subsequent significant accumulation of metabolites [35].

Once the culture was chosen, the cultivation conditions (photoautotrophic, heterotrophic, and mixotrophic), related to species, have advantages and disadvantages. The determining factor will be to offer the necessary nutrients to ensure cell growth, which be of low operational and maintenance cost, maintaining high cell productivity, controlled parameters, and high reliability [10].

Closed photobioreactors have greater profitability in growing the crop, as it guarantees a more controlled environment than those offered by open lagoons and also a

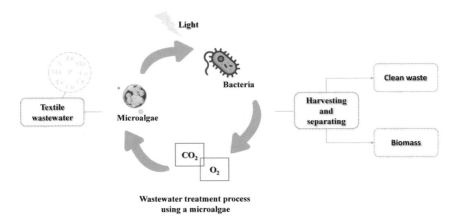

Fig. 1 Simplified schematic representation of the main stages during the treatment of textile wastewater. Nagarajan et al. Adapted from [76]

lower risk of contamination by other strains of microalgae or bacteria. This ensures high cell productivity, but at a higher cost, because this process requires CO_2 input and homogenization of nutrients. The hybrid system, on the other hand, proposes better control with less energy consumption, reducing costs. However, it has a high initial and maintenance investment, due to the fact that it is a combination of an open lagoon and a closed system. This should be evaluated on a case-by-case basis for the choice of the bioreactor and shows us how is important and various the choice of the cultivation system [32, 40, 69, 75].

To apply algae in a growth system containing wastewater, it is essential to contain the elements that are essential for the growth of the crop (carbon, nitrogen, and phosphorus), CO_2, which can be obtained from the atmosphere and from external industrial sources such as salts of soluble carbonates [21, 85]. Other media required are light and micronutrients such as Mn, Co, Cu, Fe, Mb, Zn, Cr, which textile wastewater presents in varying concentrations, facilitating cell productivity, maintenance, and the synthesis of different metabolites [10, 45, 69].

Nutrients and heavy metal ions are assimilated in the cells for their growth, consuming mainly inorganic CO_2 as a carbon source. This CO_2 is first absorbed by water forming carbonic acid, dissociating into bicarbonate and hydrogen ions, later dissociating in carbonate ions and hydrogen ions so microalgae take advantage of these ions for cell growth [95, 98].

Subsequently, through nitrification, the microalgae consume the oxygen released in their metabolic activity, reducing cultivation costs, while the phosphate macronutrients are integrated into the energy input by phosphorylation. NADP$^+$ and adenosine diphosphate (ADP) are incorporated into NADPH and adenosine triphosphate (ATP), respectively regulating the metabolic pathways of microalgae [1, 95].

The stage of harvesting and separating the biomass is very important within the process, there are several techniques, and none is considered 100% efficient and low cost. To be ideal, in addition to energy and chemical efficiency, it must have a reduced loss of secondary metabolites, but different factors influence the choice of the appropriate technique, such as species, cell concentration, and products of interest, efficiency, and cost. These techniques include isolated or combined processes and some examples are centrifugation, sedimentation, flotation, coagulation, and electrophoresis [7, 35, 64].

Among the techniques mentioned, the most suitable is flocculation, in addition to being a simpler technology and of relatively low cost, it has an efficiency of up to 99%, it will be dependent on the type of flocculant that can be divided into 3 different inorganic groups (Sulfate of aluminum, ferric sulfate), organics (cationic starch, chitosan), and bio-flocculants (fungi, enzymes). It is carried out by collecting colloidal suspension of microalgae from water treatment processes, flocculants are added to the algae suspension, where they end up settling on surfaces that are negatively charged with microalgae cells. The biomass of *Chlorella sp* and *Dunaliella tertiolecta* were recovered above 90% [7, 17, 20, 42, 99].

Sedimentation consists of a simple and economical process, where the biomass is harvested through the gravitational force of the culture medium. It is recommended when you have large, heavy cells. It can be used in combination with another technique, this being the first step [22]. Centrifugation is fast and reliable, with an efficiency of up to 90%, and the reduced operating time, however, causes excess in which it can cause damage such as cells. It is a mechanical process that requires a lot of energy, raising the cost [107].

Electrophoresis is a technology that uses the generation of ions in the algae suspension and then uses these ions to separate the cells. In electrophoresis, the separation is done through an electric field based on the zeta potential of these microalgae cells that go toward the anode for neutralization. The efficiency will depend on the ionic strength of the water, the efficiency of the methods can reach up to 95% of the microalgal cells. However, it requires a high cost of energy and electrodes [87].

The generation of the high volume of textile wastewater and the high toxicity present in it requires more advanced methods for the treatment since although there are several conventional methods available, they are not efficient in the complete removal of these pollutants. Therefore, microalgae are viable and have been shown to be an appropriate biological treatment due to the proof of effective reduction of BOD, COD, heavy metals, and complete removal of colors from the effluent, having a double role, where they can be applied simultaneously to remedy waste with high ecotoxicity and produce biomass capable of being exploited for the different secondary metabolites produced during its growth. This double role can be investigated in several dimensions, seeking to optimize the treatment of effluents and the production of fuels, for example. There are several paths to follow, which can be investigated, although there are limitations, the implementation of phycoremediation indicates that the method has great potential.

5   Biomass Production: The Main Applicability of the Biomass Microalgae After Wastewater Treatment

As previously mentioned, the research is focused on the coupling of treatment processes based on microalgae. Thus, the objective has been aimed at removing inorganic nutrients from various types of wastewater, where it is an economically viable method to save high amounts of nutrients and the water demand necessary for the cultivation of microalgae [55]. In this perspective, microalgae have been considered a potential sustainable source of biomass for countless industries in the future [56].

The biomass produced by microalgae in the treatment of effluents can be extracted and converted into a portfolio of bioproducts and biomaterials with high added value [19, 25]. Thus, microalgae grown in wastewater are a raw material rich in primary (carbohydrates, proteins, lipids) and secondary (pigments, antioxidants) metabolites. Undeniably, they have established themselves as an economically sustainable, ecological, and mainly inexhaustible source of lipids, carbohydrates, and proteins [60]. Thus, they can potentially be exploited in the bioenergy, food, and health sector [72].

Notoriously, the cultivation of wastewater is a method that in addition to being economical, has the advantage of increasing the production of biomass and modifying the lipid content and the composition of fatty acids (AF). Algae are known to accumulate approximately 30–80% of lipids, which normally consist of 90–95% of triacylglycerides [70]. It is worth mentioning that there are still substantial regulatory and technical challenges that become prohibitive for the use of microalgae biomass produced from wastewater. This is a serious bottleneck to be overcome for the expansion of microalgae-based processes because until there is a clear application for the biomass produced, the processes are unsustainable [2].

In recent years, microalgae have become a potential exploratory source of sustainable biomass raw material for the generation of biofuels [94]. In fact, wastewater demonstrates economic viability and environmental sustainability for the production of microalgae biofuels, as it is a low-cost medium [44, 91]. Consequently, microalgae biofuel has the potential to replace petrodiesel. However, the production of microalgae oil still needs to be improved [108].

From the point of view, in comparison with other oilseeds, the 3rd generation microalgae become a viable energy alternative, as it addresses all the main problems of 1st and 2nd generation biofuels [73]. Consequently, compared to maize that yields only about 560 L per ha annually, it is estimated that microalgae can produce approximately up to 94,000 L of biofuel per ha per year [109]. In addition, wastewater-based biofuel can reduce production costs by approximately 50%, making it comparable to petroleum diesel [52].

In studies carried out based on the life cycle assessment (LCA) of the cultivation of algae from wastewater for the production of biofuels, it was concluded that the ecological footprint is remarkably reduced [52]. In short, microalgae biofuel has the potential to significantly mitigate greenhouse gas emissions compared to fossil fuels [49].

Based on this understanding, microalgae biomass can produce a variety of biofuels, including liquid fuels (such as biodiesel, bioethanol) and gaseous fuels (such as biohydrogen, biomethane) [12, 71]. Biodiesel is the energy product of the biochemically converted microalgae most cited and researched by the scientific community. As such, it is considered a potential direct substitute or blend compound for petroleum diesel [30]. It is important to mention that the lipid content of microalgae grown in wastewater increased notably from 10% to 25–30%. Therefore, it can also be applied for the generation of biodiesel [23]. The important steps in the biochemical conversion of biomass into biodiesel from microalgae are composed of the cultivation of microorganisms, lipid extraction, and transesterification [24]. However, the production of biomass in biodiesel requires downstream processing steps that require high energy costs (for example, the harvest and drying steps), resulting in costly operating costs. Therefore, the pre-treatment steps must be improved, increasing productivity [100].

Microalgae can also be applied as a raw material for biogas. Thus, this process is considered simpler, as it does not require oil concentration and extraction steps, and can avoid high energy consumption [26]. At the same time, biogas produced by anaerobic digestion (AD) can be applied to generate heat or electricity in cogeneration engines, be injected into the natural gas network, or used as fuel for vehicles after undergoing a modernization process [92].

On the other hand, bioethanol is recently the most widely used alcoholic biofuel for the production of bioenergy and the main biofuel in the global market. In addition to being considered a clean fuel, it is expected to produce 160 billion liters. As a result, it experienced a rapid expansion in its global manufacturing [13]. Substantially, bioethanol can be produced from microalgae and consists of steps: the production and harvesting of biomass, the pre-treatment for the release of sugars in ethanol, the separation, and purification of ethanol, respectively [106]. Unfortunately, the production of microalgae bioethanol is reported as an expensive energy process, with relatively high costs. Thus, in today's economies, these limitations make large-scale production of this biofuel unfeasible [58].

Finally, biohydrogen is an essential fuel solution to the substantial challenges related to global warming and increasing greenhouse gas emissions, as it has high energy content and carbon-free combustion properties [76]. In fact, an encouraging approach to the generation of bioenergy arises from the biological production of hydrogen from organic substrates using phototrophic microorganisms in the treatment of wastewater [15, 43]. It is worth mentioning that algae biohydrogen is the ideal source of energy, free from pollution and global warming. Because energy consumption is interrupted with this technology and the environment is protected in an ecologically sustainable and correct way [90].

In view of these aspects, it is worth mentioning that, when combining wastewater treatment with the production of biofuels from microalgae biomass, it is an extremely promising alternative [51].

6 Conclusion and Future Perspectives

The textile industry is a sector of great importance, which contributes significantly to financial activity in several countries. One of the bottlenecks in this sector is the high production of waste. These effluents from textile industries are harmful pollutants, which end up being discarded in the environment, presenting a risk of ecotoxicity and potential bioaccumulation, which can directly affect humans through the food chain.

Today there is strict government legislation and regulation with regard to the remediation of textile effluents, making the topic widely studied by researchers. In addition, although there are several physical and chemical methods already applied in the treatment of effluents, they have disadvantages such as low removal efficiency, generation of large amounts of sludge, and high operating costs.

In this context, phycoremediation has been shown to be a viable biological treatment in several studies due to the effective reduction of BOD, COD, heavy metals, and complete removal of colors from the effluent. It is also carbon neutral and therefore an environmentally friendly alternative. Thus, it can be concluded that microalgae can be used as a low-cost alternative to assist in the sustainable remediation of textile effluents in addition to producing biomass rich in interesting compounds for the generation of diverse bioproducts.

Finally, advances in real-time monitoring of phycoremediation and technical–economic aspects in the implementation of microalgae biomass production coupled with wastewater treatment must be considered to validate and implement the results. Additionally, the effect of the mixed microbial consortium (in between autotrophic and heterotrophic populations) may provide new insights into the development of robust microbiological solutions for large-scale remediation processes.

References

1. Abinandan S, Shanthakumar S (2015) Challenges and opportunities in application of microalgae (Chlorophyta) for wastewater treatment: a review. Renew Sustain Energy Rev 52:123–132. https://doi.org/10.1016/j.rser.2015.07.086
2. Acién FG, Gómez-Serrano C, Morales-Amaral MDM, Fernández-Sevilla JM, Molina-Grima E (2016) Wastewater treatment using microalgae: how realistic a contribution might it be to significant urban wastewater treatment? Appl Microbiol Biotechnol 100(21):9013–9022. https://doi.org/10.1007/s00253-016-7835-7
3. Acuner E, Dilek FB (2004) Treatment of tectilon yellow 2G by *Chlorella vulgaris*. Process Biochem 39(5):623–631. https://doi.org/10.1016/S0032-9592(03)00138-9
4. Ahmed SF, Mofijur M, Nuzhat S, Chowdhury AT, Rafa N, Uddin MA, Show PL (2021) Recent developments in physical, biological, chemical, and hybrid treatment techniques for removing emerging contaminants from wastewater. J Hazar Mat 125912. https://doi.org/10.1016/j.jhazmat.2021.125912
5. Aragaw TA, Asmare AM (2018) Phycoremediation of textile wastewater using indigenous microalgae. Water Pract Technol 13(2):274–284. https://doi.org/10.2166/wpt.2018.037

6. Archana LKN (2013) Biological methods of dye removal from textile effluents—a review. J Biochem Technol 3:177–180
7. Barros AI, Gonçalves AL, Simões M, Pires JC (2015) Harvesting techniques applied to microalgae: a review. Renew Sustain Energy Rev 41:1489–1500. https://doi.org/10.1016/j. rser.2014.09.037
8. Behera M, Nayak J, Banerjee S, Chakrabortty S, Tripathy SK (2021) A review on the treatment of textile industry waste effluents towards the development of efficient mitigation strategy: an integrated system design approach. J Environ Chem Eng 105277. https://doi.org/10.1016/ j.jece.2021.105277
9. Behl K, Joshi M, Sharma M, Tandon S, Chaurasia AK, Bhatnagar A, Nigam S (2019) Performance evaluation of isolated electrogenic microalga coupled with graphene oxide for decolorization of textile dye wastewater and subsequent lipid production. Chem Eng J 375. https:// doi.org/10.1016/j.cej.2019.121950
10. Behl K, SeshaCharan P, Joshi M, Sharma M, Mathur A, Kareya MS, Nigam S (2020) Multifaceted applications of isolated microalgae Chlamydomonas sp. TRC-1 in wastewater remediation, lipid production and bioelectricity generation. Bioresour technol 304:122993. https:// doi.org/10.1016/j.biortech.2020.122993
11. Bhatia D, Sharma NR, Singh J, Kanwar RS (2017) Biological methods for textile dye removal from wastewater: a review. Critical Rev Environ Sci Technol 47(19):1836–1876. https://doi. org/10.1080/10643389.2017.1393263
12. Bhushan S, Kalra A, Simsek H, Kumar G, Prajapati SK (2020) Current trends and prospects in microalgae-based bioenergy production. J Environ Chem Eng 104025. https://doi.org/10. 1016/j.jece.2020.104025
13. Bibi R, Ahmad Z, Imran M, Hussain S, Ditta A, Mahmood S, Khalid A (2017) Algal bioethanol production technology: a trend towards sustainable development. Renew Sustain Energy Rev 71:976–985. https://doi.org/10.1016/j.rser.2016.12.126
14. Brar A, Kumar M, Vivekanand V, Pareek N (2019) Phycoremediation of textile effluent-contaminated water bodies employing microalgae: nutrient sequestration and biomass production studies. Int J Environ Sci Technol 16:7757–7768. https://doi.org/10.1007/s13762-018-2133-9
15. Brennan L, Owende P (2010) Biofuels from microalgae—a review of technologies for production, processing, and extractions of biofuels and co-products. Renew sustain energy rev14(2):557–577. https://doi.org/10.1016/j.rser.2009.10.009
16. Bulca Ö, Palas B, Atalay S, Ersöz G (2021) Performance investigation of the hybrid methods of adsorption or catalytic wet air oxidation subsequent to electrocoagulation in treatment of real textile wastewater and kinetic modelling. J Water Process Eng 40. https://doi.org/10.1016/ j.jwpe.2020.101821
17. Cai T, Park SY, Li Y (2013) Nutrient recovery from wastewater streams by microalgae: status and prospects. Renew Sustain Energy Rev 19:360–369. https://doi.org/10.1016/j.rser.2012. 11.030
18. Cardoso NF, Lima EC, Royer B, Bach MV, Dotto GL, Pinto LA, Calvete T (2012) Comparison of Spirulina platensis microalgae and commercial activated carbon as adsorbents for the removal of Reactive Red 120 dye from aqueous effluents. J Haz Mat 241:146–153. https:// doi.org/10.1016/j.jhazmat.2012.09.026
19. Chai WS, Tan WG, Munawaroh HSH, Gupta VK, Ho SH, Show PL (2020) Multifaceted roles of microalgae in the application of wastewater biotreatment: a review. Environ Pollut 116236. https://doi.org/10.1016/j.envpol.2020.116236
20. Chatsungnoen T, Chisti Y (2016) Harvesting microalgae by flocculation–sedimentation. Algal Res 13:271–283. https://doi.org/10.1016/j.algal.2015.12.009
21. Cheah WY, Show PL, Chang JS, Ling TC, Juan JC (2015) Biosequestration of atmospheric CO_2 and flue gas-containing $CO2$ by microalgae. Bioresour Technol 184:190–201. https:// doi.org/10.1016/j.biortech.2014.11.026
22. Cheah WY, Ling TC, Show PL, Juan JC, Chang JS, Lee DJ (2016) Cultivation in wastewaters for energy: a microalgae platform. Appl Energy 179:609–625. https://doi.org/10.1016/j.ape nergy.2016.07.015

23. Chen G, Zhao L, Qi Y (2015) Enhancing the productivity of microalgae cultivated in wastewater toward biofuel production: a critical review. Appl Energy 137:282–291. https://doi.org/10.1016/j.apenergy.2014.10.032

24. Cheng CH, Du TB, Pi HC, Jang SM, Lin YH, Lee HT (2011) Comparative study of lipid extraction from microalgae by organic solvent and supercritical CO_2. Bioresour Technol 102(21):10151–10153. https://doi.org/10.1016/j.biortech.2011.08.064

25. Chew KW, Yap JY, Show PL, Suan NH, Juan JC, Ling TC, Chang JS (2017) Microalgae biorefinery: high value products perspectives. Bioresour Technol 229:53–62. https://doi.org/10.1016/j.biortech.2017.01.006

26. Collet P, Hélias A, Lardon L, Ras M, Goy RA, Steyer JP (2011) Life-cycle assessment of microalgae culture coupled to biogas production. Bioresour Technol 102(1):207–214. https://doi.org/10.1016/j.biortech.2010.06.154

27. Cheriaa J, Bettaieb F, Denden I, Bakhrouf A (2009) Characterization of new algae isolated from textile wastewater plant. J Food Agric Environ 7(3–4):700–704

28. Chu WL, See YC, Phang SM (2009) Use of immobilised *Chlorella vulgaris* for the removal of colour from textile dyes. J Appl Phycol 21:641. https://doi.org/10.1007/s10811-008-9396-3

29. Daneshvar N, Ayazloo M, Khataee AR, Pourhassan M (2007) Biological decolorization of dye solution containing Malachite Green by microalgae *Cosmarium* sp. Biores Technol 98(6):1176–1182. https://doi.org/10.1016/j.biortech.2006.05.025

30. Daroch M, Geng S, Wang G (2013) Recent advances in liquid biofuel production from algal feedstocks. Appl Energy 102:1371–1381. https://doi.org/10.1016/j.apenergy.2012.07.031

31. Dawood S, Sen T (2014) Review on dye removal from its aqueous solution into alternative cost effective and non-conventional adsorbents. J Chem Process Eng 1(104):1–11

32. Deprá MC, Mérida LG, de Menezes CR, Zepka LQ, Jacob-Lopes E (2019) A new hybrid photobioreactor design for microalgae culture. Chem Eng Res Des 144:1–10. https://doi.org/10.1016/j.cherd.2019.01.023

33. El-Kassas HY, Mohamed LA (2014) Bioremediation of the textile waste effluent by Chlorella vulgaris. Egypt J Aquat Res 40(3):301–308. https://doi.org/10.1016/j.ejar.2014.08.003

34. Ergene A, Ada K, Tan S, Katırcıoğlu H (2009) Removal of Remazol brilliant blue r dye from aqueous solutions by adsorption onto immobilized Scenedesmus quadricauda: equilibrium and kinetic modeling studies. Desalination 249(3):1308–1314. https://doi.org/10.1016/j.desal.2009.06.027

35. Fazal T, Mushtaq A, Rehman F, Khan AU, Rashid N, Farooq W, Rehman MSU, Xu J (2018) Bioremediation of textile wastewater and successive biodiesel production using microalgae. Renew Sust Energy Revi 82:3107–3126. https://doi.org/10.1016/j.rser.2017.10.029

36. Ge S, Champagne P (2017) Cultivation of the marine macroalgae *Chaetomorpha linum* in municipal wastewater for nutrient recovery and biomass production. Environ Sci Technol 51(6):3558–3566. https://doi.org/10.1021/acs.est.6b06039

37. Ghaly A, Ananthashankar R, Alhattab M, Ramakrishnan V (2014) Production, characterization and treatment of textile effluents: a critical review. J Chem Eng Proc Technol 5:1–18. https://doi.org/10.4172/2157-7048.1000182

38. Ghazal FM, Mahdy ESM, Fattah EL, MSA, EL-Sadany AEG, Doha (2018) The use of microalgae in bioremediation of the textile wastewater effluent. Nat Sci 16(3):98–104. https://doi.org/10.7537/marsnsj160318.11

39. Gonçalves AL, Pires JCM, Simões M (2017) A review on the use of microalgal consortia for wastewater treatment. Carbon Diox Mitig Micr Systems 24:403–415. https://doi.org/10.1016/j.algal.2016.11.008

40. Gong M, Bassi A (2016) Carotenoids from microalgae: a review of recent developments. Biotechnol Adv 34(8):1396–1412. https://doi.org/10.1016/j.biotechadv.2016.10.005

41. Gosavi VD, Sharma S (2014) A general review on various treatment methods for textile wastewater. J Environ Sci Comput Sci Eng Technol 3:29–39. ISSN: 2278–179X

42. Grima EM, Belarbi EH, Fernández FA, Medina AR, Chisti Y (2003) Recovery of microalgal biomass and metabolites: process options and economics. Biotechnol Adv 20(7–8):491–515. https://doi.org/10.1016/S0734-9750(02)00050-2

43. Guo L, Li XM, Zeng GM, Zhou Y (2010) Effective hydrogen production using waste sludge and its filtrate. Energy 35(9):3557–3562. https://doi.org/10.1016/j.energy.2010.04.005
44. Gupta S, Pawar SB (2018) An integrated approach for microalgae cultivation using raw and anaerobic digested wastewaters from food processing industry. Bioresour Technol 269:571–576. https://doi.org/10.1016/j.biortech.2018.08.113
45. Hannon M, Gimpel J, Tran M, Rasala B, Mayfield S (2010) Biofuels from algae: challenges and potential. Biofuels 1(5):763–784. https://doi.org/10.4155/bfs.10.44
46. Hessel C, Allegre C, Maisseu M, Charbit F, Moulin P (2007) Guidelines and legislation for dye house effluents. J Environ Manage 983:171–180. https://doi.org/10.1016/j.jenvman.2006.02.012.
47. Holkar CR, Jadhav AJ, Pinjari DV, Mahamuni NM, Pandit AB (2016) A critical review on textile wastewater treatments: possible approaches. J Environ Manag 182:351–366. https://doi.org/10.1016/j.jenvman.2016.07.090
48. Huang J, Liu J, Kuo J, Xie W, Zhang X, Chang K, Evrendilek F (2019) Kinetics, thermodynamics, gas evolution and empirical optimization of (co-) combustion performances of spent mushroom substrate and textile dyeing sludge. Bioresour Technol 280:313–324. https://doi.org/10.1016/j.biortech.2019.02.011
49. Huang G, Chen F, Wei D, Zhang X, Chen G (2010) Biodiesel production by microalgal biotechnology. Appl Energy 87(1):38–46. https://doi.org/10.1016/j.apenergy.2009.06.016
50. Hussein FH (2013) Chemical properties of treated textile dyeing wastewater. Asian J Chem 25:9393–9400. https://doi.org/10.14233/ajchem.2013.15909A
51. Hussain F, Shah SZ, Ahmad H, Abubshait SA, Abubshait HA, Laref A, Iqbal M (2021) Microalgae an ecofriendly and sustainable wastewater treatment option: biomass application in biofuel and bio-fertilizer production. A Rev Renew Sustain Energy Rev 137. https://doi.org/10.1016/j.rser.2020.110603
52. Javed F, Aslam M, Rashid N, Shamair Z, Khan A, Yasin M, Bazmi AA (2019) Microalgae-based biofuels, resource recovery and wastewater treatment: a pathway towards sustainable biorefinery. Fuel 255. https://doi.org/10.1016/j.fuel.2019.115826
53. John J (2000) A self-sustainable remediation system for acidic mine voids. In: Proceedings of the 4th international conference of diffuse pollution, Bangkok, Thailand, pp 506–511
54. Jose S, Archanaa S (2019) Phycoremediation of textile wastewater: possibilities and constraints. In: Gupta SK, Bux F (eds) Application of microalgae in wastewater treatment. Springer Nature, Switzerland, pp 291–319. https://doi.org/10.1007/978-3-030-13913-1_14
55. Kadir WNA, Lam MK, Uemura Y, Lim JW, Lee KT (2018) Harvesting and pre-treatment of microalgae cultivated in wastewater for biodiesel production: a review. Energy Conv Manag 171:1416–1429. https://doi.org/10.1016/j.enconman.2018.06.074
56. Kalra R, Gaur S, Goel M (2021) Microalgae bioremediation: A perspective towards wastewater treatment along with industrial carotenoids production. J Water Proc Eng 40. https://doi.org/10.1016/j.jwpe.2020.101794
57. Kehinde FO, Aziz HA (2014) Textile waste water and the advanced oxidative treatment process, an overview. Int J Innovat Res Sci Eng Technol 3:15310–15317. https://doi.org/10.15680/IJIRSET.2014.0308034
58. Khan MI, Shin JH, Kim JD (2018) The promising future of microalgae: current status, challenges, and optimization of a sustainable and renewable industry for biofuels, feed, and other products. Microb Cell Fact 17, 36. https://doi.org/10.1186/s12934-018-0879-x
59. Kishor R, Purchase D, Saratale GD, Saratale RG, Ferreira LFR, Bilal M, Bharagava RN (2021) Ecotoxicological and health concerns of persistent coloring pollutants of textile industry wastewater and treatment approaches for environmental safety. J Environ Chem Eng 105012. https://doi.org/10.1016/j.jece.2020.105012
60. Kumar L, Bharadvaja N (2021) algal-based wastewater treatment and biorefinery. In: Wastewater treatment, pp 413–432. Elsevier. https://doi.org/10.1016/B978-0-12-821881-5.00020-9
61. Kurade MB, Waghmode TR, Jadhav MU, Jeon BH, Govindwar SP (2015) Bacterial–yeast consortium as an effective biocatalyst for biodegradation of sulphonated azo dye Reactive Red 198. RSC Adv 5:23046–23056. https://doi.org/10.1039/C4RA15834B

62. Lim SL, Chu WL, Phang SM (2010) Use of Chlorella vulgaris for bioremediation of textile wastewater. Bioresour Technol 101:7314–7322. https://doi.org/10.1016/j.biortech. 2010.04.092

63. Liu RR, Tian Q, Yang B, Chen J (2010) Hybrid anaerobic baffled reactor for treatment of desizing wastewater. Int J Environ Sci Technol 7:111–118. https://doi.org/10.1007/BF0332 6122

64. Majid M, Shafqat S, Inam H, Hashmi U, Kazi AG (2014) Production of algal biomass. In: Hakeem K, Jawaid M, Rashid U (eds) Biomass and bioenergy. Springer, Cham. https://doi. org/10.1007/978-3-319-07641-6_13

65. Mani S, Chowdhary P, Bharagava RN (2019) Textile wastewater dyes: toxicity profile and treatment approaches. In: Emerging and eco-friendly approaches for waste management, pp. 219–244. Springer, Singapore. https://doi.org/10.1007/978-981-10-8669-4_11

66. Maqbool Z, Hussain S, Ahmad T, Nadeem H, Imran M, Khalid A et al (2016) Use of RSM modeling for optimizing decolorization of simulated textile wastewater by Pseudomonas aeruginosa strain ZM130 capable of simultaneous removal of reactive dyes and hexavalent chromium. Environ Sci Pollut Res 23:11224–11239. https://doi.org/10.1007/s11356-016-6275-3

67. Makertihartha IGBN, Rizki Z, Zunita M, Dharmawijaya PT (2017) Dyes removal from textile wastewater using graphene based nanofiltration 110006–110006

68. Martorell MM, Pajot HF, de Figueroa LI (2017) Biological degradation of reactive black 5 dye by yeast Trichosporon akiyoshidainum. J Environ Chem Eng 5(6):5987–5993. https://doi.org/10.1016/j.jece.2017.11.012

69. Mata TM, Martins AA, Caetano NS (2010) Microalgae for biodiesel production and other applications: a review. Renew Sustain Energy Rev 14(1):217–232. https://doi.org/10.1016/j. rser.2009.07.020

70. Mehar J, Shekh A, Uthaiah Malchira N, Sarada R, Chauhan VS, Mudliar S (2019) Automation of pilot-scale open raceway pond: A case study of CO_2-fed pH control on Spirulina biomass, protein and phycocyanin production. J CO2 Utilization 33:384–393. https://doi.org/10.1016/ j.jcou.2019.07.006

71. Mishra S, Roy M, Mohanty K (2019) Microalgal bioenergy production under zero-waste biorefinery approach: recent advances and future perspectives. Bioresour Technol 292. https:// doi.org/10.1016/j.biortech.2019.122008

72. Mohan SV, Dahiya S, Amulya K, Katakojwala R, Vanitha TK (2019) Can circular bioeconomy be fueled by waste biorefineries—a closer look. Bioresour Technol Rep 7. https://doi.org/10. 1016/j.biteb.2019.100277

73. Mohammady NG, El-Sayed HS, Taha HM, Fakhry EM, Mahmoud NH, Mohamed JH, Mekawy LM (2015) Chlorella sp. as a source of biodiesel and by-products: an integral study of Med-algae Project; Part A. Int J Techno Chem Res 1(3):144–151. ISSN:2395-4248

74. Mohammad F (2015) High-energy radiation induced sustainable coloration and functional finishing of textile materials. Ind Eng Chem Res 54(15):3727–3745. https://doi.org/10.1021/ acs.iecr.5b00524

75. Mooij PR, Stouten GR, Van Loosdrecht MC, Kleerebezem R (2015) Ecology-based selective environments as solution to contamination in microalgal cultivation. Curr Op Biotechnol 33:46–51. https://doi.org/10.1016/j.copbio.2014.11.001

76. Nagarajan D, Chang JS, Lee DJ (2020) Pretreatment of microalgal biomass for efficient biohydrogen production–recent insights and future perspectives. Bioresour Technol 302. https://doi. org/10.1016/j.biortech.2020.122871

77. Nidheesh PV, Kumar A, Babu DS, Scaria J, Kumar MS (2020) Treatment of mixed industrial wastewater by electrocoagulation and indirect electrochemical oxidation. Chemosphere 251. https://doi.org/10.1016/j.chemosphere.2020.126437

78. Olguín EJ, Sánchez-Galván G (2012) Heavy metal removal in phytofiltration and phycoremediation: the need to differentiate between bioadsorption and bioaccumulation. New Biotechnol 30:3–8. https://doi.org/10.1016/j.nbt.2012.05.020

79. Oswald WJ, Gotaas HB, Golueke CG, Kellen WR (1957) Algae in wastewater treatment. Sew Ind Waste 29:437–455
80. Pal P (2017) Industry-specific water treatment: case studies. Ind Water Treat Process Technol 243–511. https://doi.org/10.1016/B978-0-12-810391-3.00006-0
81. Pang YL, Abdullah AZ (2013) Current status of textile industry wastewater management and research progress in Malaysia: a review. Clean-Soil, Air, Water 41(8):751–764. https://doi.org/10.1002/clen.201000318
82. Paz A, Carballo J, Perez MJ, Domínguez JM (2017) Biological treatment of model dyes and textile wastewaters. Chemosphere 181:168–177. https://doi.org/10.1016/j.chemosphere.2017.04.046
83. Paździor K, Bilińska L, Ledakowicz S (2019) A review of the existing and emerging technologies in the combination of AOPs and biological processes in industrial textile wastewater treatment. Chem Eng J 376. https://doi.org/10.1016/j.cej.2018.12.057
84. Pereira L, Alves M (2012) Dyes—environmental impact and remediation. In: Environmental protection strategies for sustainable development, pp. 111–162. Springer, Dordrecht. https://doi.org/10.1007/978-94-007-1591-2_4
85. Pires JC (2017) COP21: the algae opportunity? Renew Sustain Energy Rev 79:867–877. https://doi.org/10.1016/j.rser.2017.05.197
86. Popli S, Patel UD (2015) Destruction of azo dyes by anaerobic–aerobic sequential biological treatment: a review. Int J Environ Sci Technol 12:405–420. https://doi.org/10.1007/s13762-014-0499-x
87. Pragya N, Pandey KK, Sahoo PK (2013) A review on harvesting, oil extraction and biofuels production technologies from microalgae. Renew Sustain Energy Rev 24:159–171. https://doi.org/10.1016/j.rser.2013.03.034
88. Qie F, Zhu J, Rong J, Zong B (2019) Biological removal of nitrogen oxides by microalgae, a promising strategy from nitrogen oxides to protein production. Bioresour Technol 292. https://doi.org/10.1016/j.biortech.2019.122037
89. Rahman BMA, Rahman MM, Shaid A, Bashar MM, Khan MA (2016) Scope of reusing and recycling the textile wastewater after treatment with gamma radiation. J Clean Prod 112:3063–3071. https://doi.org/10.1016/j.jclepro.2015.10.029
90. Rathore D, Singh A (2013) Biohydrogen production from microalgae. In: Biofuel technologies, pp 317–333. Springer, Berlin, Heidelberg. https://doi.org/10.1007/978-3-642-34519-7_13
91. Rawat I, Kumar RR, Mutanda T, Bux F (2011) Dual role of microalgae: phycoremediation of domestic wastewater and biomass production for sustainable biofuels production. Appl Energy 88(10):3411–3424. https://doi.org/10.1016/j.apenergy.2010.11.025
92. Rincón B (2020) Biogas from microalgae. In: Handbook of microalgae-based processes and products, pp 311–328. Academic Press. https://doi.org/10.1016/B978-0-12-818536-0.00012-9
93. Robinson T, Mcmullan G, Marchant R, Nigam P (2001) Remediation of dyes in textile effluent: a critical review on current treatment technologies with a proposed alternative. Bioresour Technol 77:247–255. https://doi.org/10.1016/S0960-8524(00)00080-8
94. Salama ES, Kurade MB, Abou-Shanab RA, El-Dalatony MM, Yang IS, Min B, Jeon BH (2017) Recent progress in microalgal biomass production coupled with wastewater treatment for biofuel generation. Renew Sustain Energy Rev 79:1189–1211. https://doi.org/10.1016/j.rser.2017.05.091
95. Salgueiro JL, Perez L, Maceiras R, Sanchez A, Cancela A (2016) Bioremediation of wastewater using Chlorella vulgaris microalgae: Phosphorus and organic matter. Int J Environ Res 10(3):465–470. https://doi.org/10.22059/IJER.2016.58766
96. Sarayu K, Sandhya S (2012) Current Technologies Saini RD (2017) Textile organic dyes: polluting effects and elimination methods from textile waste water. Int J Chem Eng Res 9(1):121–136. ISSN: 0975-6442.
97. Samsami S, Mohamadi M, Sarrafzadeh MH, Rene ER, Firoozbahr M (2020) Recent advances in the treatment of dye-containing wastewater from textile industries: overview and perspectives. Proc Saf Environ Prot. https://doi.org/10.1016/j.psep.2020.05.034

98. Sekomo CB, Rousseau DP, Saleh SA, Lens PN (2012) Heavy metal removal in duckweed and algae ponds as a polishing step for textile wastewater treatment. Ecological Eng 44:102–110. https://doi.org/10.1016/j.ecoleng.2012.03.003

99. Seo YH, Park D, Oh YK, Yoon S, Han JI (2015) Harvesting of microalgae cell using oxidized dye wastewater. Bioresour Technol 192:802–806. https://doi.org/10.1016/j.biortech.2015.05.074

100. Severo IA, Siqueira SF, Deprá MC, Maroneze MM, Zepka LQ, Jacob-Lopes E (2019) Biodiesel facilities: what can we address to make biorefineries commercially competitive? Renew Sustain Energy Rev 112:686–705. https://doi.org/10.1016/j.rser.2019.06.020

101. Seow TW, Lim CK (2016) Removal of dye by adsorption: a review. Int J Appl Eng Res 11:2675–2679

102. Singh K, Arora S (2011) Removal of synthetic textile dyes from wastewaters: a critical review on present treatment technologies. Crit Rev Environ Sci Technol 41(9):807–878. https://doi.org/10.1080/10643380903218376

103. Singh PK, Singh RL (2017) Bio-removal of azo dyes: a review. Int J App Sci Biotechnol 5(2):108–126. https://doi.org/10.3126/ijasbt.v5i2.16881

104. Singh RP, Singh PK, Gupta R, Singh RL (2019) Treatment and recycling of wastewater from textile industry. In: Advances in biological treatment of industrial waste water and their recycling for a sustainable future, pp 225–266. Springer, Singapore. https://doi.org/10.1007/978-981-13-1468-1_8

105. Sinha S, Singh R, Chaurasia AK, Nigam S (2016) Self-sustainable Chlorella pyrenoidosa strain NCIM 2738 based photobioreactor for removal of direct Red-31 dye along with other industrial pollutants to improve the water-quality. J Hazard Mater 306:386–394. https://doi.org/10.1016/j.jhazmat.2015.12.011

106. Suali E, Sarbatly R (2012) Conversion of microalgae to biofuel. Renew Sustaina Energy Rev 16(6):4316–4342. https://doi.org/10.1016/j.rser.2012.03.047

107. Teixeira CMLL, Kirsten FV, Teixeira PCN (2012) Evaluation of Moringa oleifera seed flour as a flocculating agent for potential biodiesel producer microalgae. J Appl Phycol 24:557–563. https://doi.org/10.1007/s10811-011-9773-1

108. Udaiyappan AFM, Hasan HA, Takriff MS, Abdullah SRS (2017) A review of the potentials, challenges and current status of microalgae biomass applications in industrial wastewater treatment. J Water Proc Eng 20:8–21. https://doi.org/10.1016/j.jwpe.2017.09.006

109. Ullah K, Ahmad M, Sharma VK, Lu P, Harvey A, Zafar M, Anyanwu CN (2014) Algal biomass as a global source of transport fuels: overview and development perspectives. Prog Natural Sci: Mat Int 24(4):329–339. https://doi.org/10.1016/j.pnsc.2014.06.008

110. USEPA (1996) Manual: best management practices for pollution prevention in the textile industry. https://nepis.epa.gov/EPA/html/DLwait.htm?url=/Exe/ZyPDF.cgi/30004Q2U/?Dockey=30004Q2U. Accessed 18 may 2021

111. Yaseen DA, Scholz M (2019) Textile dye wastewater characteristics and constituents of synthetic effluents: a critical review. Int J Environ Sci Technol 16:1193–1226. https://doi.org/10.1007/s13762-018-2130-z

112. Zhang ZB, Liu ZZ, Wu L, Li PF (2006) Detection of nitric oxide in culture media and studies on nitric oxide formation by marine microalgae. Med Sci Monit 12:75–85

Potential for Constructed Wetlands Aimed at Sustainable Wastewater Treatment, Reuse, and Disposal in Dyestuff and Textile Sectors

Shardul D. Prabhu, B. Lekshmi, and Shyam R. Asolekar

Abstract It is envisaged in this research that dyestuff and textile processing sectors can be mutually inspired by the strategies developed by them for treatment and reuse of dye-bearing wastewaters because their respective environmental concerns are comparable. It is well known that decolourization and degradation of mixture of dyes is difficult with conventional biological wastewater treatment. Also, it is rather expensive using the advanced physico-chemical treatment processes—which are typically resource-intensive and have large carbon footprint and electrical power requirements. To break this gridlock, a logical alternative could be to segregate wastewater streams in production plants and render appropriate treatment to them separately and make efforts to minimize the costs of environmental protection. This proposed integrated wastewater treatment framework also minimizes water footprint through the production of reuse-quality water. This research highlights the significance of incorporation of the Natural Treatment Systems (i.e. Nature-based Solutions) making the treatment strategy eco-centric and innovative.

Keywords Textile · Dyestuff · Dye-bearing wastewater · Constructed Wetlands · Natural treatment system · Nature-based solution · Biological · Physico-chemical · Reuse water · Eco-centric technology · Recycling · Effluent

1 The Context of Dyestuff and Textile Sectors in India

Dyes and dyestuff have been around for centuries, but the production of the first synthetic dye "Mauveine" by Perkin in 1856 laid the foundation for the present-day synthetic dyestuff industry. Thus, it is one of the oldest and most well-established industries. Today, we can find numerous industries from textiles to pharmaceuticals to food which make use of dyes at some point in their respective production processes or applications. The global dye industry is tremendous with more than 100,000 commercially manufactured organic dyes totalling over 7×10^5 Tons produced

S. D. Prabhu · B. Lekshmi · S. R. Asolekar (✉)
Environmental Science & Engineering Department, Indian Institute of Technology Bombay, Mumbai 400076, India

annually [119]. As one would reckon, the largest contribution to the global synthetic organic dye production and demand is by the textile industry. The global demand in the dyes and pigment market was forecast to be $19.5 billion in 2019 and about half of it is contributed by the textile industry [89].

With dyestuffs being used by a variety of industrial applications, it is no wonder that the dyestuff industry has developed exponentially since its inception and now the dyestuffs can be classified into about 15 categories and chemistries based on different functions and applications. Every type of dye, based on its chemistry, has a different range of colours, fixation efficiency, and different type of cloth [37]. It can be seen that the dyestuff and textile industries are closely intertwined.

Efforts are made in this chapter to articulate and highlight the opportunity for using a natural treatment method like Constructed Wetlands for the treatment, reuse, and disposal of wastewaters generated in the dyestuff and textile sectors. There is a remarkable similarity between the physical and chemical characteristics of wastewaters from the dyestuff and textile sectors. Hence, the issues associated with biodegradability of the pollutant molecules, too, is comparable in both sectors.

1.1 Environmental Challenges Faced in Dyestuffs and Textile Sector

India is a key player globally in, both, the production of dyes as well as dye intermediates. Nearly all the dyestuff and dye intermediate giant multinational corporations in the world have collaborations for large-scale production in India. Also, there are several home-grown small and medium-scale dyestuff industries in India. No wonder the country has a significant share of dye and pigment market in the world. As a result, in 2019, India produced approx. 382,000 Tons of dyes and pigments [107]. India exported dyes worth USD 2,538 million and dye intermediates worth USD 270 million in 2019 [106].

The dyestuff and dye intermediates as well as textile sectors have enormous fluctuations in monthly production largely because these sectors are internationally influenced by the volatile markets. Also, typically they have campaign-based batch manufacturing processes. Consequently, these sectors have an immense variability in terms of, both, quantities and nature of emissions. As a result, the dyestuff and dye intermediates as well as textile sectors are being classified in the so-called "red" category in India and are monitored rather closely [14].

The regulatory agency responsible for monitoring and enforcing environmental norms for these industrial sectors at the federal level is the Central Pollution Control Board (CPCB), Ministry of Environment, Forest and Climate Change (MoEF&CC), Government of India (GoI), New Delhi. In addition, every State in the Union of India has the State's Pollution Control Board—which is responsible for enforcing the pollution control norms and have authority for making the norms even more stringent for the zones having rather excessive pollution or for protection of biodiversity, water bodies and forests.

Table 1 presents the regulatory limits for disposal of treated waters from dyestuff and dye intermediates and textile industries set by the CPCB [15, 16]. It is evident that the disposal standards prescribed by the regulatory agency are different for dyestuff and dye intermediates as well as textile sectors. For example, the permissible emission limit for colour in dyestuff and dye intermediates sector is 400 Pt–Co units and for the textile sector it is 150 Pt–Co units.

Unless serviced at the cluster-level, the tiny industries can rarely afford to construct and operate individual Effluent Treatment Plants (ETPs) because it is infeasible for them to maintain skilled ETP operators capable of operating the complex treatment technologies. Therefore, the Government of India has been zoning the small and medium-scale industries (dye and dye intermediates as well as textile sector) and dozens of Common Effluent Treatment Plants (CETPs) have been established for nearly two decades. The capital costs for establishing CETPs are mostly borne jointly by the Central Government and the respective State Government. The operation and maintenance costs for the given CETP are collectively borne by the member industries in the cluster.

Table 1 The regulatory limits for disposal of treated waters from dyestuff and dye intermediates and textile industries as prescribed by the CPCB (Central Pollution Control Board [15, 16])

Sr. no	Water quality parameter	Dyestuff and dye-intermediates industry[a]			Textile industry[b]
		Disposal in surface water	Marine disposal	Disposal on land for irrigation	
1	pH	6–8.5	5.5–9	5.5–9	6.5–8.5
2	Suspended solids	100 mg/L	–	200 mg/L	100 mg/L
3	Colour	400 Pt–Co	–	–	150 Pt–Co
4	Biochemical oxygen demand (BOD$_3$)	30 mg/L	100 mg/L	100 mg/L	30 mg/L
5	Chemical oxygen demand (COD)	250 mg/L	250 mg/L	–	250 mg/L
6	Ammoniacal nitrogen (as N)	50 mg/L	50 mg/L	–	50 mg/L
7	Total chromium	2 mg/L	2 mg/L	–	2 mg/L
8	Total dissolved Solids (TDS)	–	–	–	2,100 mg/L[*]
9	Chlorides	1,000 mg/L	–	–	–
10	Sulphates	1,000 mg/L	–	–	–

[*] May be made more stringent by CPCB or respective SPCB depending on the quality of local receiving water body
[a]Central Pollution Control Board [15]
[b]Central Pollution Control Board [16]

However, despite the sustained efforts, all the smaller units (textile artists and dyers) have not been connected yet to the respective CETPs in their vicinities. As a result, co-disposal of municipal wastewaters (sewages) and the dye-bearing wastewaters from some small-scale textile and dyestuff industries, scattered in remote rural locations or residential zones of peri-urban communities, continue to potentially pose threats to the wastewater treatment plants of the respective communities.

It is well known that the main raw materials in dyestuff manufacturing are petrochemical-based organic molecules—which undergo a series of unit operations and unit processes aimed at constructing large, typically non-biodegradable complex dye molecules. A few illustrative dye structures are shown in Fig. 1.

It is no wonder that dye molecules are difficult to degrade and break down owing to their large molecular size and complex aromatic structure. Decolourization of dye-bearing wastewater streams is typically achieved through the breaking of bonds of the dye molecule with the respective chromophore(s)—which are the colour-imparting substitutions. Although it may appear straightforward, in reality, this process is rather complex because there could be multiple chromophores having different bond energies. In addition, the interferences caused by the co-existing background ionic strength and other solvents, intermediates, and by-products do make the selective bond cleaving nearly impossible. It is well known that the dye molecules are further used in textile sector (dyeing and printing) and eventually the unused (unfixed) dyes end up in the emitted wastewaters.

Studies have shown the harmful impacts of discharging untreated textile dyeing effluent mixed with sewage into ponds. Apart from the colour, COD, and BOD, a major source of pollution from textile dyeing effluents are heavy metals. Heavy metals such as Lead (Pb), Cadmium (Cd), Iron (Fe), Manganese (Mn), Nickel (Ni), Copper (Cu), and Zinc (Zn) find their way into the crops when this polluted water from the pond is used for irrigation.

One study conducted by [27] showed that in Haryana, India, where untreated textile effluent was being mixed with sewage and let into a nearby pond, the bioaccumulation of Pb was dangerously over safe limits. Further, the impact of heavy metals on ecological and human health is exacerbated when these metals leach into the groundwater, where they can perpetually stay. It is well known that if such waters are used for drinking or irrigation, the heavy metals would tend to bioaccumulate in crops and human tissues. For example, the highest concentration of Pb (28 mg/Kg) was reported in the roots of Cauliflower—which was about 10 times higher than the permissible safe limit of Pb (2.5 mg/Kg) in raw vegetables.

1.2   The Scope and Specific Objectives

The industrial sector of dyestuffs and dye intermediates happens to be one of the strong sectors in India because the country has a significant share of the dye and pigment market in the world. Also, India has a robust textile industrial sector—which is the ready market for dyestuff sector. As a result, nearly all the dyestuff and

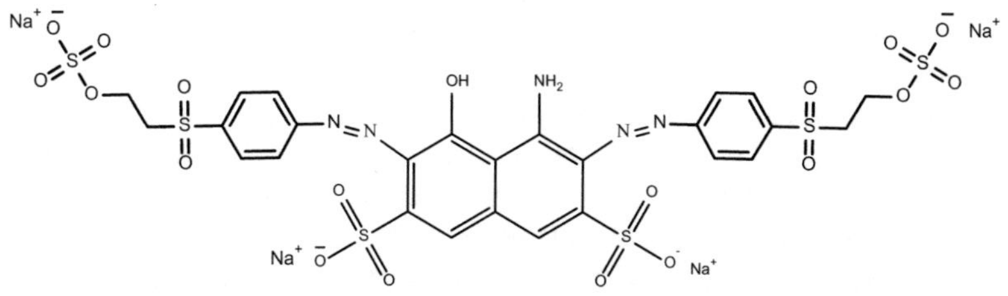

Reactive Black 5 Dye
Chemical Formula: $C_{26}H_{22}N_5O_{19}S_6Na_3$
Molecular Weight: 991.79

(a)

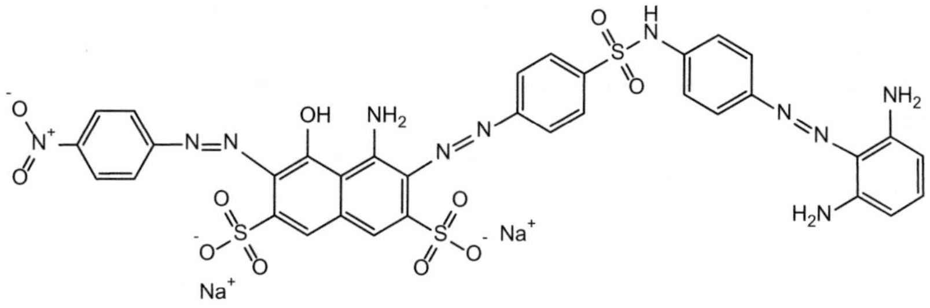

Acid Black 210
Chemical Formula: $C_{34}H_{25}N_{11}Na_2O_{11}S_3$
Molecular Weight: 905.80

(b)

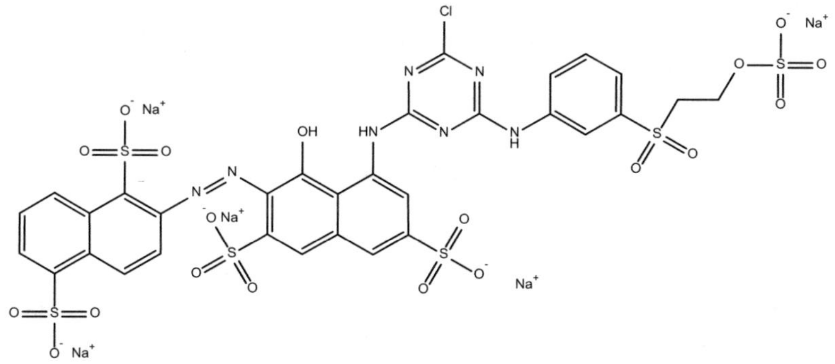

Reactive Red 195
Chemical Formula: $C_{31}H_{19}ClN_7O_{19}S_6Na_5$
Molecular Weight: 1134.82

(c)

Fig. 1 Structures of typical dye molecules **a** Reactive Black 5, **b** Acid Black 210, and **c** Reactive Red 195

dye intermediate giant manufacturers have collaborations in India and traditionally the country has had home-grown dyestuff manufacturers over the past seven decades.

Although it is obvious that the dyestuff and textile sectors are partners in the same industrial ecosystem having close market relations, it is rarely appreciated that their respective environmental concerns, too, are comparable. The presence of colour-causing dyes and dye intermediates in wastewaters from dyestuff sectors have been a major concern for the environmentally conscientious dyestuff manufacturers as well as the environmental protection regulatory agencies in India (i.e. the Central Pollution Control Board, Government of India, New Delhi and the State Pollution Control Boards in the various States in Union of India). Owing to the regulatory pressure and the escalating socio-political leveraging by environmental activists, the Indian dyestuff industry has implemented several innovations while combining the conventional biological and physico-chemical treatment technologies to meet the regulatory expectations.

The scope of the study presented in this Chapter encompasses the lessons and experiences from treatment of wastewaters bearing dyes and the associated pollutants found in textile and in dyestuff sectors. With the backdrop of the context and scope described above, the following objectives are articulated for this Chapter:

(1)   To map the steps in typical production processes in textile and dyestuff sectors and to flag the nature of the corresponding dye-bearing wastewater streams. Efforts will also be made to present the shortcomings of conventional treatment strategies for the degradation of dyes in wastewater treatment.

(2)   To evaluate the potential benefits of careful segregation and targeted treatment of the intermediate streams in the textile sector, especially those which are not very highly polluted (e.g. filter cake wash water, various wastewaters from cloth washing in textile processing, and vessel washing water). This could be a winning strategy for resource conservation and waste minimization. It is aimed in this study to establish that the treatment and reuse of source separated streams may prove to be, both, economic as well as environmentally sustainable.

(3)   To present the elements of the process, configurations and variants of engineered Constructed Wetlands (CWs) for rendering wastewater treatment in, both, the textile as well as dyestuff sectors.

(4)   To articulate the possibilities of enhancing the potential for recycling and reusing treated water by employing engineered constructed wetlands and also incorporating the integrated approach to dye wastewater management. The present state of the art of wastewater treatment in textile and dyestuff sectors have not yet implemented the sunrise technologies referred to as the Natural Treatment Systems or sometimes the Nature-based Solutions.

2 The Genesis of Pollution in Dyestuff & Textile Sectors

It is well known that decolourization and degradation of dyes is a formidable task and no modern archetypal treatment technology is capable of achieving this feat to the desired extent. In this section, efforts are made to map the steps in production processes in the textile and dyestuff sectors. The comparative accounts of these sectors are presented with the hope that the similarity of environmental challenges can be appreciated on one hand. On the other hand, it will be instructive to appreciate the efforts made by these sectors to resolve their respective challenges through modernizing their production processes and also by segregating certain critical waste streams.

2.1 Pollution Causing Steps in Production of Dyestuffs

The process flow diagram for production of synthetic dyestuffs comprising of typical unit operations and unit processes is presented in Fig. 2. The process comprises of Raw Material Unloading and Storage, Chemical Synthesis of Dyestuff, Isolation of Dyestuff, Filtration, Drying, and Quality Checks and Packaging. The Chemical Synthesis consists of multiple steps such as Diazotization, Amination, Sulphonation, Coupling, etc. The Isolation of Dyestuff (precipitation) is carried out by the addition of a salt with a common ion with the dyestuff molecule to separate the dye salt by employing the common ion effect. This process is also known in the industry as "salting out the dye". Conventionally, Filtration is carried out to separate the dye from the reaction liquor by employing equipment like Filter Press, Belt Filters or

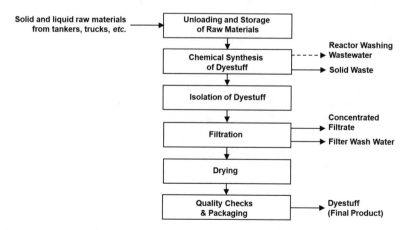

Fig. 2 Process flow diagram for production of synthetic dyestuffs comprising of typical unit operations and unit processes including Raw Material Unloading and Storage, Chemical Synthesis of Dyestuff, Isolation of Dyestuff, Filtration, Drying, and Quality Checks and Packaging

Centrifuges. The modern methods make use of membrane separation techniques for the separation of the dye molecule [58].

The filtered solid mass contains a considerable amount of moisture, and hence, it must be dried to further decrease the moisture content. Nowadays, advanced dyestuff manufacturing processes typically use Spray Drying to vaporize water and achieve the desired level of moisture content in the final product. Heat sensitive dyes are typically spray-dried at about 130 °C, whereas temperatures for other dyes can go as high as 450 °C [75]. The final step is the Quality Check and Packaging where a sample of dye from the batch is tested against the standard to check for desirable quality.

Chemical Synthesis of dyestuffs is comprised of multiple unit processes, and naturally, several raw materials are added to the reactors. The products of reaction thus become part of the complex reaction mixture. The dyestuff reaction mixture mostly consists of the unreacted organic reactants (dye intermediates) which did not couple and salt like NaCl which are generated as a by-product of the synthesis reactions. The filtration process traps the dyestuff product along with some impurities. Everything else becomes part of the wastewater i.e. the filtrate from the Filtration step (or Permeate stream from the membrane filtration unit). It is estimated that the wastewater generation from this step is anywhere between 1 and 700 L per Kg of product for most dyes. For Vat dyes, this value could be up to 8,000 L per Kg of product [121]! The wastewater from Chemical Synthesis is mainly from the washing of reactor vessels after completion of the batch. Usually, industries will manufacture multiple dyes in the same reactors as they usually require the same machinery and vessels.

2.2 Pollution Causing Steps in Textile Processing

A generalized process flow diagram for wet processing of textile fabrics is presented in Fig. 3. The process typically comprises of Desizing, Scouring, Bleaching, Mercerizing, Dyeing, and Finishing. It should be noted that some of the steps depicted here may not be performed for certain types of fabrics. The Desizing process is the preprocessing step which includes the addition of sizing and desizing chemicals to increase the workability of the fabric. Scouring is a "cleaning" step that consists of washing away the unwanted non-cellulosic material from the fabric.

The next step is Bleaching, wherein all colour is removed to enhance the appearance of various shades on the fabric. Mercerizing involves treating the fabric with Caustic Soda (Sodium Hydroxide, NaOH) to strengthen the fabric, provide a lustrous finish, and to better facilitate dye uptake. The subsequent step is the core of the textile wet processing i.e. Dyeing. The dyeing process consists of contacting the fabric with the dye solution in the dye bath. Once the colour is imparted, there may be an additional step of Printing on the fabric. Finally, the dyed textile is finished, during which it undergoes multiple smaller steps like disinfection and cleaning [73].

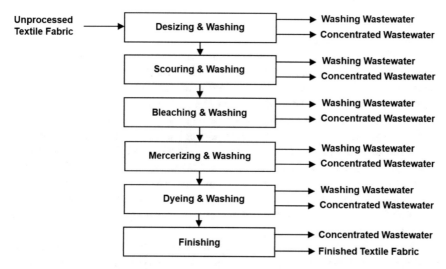

Fig. 3 Process flow diagram for wet processing of textile fabrics comprising of typical unit operations and unit processes including Desizing, Scouring, Bleaching, Mercerizing, Dyeing, and Finishing. Some of the steps depicted here may not be performed for certain types of fabrics

As seen in Fig. 3, the fabric is washed after almost every step in the textile wet processing and every step contributes to liquid wastewater [19]. The wastewater in textile industry is generated mainly in the dye baths, cloth-washing stages, and in other maintenance and cleaning operations [112]. The highest contribution of wastewater is from the Desizing step which is about 50% of the total wastewater contribution and has a considerable amount of BOD [73]. The wastewaters from the multiple steps in textile sector are variable in their quality and quantity depending on the fabric and the process employed; but they are generally characterized by a deep colour, high Chemical Oxygen Demand (COD), high Total Dissolved Salts (TDS), high metal concentrations, and in some cases, they may also have high Nitrogen and Phosphorus concentrations.

3 Shortcomings of Conventional Treatment Strategies for Removal of Dyes in Wastewater Treatment

Degradation and decolourization of dyes encountered in wastewater treatment of dyestuff or textile manufacturing industries usually consists of a combination of certain physico-chemical and biological treatment steps. It is almost impossible to achieve complete dye wastewater treatment using just one approach. The most commonly used pathway for biological treatment is the Activated Sludge Process (ASP). Some conventional physico-chemical approaches to dye wastewater treatment include: Chemical Adsorption, Membrane Separation, Ion Exchange,

Fenton's Reagent Treatment, Ozonation, Photochemical Treatment, or Coagulation and Flocculation [104, 112].

The conventionally employed indicator for the success of dye wastewater treatment, and rightly so, is colour removal. But, colour removal alone does address the entire problem. It is imperative to take into consideration the other water quality parameters—especially Chemical Oxygen Demand (COD), Biochemical Oxygen Demand (BOD), or Dissolved Organic Content (DOC). In fact, these parameters are believed to be shedding more light on the health of the environment and ecosystem—where the treated waters would be eventually disposed of. The BOD/COD ratio of dye wastewater is usually extremely low. For example, the BOD/COD ratio of textile dyeing effluent containing the reactive dye Drimaren HF was 0.35 in the study conducted by [2]. In yet another study, it was determined using three dyes that the BOD/COD ratio of the mixture was 0.26, which suggested that the dyes were recalcitrant to biodegradation [13].

The Aim Should be the Removal of Colour, COD, and Nutrients: In reality, the ideal elemental composition of wastewater containing dyes and textile processing-related chemicals is much different than the desired composition of Carbon, Hydrogen, Nitrogen, and Oxygen for conducting pollutants. This deviation from desirable composition eventually leads to making dye intermediates, dyestuff and textile effluent difficult for degradation and removal. Therefore, one should never aim at the removal of colour as the primary objective of wastewater treatment in textile or dyestuff sectors. Instead, the removal of colour, COD, and nutrients must be seen as the ultimate objective of the wastewater treatment.

Clearly, it is difficult to conduct aerobic or anaerobic dye degradation. The remedy, therefore, often employed in environmental technologies, focusses on the development of a specific strain of microorganisms or enzymes to enhance the effectiveness of biological treatment (often anaerobic or anoxic). In such events, the first casualty is of achieving "the desirable eco-friendly" wastewater quality at the end of treatment that is safe to be disposed of into the environment or amenable for further reuse. Recall the case study presented in Sect. 1.1, showing the adverse effects of using seemingly safe-to-use water which in reality was contaminated with textile dyeing wastewater on the ecosystem and human health.

Difficulty in Treating the Mixed Dye Wastewaters: The problem worsens when there is a mixture of dyes, as will be the case in any industry based on the ongoing production campaigns. The extent of decolourization and subsequent mineralization of dyes varies greatly with the class of dyes as some bonds may be exceptionally tough to break using biological treatment. Clearly, different dyes will degrade under different redox conditions. This fact is highlighted rather effectively in a study published by [87], where only 44 of the 87 dyes tested in the experiment appeared to be removed to reasonable extent. The proxy parameters used in this study were Dissolved Organic Content (DOC) reduction and decolourization. Thus, the colour removal indeed cannot sufficiently guarantee the destruction of organic content associated with the dye molecules present in wastewater. The mechanism of decolourization in conventional ASP, however, could not be attributed with certainty to biodegradation and biosorption onto the suspended microbial cells in the sludge was conjectured to be a major contributor [87].

A similar observation was made by [23] confirming the relatively significant non-biodegradability of a mixture of dyes. In their study, the textile industry effluents containing mixture of dyes were not completely degraded—even after a long retention time (over 10 days in aeration tank) in aerobic bioreactor during conventional biological wastewater treatment.

Challenges Posed by the Complex Structures of Dye Molecules: Dye compounds are large organic molecules with high molecular weight and multiple bonds at many locations within their structure. This problem is worsened when a mixture of dye molecules is present in high concentrations such as in the case of textile dyeing effluent or effluent from a dye manufacturing industry. It is well known that dye molecules are recalcitrant to biological degradation. Textile effluent containing a mixture of dye compounds does not completely degrade even after 10 days of treatment in aeration tank in the conventional wastewater treatment scheme [23].

It was mentioned before that a single treatment approach may not work for the treatment and decolourization of dye wastewater. In a study reported by [77], the performance of an anaerobic–aerobic treatment scheme for the degradation of Reactive Black 5 (an azo dye) was evaluated. The study concluded that the destruction of the dye was favoured in low Dissolved Oxygen (DO) concentrations and with the addition of external substrate (glucose). The maximum dye decolourization of 96% was observed at a Dissolved Oxygen concentration of 0.5 mg/L and glucose concentration of 2 mg/L with a hydraulic retention time of 48 h. Thus, the action of the enzyme *azoreductase*, led to fragmentation of dye molecule and thereby resulted in decolourization.

It was also observed by [77] that the partially anaerobically treated dye wastewater was prone to "re-colourization" when exposed to air in laboratory studies. Accordingly, it was envisaged that the "re-colourization" occurred due to oxidation of amine group (–HN–NH–) eventually leading to re-formation of the azo group (–N = N–) resulting from dissolution of oxygen from the air. Further, [77] reported that the fragmented and decolourized azo dye metabolites were readily degraded in aerobic laboratory reactor.

The Problem of High TDS: During the process of textile dyeing, inorganic salts like Sodium Chloride or Sodium Sulphate are added to the dye bath. The addition of inorganic salts like Sodium Chloride (NaCl) or Sodium Sulphate (Na_2SO_4) to the dye bath is critical in order to promote the dye fixation onto the fibre. The inorganic salt acts only as a catalyst and is not consumed in the dyeing process. The resulting effect is a high Total Dissolved Solids (TDS) concentration in the dye bath wastewater. The concentration of TDS in the dye bath effluent will vary with multiple factors such as type of dye, dyeing technique, type of fibre, etc. It has been demonstrated by Mirbolooki et al. that the high salinity textile wastewaters hindered the growth of microbes in activated sludge. They performed experiments at a high concentration of dyes of 500 mg/L. For these conditions, the COD removal was 80.71%, 59.44%, and 14.92% for a TDS concentration of 1,000 mg/L, 5,000 mg/L, and 10,000 mg/L respectively. The point of inflection was at the 5,000 mg/L TDS mark, after which, the Mixed Liquor Suspended Solids (MLSS) of the activated sludge kept decreasing—indicating a decrease in the microbial survival [76].

This problem is exacerbated when there are several dyeing operations in progress and their mixed effluent might lead to a high dye concentration and a high TDS concentration in the final mixed wastewater at the biological treatment stage. In general, the activated sludge can tolerate shock loads to a certain extent and the resilience varies with different microbial populations and different dyes. Hence, it is a prudent and reasonable idea to have a sort of "treatment completion allowance after the microbial biodegradation step by making use of CWs". The CWs can function as a polishing step which will only enhance the treatment capabilities of the overall system. The increasing application of nature-based water treatment solutions not only enhances the overall technological effectiveness, but it is also the environmentally defendable course of action.

4   The Horizon of Eco-centric Treatment Technologies

Several research publications, review papers and books, including Arceivala and Asolekar [4, 5] and the *Saph Pani* Handbook [120], have addressed a class of decentralized treatment systems generally referred to as Natural Treatment Systems (NTSs). As described by these authors, there are several kinds of NTSs including variants of constructed wetlands, anaerobic, facultative and aerobic waste stabilization ponds, maturation ponds, fish ponds, water hyacinth ponds, duckweed ponds, and polishing ponds. Typically, NTS is one of the most reliable alternatives to conventional wastewater treatment technology as it is relatively more eco-centric in terms of low capital and operating costs, extremely small carbon footprint and nearly negligible power and chemical requirements [5, 17, 62]. Nature-friendly and pro-society virtues of engineered Constructed Wetland were highlighted by [61, 65, 66, 111]—which is one of the noteworthy examples of the so-called NTSs.

4.1   Significance of Natural Treatment Systems for Sustainable Wastewater Treatment

Wastewater treatment systems incorporating NTSs typically have a line-up of a series of unit operations and unit processes, with the output of one unit subjected to the next unit for rendering a step-wise incremental treatment to wastewater. The first stage generally comprises of the physico-chemical removal of pollutants—which is referred to as the "primary treatment". The second stage is the so-called "secondary treatment" in which a suitable NTS could be employed to achieve the degradation of pollutants via biological pathways. There are instances wherein an advanced physico-chemical treatment process, too, is employed to degrade or fragment rather complex pollutant molecules to enable biological transformations to be effective.

These intricate treatment strategies are especially needed while treating a mixture of complex organic pollutants in industrial wastewaters, so that the pollutants can eventually be removed as sludges, biosolids or altered into transformation products that are hopefully less harmful than the parent compounds [62]. More likely than not, a suitable pre-treatment is recommended for improving the effectiveness of the primary and secondary treatment strategies. It is needless to emphasize that the care exercised in pre-treatment and segregation of waste streams at source goes long way in terms of optimizing capital and operating costs of the unit operations and unit processes. The lives of filtration and RO membranes, catalysts, activated carbon and ion exchange resins can be prolonged with these strategies—which is often overlooked while working single-mindedly towards cost cutting.

In the recent years, the Nature-based Solutions (NbS) is a newer terminology used for eco-centric technologies—which is defined by the International Union for Conservation of Nature [49] as the *"actions to protect, sustainably manage, and restore natural or modified ecosystems, that address societal challenges effectively and adaptively, simultaneously providing human well-being and biodiversity bene-fits"*. Typically, climate change mitigation, pollution abatement in air, land, and water, protection of biodiversity, renaturing the urban environment, and thereby transitioning to resilient landscapes are considered to be some important benefits of the application of NbS [49]. It appears that the NTSs in general and CWs in particular can be considered to be a sub-set of a larger class of decentralized technologies inspired by nature that is referred to as the Nature-based Solutions.

Figure 4 depicts a conceptual flow diagram of wastewater treatment strategy for rendering reuse-oriented "sustainable wastewater treatment" to effluents from production plants and sanitation fractions of industrial wastewater. One of the highlights of the proposed treatment process in Fig. 4 happens to be the pre-treatment given to some selected segregated streams within the production plants (pre-treatment at source). It is envisaged that the segregated streams could be subjected to the primary treatment in conjunction with the mixed effluents generated during production. Further, a combination of treatment using NTS/NbS and Physico-chemical treatment can be given based on the wastewater characteristics to effectively degrade the pollutants.

For sanitation wastewater (sewages generated in toilets, baths, and kitchen), treatment can be rendered by a primary treatment unit followed by secondary treatment using NTS/NbS. Finally, the tertiary treatment or polishing treatment can be given to both, secondary treated domestic as well as secondary treated industrial wastewater by employing another NTS/NbS such as Constructed Wetland. The tertiary treated water can be either reused in the industry or it can be disposed of in the legally designated receiving body.

In sum, a noteworthy feature of NTSs (as well as NbSs) is that they can very effectively be integrated with other biological and physico-chemical treatment methods to provide a sustainable solution. This proposed treatment strategy could have a

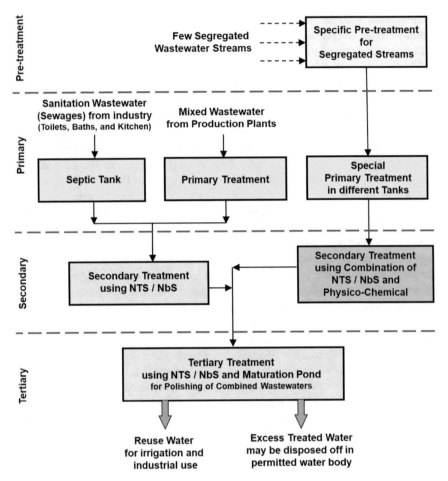

Fig. 4 A conceptual flow diagram of wastewater treatment strategy for rendering reuse-oriented "sustainable wastewater treatment" to effluents from production plants and sanitation fractions of industrial wastewater

lower carbon footprint, electrical energy requirement, and chemical costs. Thus, the proposed industrial effluent treatment strategy could help the industry render wastewater treatment in an eco-friendly manner.

4.2 Use of Engineered CWs for Treatment of Domestic Wastewaters

Broadly, an engineered Constructed Wetland (CW) consists of an open box-like shallow bed filled with river sand (packing media) having suitable planted vegetation in the sand bed. The Horizontal Sub-surface Flow Constructed Wetland (HSSF-CW) is the most widely employed configuration in India, where the wastewater is meant to stay below a layer of unsaturated packing. Prof Asolekar's research group (the senior author of this Chapter) has researched and implemented NTSs including this variant of engineered CWs widely in India [5, 7, 61–66, 111]. [8, 39, 52, 94], too, opined that the HSSF-CWs have a better potential for removing biodegradable contaminants.

Figure 5 presents the schematic representations of the major features of engineered Horizontal Sub-surface Flow Constructed Wetland (HSSF-CW) including vegetation, media, and direction of water flow for three types of media, namely: (a) single media like coarse sand or fine gravel, (b) dual media like coarse gravel at the

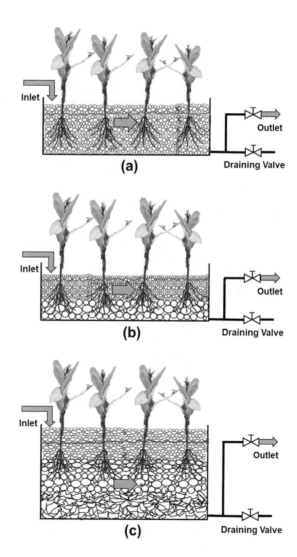

Fig. 5 The schematic representations of the major features of engineered Horizontal Sub-surface Flow Constructed Wetland (HSSF-CW) including vegetation, media, and direction of water flow for three types of media, namely: **a** single media like coarse sand or fine gravel, **b** dual media like coarse gravel at the bottom and smaller media like natural or manufactured sand at the top, and **c** triple media with coarse well-draining media like gravel or coconut shavings at the bottom, medium-sized alumina or gravel or pebbles in the middle, and smaller media like natural or manufactured sand at the top

bottom and smaller media like natural or manufactured sand at the top, and (c) triple media with coarse well-draining media like gravel or coconut shavings at the bottom, medium-sized alumina or gravel or pebbles in the middle, and smaller media like natural or manufactured sand at the top.

In India, several decentralized domestic wastewater treatment facilities in small towns and peri-urban locations employ engineered HSSF-CWs to render secondary biological wastewater treatment or at times for tertiary treatment to enhance the reuse of water. For example, the full-scale treatment facilities incorporating HSSF-CWs were demonstrated in (a) the Town of Mhaswad, State of Maharashtra, having a capacity of 250 m^3/day [65, 66]; (b) the City of Jaipur, State of Rajasthan having capacities of 40 and 60 m^3/d [92], (c) the Town of Katel, State of Maharashtra having a capacity of 18 m^3/d [111], and (d) Kothapally Village, State of Telangana having a capacity of 20 m^3/d [22].

Over the years, several researchers have also investigated CWs in India using HSSF-CWs in pilot-scale facilities [3, 54, 61, 91, 103] and in laboratory-scale reactors [10, 40, 61, 105]. The effectiveness of removal of COD, BOD, TN, and TP from domestic wastewaters in these full-scale, pilot-scale as well as laboratory-scale studies were reported to be in the range 60–90%, 80–95%, 25–75%, and 20–60%, respectively.

Some of the other important variants of engineered constructed wetlands could be the vertical flow constructed wetlands (VF-CWs), free-floating constructed wetlands (FF-CWs) as well as hybrid systems (Hybrid-CWs) [63]. The use of CWs to remove organic contaminants and nutrients from wastewaters (especially domestic wastewaters) is being explored extensively by researchers and also implemented around the world [117].

Figure 6 depicts the schematic representations of the major features of engineered Free Water Surface-flow CW (FWS-CW) and Vertical Flow CW (VF-CW) including vegetation, media, and direction of water flow for three types of media, namely: (a) FWS-CW having a single media like coarse sand or fine gravel and (b) VF-CW having dual media like coarse gravel at the bottom and smaller media like natural or manufactured sand at the top. Further, Fig. 7 presents the schematic representations of the major features of engineered Floating CW (FCW) including vegetation, media, and direction of water flow for three types of media, namely: (a) FCW having a floating bed or mat consisting of single media like lightweight granules of pearlite or coir or synthetic felt and (b) FCW having hydroponic vegetation tied to a floating raft or frame of bamboo or PVC pipes.

4.3 Kinetics of Removal of COD, BOD and Nutrients in CWs

Various researchers have reported kinetic studies and its importance in evidencing mechanisms of pollutant removal processes from wastewater using constructed wetlands [60, 61, 97, 98, 110]. The first-order and pseudo-first-order degradation kinetic models were used to predict the removal performance for pollutants

Fig. 6 The schematic representations of the major features of engineered Free Water Surface-flow CW (FWS-CW) and Vertical Flow CW (VF-CW) including vegetation, media, and direction of water flow for three types of media, namely: **a** FWS-CW having a single media like coarse sand or fine gravel and **b** VF-CW having dual media like coarse gravel at the bottom and smaller media like natural or manufactured sand at the top

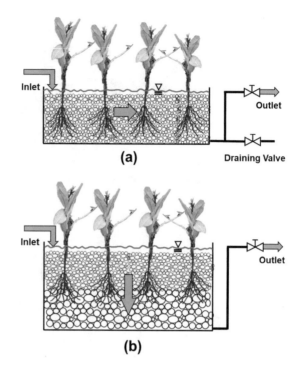

in wastewater using constructed wetlands in laboratory-scale reactors [61, 108, 126], pilot-scale systems [50, 60, 61, 70, 98], and full-scale treatment plants [20, 31, 42, 66, 109].

The Critical Role Played by Plants and Microbes: In constructed wetlands, there is a heterogeneous environment and several microenvironments coexist. Biogeochemical reactions occur, and that favours various metabolic pathways to degrade recalcitrant compounds, and varieties of microbiological communities perform the enzymatic activity to achieve the same [114]. Ramified roots in the constructed wetland and the presence of microorganisms help in creating a microenvironment inside. Aerobic zones near anoxic or anaerobic ones establish oxic or anoxic interfaces in the media through active transport of oxygen through the rhizosphere via plant tissues [43]. Other factors that influence removal are biodegradation, sedimentation, plant exudates, uptake, and sorption into media [47].

Sorption mainly occurs on the surface of media, sediment, or particulate organic matter. A compound's affinity is determined by environmental pH, redox potential, stereochemical structure, and the chemical nature of the sorbent and sorbed molecule. Furthermore, in wetland beds, recalcitrant compounds with a planar aromatic structure show intercalation between the layers of solids [114]. One other mechanism is

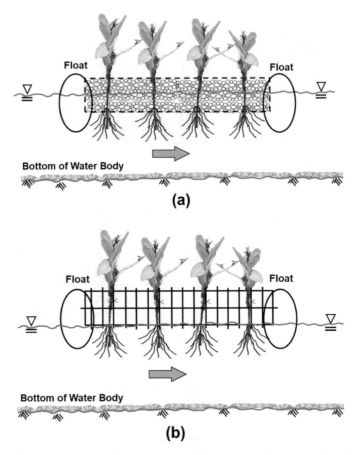

Fig. 7 The schematic representations of the major features of engineered Floating CW (FCW) including vegetation, media, and direction of water flow for three types of media, namely: **a** FCW having a floating bed or mat consisting of single media like lightweight granules of pearlite or coir or synthetic felt and **b** FCW having hydroponic vegetation tied to a floating raft or frame of bamboo or PVC pipes

the lipid portion of suspended solids or the lipophilic cell membrane of microorganisms which aids in the sorption of refractory substances by hydrophobic interactions between aliphatic and aromatic groups. The electrostatic interaction of negatively charged microbe surfaces with positively charged pharma chemical groups also contributes in their elimination [114].

Oxygen exudation into the rhizosphere by the macrophytes in the wetland bed facilitates enhanced removal through the prevalence of aerobic conditions in the subsurface by promoting biochemical pathways [43]. Furthermore, microbial activity in the vicinity of the root zone is significantly higher than that of other parts in the wetland bed [12]. It was also reported the fate of recalcitrant compounds in CWs could be directly or indirectly affected by active plant uptake [95].

The Kinetics of Pollutant Degradation: The kinetic studies are significant to evidence the mechanisms of pollutant removal in CWs [5, 110]. Kinetic rate constants reported for the removal of organics and nutrients in full-scale HSSF-CW are very limited. First-order kinetic rate constants for chemical oxygen demand (COD) and total nitrogen (TN) removal in full-scale HSSF-CW treating domestic wastewater (95 m^3/d) in Nunavut, Canada was studied by [42]. Rate constants for COD and TN were reported as 106 m/y and 20 m/y, respectively. A variety of native vegetation was used in this full-scale treatment plant, including Bryophyte spp., Hippuris vulgaris, Carex spp., and Salix spp as reported. Further, investigation of full-scale HSSF-CW planted with *Typha latifolia* and *Schoenoplectus americanus* in Greensboro, North Carolina, reported rate constants of 4.45 m/y and 1.67 m/y for TN and total phosphorous (TP), respectively [109]. Nitrate nitrogen and ammonia nitrogen rate constants reported in full-scale CW in Beijing, China, were 27 m/y and 14 m/y, respectively [20].

Further, the kinetics of organics and nutrient removal in laboratory-scale or batch scale CW were extensively investigated around the globe. Rugaika et al. [98] reported COD rate constant of 0.692 d^{-1} for pilot-scale HSSF-CW using *Typha latifolia* vegetation at Tanzania. Rate constants of 7 m/y, 4.1 m/y, and 1.6 m/y for Total Kjeldahl Nitrogen and Total Phosphorous, respectively, as found in the study conducted in pilot-scale HSSF-CWs at Nova Scotia, Canada [50]. Another study reports ammonia nitrogen (0.17 d^{-1}) and orthophosphate (0.215 d^{-1}) rate constants in pilot-scale HSSF-CWs with *Canna* vegetation in Taiwan [70].

Batch and continuous mode operation of lab-scale hybrid flow constructed wetland treating municipal wastewater was studied by [41]. *Pistia stratiotes*, *Typha angustifolia* and *Water Hyacinth* were the macrophytes used in the study and the first-order volumetric rate constant and areal rate constant for BOD removal in HSSF-CW were reported to be 0.28–0.31 d^{-1} and 0.11–0.12 m/d, respectively for the retention period of 4 days. Similarly, the volumetric rate constants for COD, TKN, and TP were 0.05–0.06 d^{-1}, 0.13⁻0.14 d^{-1}, and 0.11⁻0.14 d^{-1}, respectively. Further, the areal rate constants for COD, TKN and TP were 0.02 m/d, 0.05–0.06 m/d, and 0.05–0.06 m/d, respectively [41].

In India, few studies have reported rate constants for full-scale constructed wetlands treating domestic wastewaters. Kinetic studies conducted in "CW4Reuse" technology, a HSSF-CW based decentralized domestic wastewater treatment system (250 m^3/d), in Mhaswad, India by Lekshmi et al. [66] reported the rate constants of organics and nutrients removal. The volumetric rate constants for COD, BOD, TKN and TP were observed to be 0.0553 ± 0.0071 h^{-1}, 0.0802 ± 0.017 h^{-1}, 0.0485 ± 0.019 h^{-1}, and 0.0275 ± 0.0197 h^{-1}, respectively for the retention period of 24 h [66]. Organics and nutrients removal was evaluated in HSSF-CW by [51] in southern India employed for treatment of domestic wastewater generated from a school complex (350 L/d). Author reports first-order removal rate constants for COD, BOD, TN and TP as 0.24 d^{-1}, 0.30 d^{-1}, 0.16 d^{-1}, and 0.07 d^{-1}, respectively.

In sum, very high rates of removal of organics and nutrients were witnessed by employing constructed wetlands for the treatment of domestic wastewaters. Variations of the rate of removal in these systems may be attributed to the different operational conditions, the difference in vegetation, substrate, etc. and other system dynamics.

5   Significance of CWs for Treatment of Industrial Wastewaters

Global applications of subsurface flow, surface flow, vertical flow, and hybrid constructed wetlands systems for treatment of wastewaters from industries such as textile, refinery, pulp and paper, tannery, aquaculture, winery, distillery, brewery, abattoir, dairy and cheese, olive mill, seafood, sugar, potato, laundry, chemical industry, explosives, steel, wood waste, coke plant, coal gasification, tool industry, and floriculture farm were critically reviewed by [118]. However, wastewaters from certain industrial sectors including textile, tannery, oil and refinery, food industry, pulp and paper, as well as acid mine drainage could have typically rather high concentrations of organic matter (COD ranging 5,000 mg/L–1,20,000 mg/L), oil and grease (up to 1,800 mg/L), salinity (even up to 80,000 mg/L), or highly acidic or alkaline conditions and presence of recalcitrant and toxic pollutants. This can indeed pose a significant challenge for the treatment of such wastewaters with the help of Natural Treatment Systems in general and CWs in particular. As pointed out by [122], such extreme conditions, too, can be treated using engineered CWs, provided a suitable primary treatment and a secondary treatment comprising of appropriate advanced physico-chemical treatment or anaerobic biological treatment is given before employing engineered CWs to give an incremental treatment to secondary treated wastewaters.

Passive operation and strong buffering capacity to high organic load are some of the features of CWs for application in industrial effluents. Pre-treatment, including settling, dissolved air floatation, and pH adjustments, was recommended by [122] for improvement of the performance of CWs in the instances of applications for industrial wastewaters. They also reported the significant role played by other operational modifications including the addition of external carbon source and artificial aeration to mitigate poor oxygen transfer for ensuring smooth operation of engineered CWs because there are certain deficiencies in the case of industrial effluents and they need to be supplemented and complimented before the wastewaters are subjected to CWs for treatment of industrial effluents for better performance. Augmentation of engineered CW with bacterial endophytes for treatment of wastewater from textile industry [44], CW augmented with microbial fuel cell (CW-MC) for treating saline wastewater [123], hybrid-constructed wetlands for fertilizer wastewater treatment [74] are the recent advancements in CWs for treating industrial effluents.

5.1   Removal of Industrial Pollutants Using Constructed Wetlands

In recent years, CWs have been explored for their application in the removal of industrial pollutants. However, these studies were mostly narrowed to laboratory-scale and pilot-scale. Few significant studies on the removal of industrial pollutants using constructed wetlands have been discussed below.

Industrial cork boiling wastewater was effectively treated using HSSF-CW, and the average removal was achieved for COD—75%, BOD_5—91.7%, and Total Phenols—69.1%, as reported by [36]. However, 5 ± 1 days of HRT was required for treating the wastewater in laboratory-scale reactors attributed to the low biodegradability of cork boiling effluent. Though low colour removal (35%) was reported in this study, higher removal of total phenolic compounds was established using HSSF-CWs when compared with the conventional biological treatment methods such as activated sludge process or anaerobic digestion.

Concentrated livestock wastewaters when treated with CWs were reported to achieve 65%, 53%, 42%, and 42% for BOD, TSS, TN, and TP [59]. CWs also achieved significant removal of organics and nutrients in dairy wastewater. The average removal of COD, BOD, TKN, and TP for dairy wastewater is reported to be 68.7%, 57.9%, 25.7%, and 29.9% [101]. Further, tannery industrial wastewater was studied for its treatment in HSSF-CW, and the performance was reported to be 99.83% removal for chromium [53]. García-Valero et al. [32] have also demonstrated the removal of 78% TP and 48% Cr from tannery wastewater using HSSF-CWs. [93] revealed pulp and paper mill wastewater treatment using constructed wetland with an average efficiency of removal of 86.6% and 80% for COD and BOD in the summer season. These systems could achieve significant removal with higher HRTs (>3 day). Moreover, industrial wastewaters often pose a threat to the health of wetland biota, therefore, adequate pre-treatment is recommended while treating industrial effluents using CWs.

5.2   Case Studies on Co-disposal of Industrial and Domestic Wastewaters

Traditionally, in the processes of urbanization, several small and cottage-scale industries have congregated in cities. As a result, urban domestic wastewaters got mixed with partially treated wastewaters from industries within the city limits. This phenomenon is observed all over the world. Field-scale applications of engineered constructed wetlands in full-scale wastewater treatment facilities (for a mixture of industrial and domestic wastewaters) are increasingly being practised globally as well as in India. In this context, few case studies that have demonstrated the treatment of urban effluents on a large scale in India are discussed below.

Karnal type—constructed wetland for treatment of wastewater at Ujjain, Madhya Pradesh, Central India: A soil–plant bed is treating 1.79 MLD wastewater at Ujjain, Madhya Pradesh. Reportedly, the performance of the system is reported to be satisfactory. This technology has also gained attention due to the feedstock production for the pulp and paper industry through waste management [62, 64].

Constructed wetland for treatment of 0.05 MLD wastewater at Katchpua slum, City of Agra, Uttar Pradesh, India: HSSF-CW with a capacity of 0.05 MLD treating urban wastewater from the Katchpua slum was studied by Kumar et al. [63]. Annual performance of the treatment system was reportedly 61%, 64%, 90–99% removal for BOD, COD, and FC, respectively.

Constructed wetland for treatment of 7.8 MLD wastewater at Mansagar Lake, City of Jaipur, Rajasthan, India: HSSF-CW was employed for tertiary treatment for the 7,800 m^3/day wastewater generated in the metropolitan City of Jaipur. As reported by Asolekar et al. [7] and Kumar et al. [63], the maximum performance of the system for Mansagar Lake rejuvenation in Jaipur is 96% BOD removal, 87% COD removal, and 99.9% FC removal.

6   Use of Constructed Wetlands in Dye Wastewater Management

Constructed Wetlands (CWs) have conventionally found their application in treatment of highly biodegradable wastewaters like municipal wastewaters (as discussed in Sect. 4). As discussed in Sect. 5, engineered Constructed Wetlands can also be incorporated in industrial wastewater treatment systems receiving wastewaters from refinery, pulp and paper, tannery, textile, aquaculture, winery, distillery, brewery, abattoir, dairy and cheese, olive mill, seafood, sugar, potato, laundry, chemical industry, explosives, steel, wood waste, coke plant, coal gasification, tool industry, and flower farms. In recent times, CWs are being explored for the treatment of wastewaters from textile industry or dye manufacturing industry [55, 56, 113, 116, 118, 122].

These industrial applications of CWs have become possible due to certain innate virtues of the "ecosystems" present in wetlands including sedimentation, adsorption of pollutants at the media-water interface as well as interfaces of biofilms and plant roots, microbial degradation, and plant uptake. Further, the engineered CWs are considered to be versatile and are amenable to the improvement of treatment efficiencies by incorporating a variety of wastewater flow regimes within the wetland beds in conjunction with a range of geometric shapes and depths of the beds. In summary, the unique possibilities are offered by incorporating the engineered CWs in the treatment systems for industrial wastewaters by carefully employing the suitable degradation kinetic regimes achieved by the consortium of plants and microorganisms in combination with the appropriate flow regimes within the wetland beds.

It can be seen from the multiple studies presented in the subsequent sections that CWs possess commendable dye decolourization and degradation capabilities. Among the several bioremediation technologies, the unique feature of CWs is the use of combination of vegetation and microorganisms. Vegetation and the associated rhizospheric microorganisms play a critical role in pollutant degradation with the help of microbially secreted enzymes facilitating the uptake and metabolic processing of fragments of the complex molecules. For example, [80] reported that extent of decolourization of dye wastewaters via uptake by vegetation is greatly affected by the size of molecule. Among other performance indicators, the studies reported excellent nutrient removal, especially Nitrogen, which can also be attributed to the uptake by vegetation. It has now been unequivocally established that vegetation plays a considerable role in dye wastewater treatment via the uptake of dyes [29].

6.1 Constructed Wetlands for Secondary Treatment

Various experimental researchers have investigated the kinetics of removal of dyes from synthetic wastewaters or real industrial effluents with the help of laboratory reactors, pilot plants, and also with the help of real life engineered CWs installed for industrial wastewater treatment. Accordingly, Table 2 presents a summary of the recent literature on use of engineered Constructed Wetlands for the treatment of dye-bearing wastewaters. The Table organizes the categorical findings for the removal of a variety of dyes with the help of different types of CWs, vegetation and media. It also presents the salient observations and results on hydraulic retention time, decolourization efficiencies as well as other pollutant removal efficiencies.

It can be observed from Table 2 that there are numerous studies in the literature depicting the use of CWs for dye wastewater treatment. CWs are a complex micro-ecosystem and several mechanisms contribute to the degradation of pollutant molecules. It is well understood by now that the treatment performance of CWs is governed by interplay of numerous system parameters and environmental factors.

Although the main component of these studies is examining the decolourization and degradation of dyes, most studies have taken a holistic approach and also monitored the progressive changes in other associated water quality indicators such as aggregated organic parameters like Chemical Oxygen Demand (COD), Biochemical Oxygen Demand (BOD), Total Organic Carbon (TOC), physical parameters like Total Suspended Solids (TSS), Total Dissolved Solids (TDS), and nutrients such as Ammoniacal Nitrogen (NH_4-N), Phosphates (PO_4-P), and Sulphates (SO_4^{2-}).

Difficulty in Treating Mixture of Dyes: Most studies highlighted in Table 2 have examined the performance of decolourization and degradation of simulated dye wastewaters containing a single type of dye molecule. The highlighted studies have covered major types of dyes such as Reactive dyes, Disperse dyes, Acid dyes, Basic dyes, and Vat dyes. Some studies have also examined the performance of CW systems for real dye wastewater [57, 78, 99, 102]. It is well known that the treatment efficiency of dye wastewaters in CWs decreases with increasing complexity of the

Table 2 Summary of recent literature on use of Constructed Wetlands for treatment of dye-bearing wastewaters

Sr. no	Dye (s) treated	Source of Wastewater	Configuration and type of wastewater treatment units	Vegetation in CW	Media in CW	Hydraulic retention time (hour)	Colour removal efficiency	Pollutant removal	References
1	Reactive Red 22, Vat Red 13, and Reactive Black 5	Synthetic	VFCW	Unspecified	Sand, Gravel, Peat, and Zeolite	24	Absorbance: 40–70%	COD: ≈ 88%	[81]
2	Reactive Black 5, Disperse Yellow 211, and Vat Yellow 46	Synthetic dye-bath wastewater (0.03 g/L each)	2 VFCW + 1 HSSF CW	*Phragmites australis*	Gravel, Coarse Sand, and Fine Sand	24	Absorbance: 89–93%	COD: ≈ 84% TOC: ≈ 89% SO_4^{2-}: ≈ 88% T-N: ≈ 52% Org-N: ≈ 87%	[13]
3	Acid Orange 7 (AO7)	Synthetic (50–100 mg/L)	Aerated and Non-aerated Up-flow CWs	*Phragmites australis*	Gravel supported on glass beads	72–144	Absorbance: 90–95% % w/w AO7: 94–98%	COD: 74–90% T-N: 44–67% T-P: 12–28% NH_4-N: 24–99% NO_3-N: 21–99%	[83] and [82]
4	Reactive Red 2, (RR2) Reactive Red 120 (RR120), and Reactive Red 141 (RR141)	Synthetic (20 mg/L each dye)	Hydroponic FCW	*Echinodorus cordifolius L*	(1) with clay soil base and (2) no media	168	% w/w Dyes: RR2: ≈ 97% RR120: ≈ 92% RR141: ≈ 88%	pH: 7.3–7.7 TDS: 42%	[80]
5	Unspecified Azo dye	Synthetic (11.5 mg/L)	FWS CW + HSSF	*Phragmites australis*	Shale	96	Absorbance: 90–98%	COD: 82–92%	[21]

(continued)

Table 2 (continued)

Sr. no	Dye (s) treated	Source of Wastewater	Configuration and type of wastewater treatment units	Vegetation in CW	Media in CW	Hydraulic retention time (hour)	Colour removal efficiency	Pollutant removal	References
6	Disperse Red BF, Disperse Yellow G, Disperse Bryal Blue, Rubine GFL, and Brown REL	Real textile effluent and Synthetic (50 mg/L)	VFCW	*Portulaca grandiflora*	Fine Gravel, Coconut Shavings, and Soil	48–72	Absorbance: 73–89%	COD: ≈ 73% BOD: ≈ 54% TOC: ≈ 52% Turbidity: ≈ 57% TDS: ≈83% TSS: ≈ 71%	[57]
7	Mixture of dyes	Real textile wastewater	VFCW + HSSF CW	*Phragmites australis, Dracaena sanderiana, and Asplenium platyneuron*	Sugarcane Bagasse and Sylhet Sand	211	Absorbance: 61–79%	Turbidity: 83–90% TSS: 38–65% NH_4-N: 70–80% NO_3-N: 72–76% COD: ≈ 89% BOD: 95–96%	[99]
8	Mixture of dyes	Real textile effluent	VFCW	*Typha domingensis*	Coconut Shavings, Gravel, Sand, Soil	72	Colour 68–92%	COD: ≈ 80% BOD: ≈ 77% TSS: 13–27%	[102]
9	Active Brilliant Red X-3B (ABRX3)	Synthetic (100–500 mg/L)	VFCW + MFC with GAC as electrodes	*Ipomoea aquatic*	Gravel, scree, glass beads, and biological ceramics	36–96	Absorbance: ≈ 93%	COD: ≈ 86%	[28]
10	Indigo Vat dye	Synthetic	VFCW	*Canna indica and Typha angustifolia*	Sand, Gravel, and Zeolite	270	Absorbance: 85–98%	COD: ≈ 62%	[25]
11	Yellow 2G	Synthetic	VFCW	*Canna indica and Typha angustifolia*	Sand, Gravel, and Zeolite	270	Absorbance: 94–99%	COD: 47–76% NH_4-N: 23–54% PO_4-P: 80–95% SO_4^{2-}: 38–58%	[24]

(continued)

Table 2 (continued)

Sr. no	Dye (s) treated	Source of Wastewater	Configuration and type of wastewater treatment units	Vegetation in CW	Media in CW	Hydraulic retention time (hour)	Colour removal efficiency	Pollutant removal	References
12	Acid Blue 113 (AB113) and Basic Red 46 (BR46)	Synthetic (7 mg/L and 218 mg/L)	VFCW	*Phragmites australis*	Pea Gravel and Large Gravel	48–96	BR46: 82–97% AB113: 68–80% Mixture of BR46 & AB113: $\approx$ 68%	COD: 50–89% NH_4-N: 11–66% NO_3-N: 45–75% PO_4^{3-}: 27–89%	[45, 46]
13	Mixture of dyes	Discharge from dye manufacturing	FWS (11 ponds)	*Sp. Lemna, Typha latifolia, E. crassipes,* and *Phragmites*	NA	192	NA	COD: $\approx$ 53% SO_4^{2-}: $\approx$ 88% S^{2-}: $\approx$ 63% TSS: $\approx$ 28% TDS: $\approx$ 70%	[78]
14	Acid Red 18 (AR 18), Acid Orange (AO 7), and Congo Red (CR)	Synthetic (200 mg/L)	VFCW + MFC with Carbon felt as electrodes	*Typha latifolia*	Glass Beads, Gravel,	768–1,080	AR 18: $\approx$ 96% AO 7: $\approx$ 67% CR: $\approx$ 60%	COD: $\approx$ 75%	[84]

COD: Chemical Oxygen Demand; BOD: Biochemical Oxygen Demand; TSS: Total Suspended Solids; T-N: Total Nitrogen; TDS: Total Dissolved Solids; NH_4-N: Ammoniacal Nitrogen; PO_4-P: Phosphates; S^{2-}: Sulphides; SO_4^{2-}: Sulphates; TOC: Total Organic Carbon; VFCW: Vertical Flow Constructed Wetland; HSSF CW: Horizontal Sub-Surface Flow Constructed Wetland; FWS CW: Free Water Sub-Surface Constructed Wetland; FCW: Floating Constructed Wetland; MFC: Microbial Fuel Cell

mixture of dyes as different dye molecules will exhibit different levels and types of pollution loads. For example, [13] carried out experiments using simulated dye bath wastewaters with two azo dyes (Reactive Black 5 and Disperse Yellow 211) and an anthraquinone dye (Vat Yellow 46) in a laboratory-scale vertical flow CW. Although the dye concentrations in all three simulated wastewaters were 0.03 g/L, the COD exhibited by Disperse Yellow 211 and Vat Yellow 46 were significantly higher than Reactive Black 5, and consequently, the COD removal efficiency for the two dyes was also lower than that for Reactive Black 5. Thus, focused and customized treatment for dye wastewaters may prove to be more efficient.

Role of Vegetation and Media: As seen in Table 2, numerous combinations of various species of vegetation and CW bed media have been explored with *Phragmites* and *Typha* being the predominantly employed vegetation species and sand and gravel as the predominantly employed CW bed media. Studies such as Noonpui and Thiravetyan [80], Dogdu and Yalcuk [24, 25], Khandare et al. [57], and Nawab et al. [78] have studied the use of other species such as Sp. Lemna, E. crassipes, *Echinodorus cordifolius L, Canna Indica*, and *Portulaca grandiflora* as well.

A study on the treatment of Reactive Red 141 containing wastewater by [79] has found that the use of emergent species like narrow-leaved cattails (*Typha angustifolia* Linn) for dye wastewater treatment is relatively unexplored. They found that the CW system planted with *Typha angustifolia* Linn supported on sand and gravel bed gave adequate removal of COD, TDS, and colour as 59%, 86%, and 58%, respectively.

The Issue of high TDS: In the above study, the TDS decreased from 1,328 mg/L to 453 mg/L in the course of the 15 days HRT. The surprisingly high removal of TDS is attributed to uptake by plants. Real-life textile effluents also have high TDS because of the salts used in the dye baths. Thus, the CW system in this study has effectively proven its capability in dealing with salinity. However, prolonged exposure to high salinity wastewaters may be detrimental for the plants as evidenced by physical observations like wilting and necrosis in the study.

Issues Associated with Removal of COD and Nutrients: The decrease in COD can be attributed to the fact that the large-sized and high molecular weight dye molecules get entrapped within the interstitial vascular spaces of plants, adsorbed on the media, and also degraded by microbial action [13, 21, 78, 79, 81]. In addition to COD and colour removal, the CW systems also showed remarkable efficiency in nutrient removal whose major mechanism is adsorption onto bed media and plant uptake [45, 46, 78, 82, 83]. One of the other advantages of using CWs is their superior pH-buffering capabilities [13, 24, 45, 79, 81].

Degradation of Azo Dyes in CWs: It is well known that azo dyes are the most common type of dyestuff used in the industry for dyeing accounting for more than 50% of the worldwide dye production every year [67]. It is not surprising that most studies on dye degradation using engineered Constructed Wetlands have been performed with azo dyes as the candidate dye molecule. The performance of CWs for treatment of a wide range of input dye concentrations have been examined from as low as 7 mg/L to as high as 784 mg/L. It should also be noted that most studies use deeper Vertical Flow Constructed Wetland (VFCW) beds because the lower portions of the beds develop facultative and anaerobic microbial ecosystem. It was discussed earlier in Sect. 3 that azo dye degradation is favoured in anaerobic conditions through the action of the enzyme *azoreductase*. It is the considered opinion of the authors of this Chapter that such facultative and anaerobic microbial ecosystems will also get developed in case of deeper HSSF CW beds.

6.2   Constructed Wetlands for Tertiary Treatment

The use of engineered CWs for treatment of dye-bearing wastewaters has been explored by several researchers as seen in Sect. 6.1 (Table 2). Some researchers have also explored the use of engineered CWs for the tertiary treatment of dye-bearing wastewaters i.e. incremental treatment imparted after secondary treatment using either biological or physico-chemical methods.

The use of engineered CWs for tertiary treatment addresses the concerns of dye wastewater treatment using conventional methods (refer to Sect. 3). Since no single treatment method is capable of complete treatment of dye wastewater, there is a possibility of the presence of residual dyes and other associated pollutants in the secondary treated water. The use of engineered CWs for further treatment of this wastewater provides the wastewater with adequate residence time within the biological system and enables further removal of the pollutants.

FCWs for Tertiary Treatment: A study by [124] explored the applicability of employing Floating Constructed Wetlands with *Lemna minor* as the vegetation for tertiary treatment of simulated dye wastewater. To examine the role of vegetation, the study compared the performance of an unplanted control FCW and a planted FCW. The planted system gave significantly higher performance in decolourization than the unplanted system, highlighting the major role of vegetation in treatment. The highest reported dye removal for a mixture containing Reactive Black 198 and Basic Red 46 was reported to be about 50%. The planted system was able to effectively reduce the COD, Suspended Solids, Total Dissolved Solids and nutrients like NH_4-N and NO_3-N. Thus, FCWs could be implemented for treatment and removal of residual dye as well as other pollutants.

Tertiary Treatment for Heavy Metal Removal: The dyestuff compounds used in textile dyeing are usually bound with metals forming dyeing complexes. These metals are an integral part for the dyeing process in the textile industry. Copper, Lead, Zinc, Cobalt, and Chromium are among the commonly used metals in textile dyeing complexes. The presence of these metals in the textile effluent will of course vary with the type of dye used and the seasonal demands, but of these metals, Chromium (Cr) is the most prevalent in textile effluent. This is because Basic dyes, Direct dyes, and Reactive Dyes use Cr as a metal complexing agent, and the share of these dyes in the textile dyeing industry is large [115].

The most commonly used biological treatment for dyestuff and textile wastewater is the Activated Sludge Process, which may not be capable of completely removing the heavy metals such as Cr. The remediation of Cr in biological treatment takes place via biosorption, bioaccumulation, and biodegradation [34].

As per the study conducted by [30], engineered CWs can be used to remove the residual Cr from the ASP system treating textile wastewater. They used a Horizontal Sub-Surface Flow Constructed Wetland (HSSF CW) with *P. australis* as the vegetation. The system had gravel as the media and had a hydraulic retention time of 1.8 days. The textile wastewater was subjected to a treatment plant consisting of primary settling, activated sludge process, secondary treatment, clariflocculation, and ozonation as the tertiary treatment before sending it to the engineered CW. It was reported by the authors that hexavalent Chromium (Cr^{6+}) concentration is higher in the presence of Ozonation along with the ASP. The study concludes that the HSSF CW system is capable of removing about 70% of Cr^{6+} and 40–50% Total Chromium.

In sum, the existing literature supports the case of using CWs for dye wastewater treatment but its full-scale application to the treatment of real dye wastewater is still a work in progress. Clearly, there is tremendous scope for integrating engineered CWs in industrial dye wastewater treatment strategies with suitable pre-treatment and primary treatment.

7 Integration of Constructed Wetlands in Dye Wastewater Management

It can be gathered from Sects. 5 and 6 that engineered Constructed Wetlands are a robust and sustainable wastewater treatment technology that can be employed in treatment of dye-bearing wastewaters. For successful industrial application, however, there is a need for integrating engineered CWs with the suitable pre-treatment, primary treatment as well as secondary physico-chemical advanced treatment. The crux of treatment of dye-bearing wastewaters in CWs is in harmonizing the interplay between the dyes and associated pollutants with the physio-biological possibilities through selection of appropriate vegetation, media, retention time, and microbial consortium. This Section aims to present a conceptual strategy for rendering targeted treatment to separated dilute wastewater streams.

7.1 Significance of Wastewater Segregation in Textile Sector

As seen in Fig. 3 in Sect. 2.2, the major processes in textile dyeing operations are desizing, scouring, bleaching, mercerizing, dyeing, printing, and finishing. The textile cloth is also given a washing after particular steps in order to remove any excess chemical compounds from that step in the process. Usually, in small-scale industries, the wet processing of textiles are batch processes. As a result, the process vessels must be cleaned after every batch so as to not carry forward any contamination which leads to generation of vessel washing effluent. Both these streams, viz. textile washing and vessel washing, together constitute the "wash effluent" from the wet processing of textile. Since these are simply wash waters, it is expected that they are considerably low pollutant concentration wastewater streams.

Multiple studies have underlined the importance of segregation of various wastewater streams from different steps in wet processing of textiles with the aim of enhancing the subsequent reuse of certain streams within the process [19, 38, 85, 96, 125]. It was also argued that this strategy will, in fact, lead to appreciable water saving. These studies employ the Best Available Techniques (BAT) approach to analyse the reuse potential of all the wastewater streams based on their chemical characteristics.

For example, Ozturk et al. [86] stated that through reuse of the segregated wastewater streams in the textile mill, the savings in total water consumption could be to the extent of 46–50% after implementing the so-called "BAT approach". They also reported that recycling of the washing and rinsing wastewater (direct recycle + recycle after treatment) could additionally yield a saving of 15–16% in freshwater consumption. Thus, the segregation of wastewater streams from textile wet processing is advantageous as it makes the dilute fractions of wastewater easier to treat and subsequently reuse or recycle.

It is well known that the water consumption in textile wet processing depends on the type of cloth as well as the equipment employed and the variety of operations involved for the production of the respective product. For example, if the finished fabric involves, both, dyeing and printing, the amount of wastewater and the pollution loads would be more when compared with the product requiring only dyeing or only printing.

In any case, for a given technology and the type of fabric, the primary parameter that influences water consumption is the Material to Liquor Ratio (MLR) as reasoned by Raja et al. [90]. From the operational point of view, every stage in the processing of textile has a specified MLR corresponding to the type of equipment and fabric. A similar analysis has been presented by [9], who also estimate that approx. 70% of the total water is consumed in the multiple washing steps in textile processing.

Illustrative Estimates for Washings Segregated in Dyeing of Woven Fabric: An illustrative example has been worked out here to highlight the significance of segregating wastewaters from various washing operations in the processing of woven fabric. Table 3 depicts the indicative quantities of the pollution loads corresponding to 1,000 kg batch of wet processing of fabric. The volumes of wastewater and the corresponding pollution concentrations and loads are estimated in this study to gain the appreciation of typical pollution generated during washing steps performed at the end of each stage in wet processing of textile dyeing and printing. The authors of this chapter gratefully acknowledge the valuable contribution of the data on Material to Liquor Ratio (MLR) in various steps of textile wet processing sourced from Raja et al. [90]. Also, the reference data on emissions corresponding to the various washing steps are sourced from [33]. Table 3 is inspired by Table 26 in [33].

The results presented based on estimates made by the authors of this chapter will enable help in evaluating the pollution loads emerging from washing operations performed after each step during wet processing of fabrics. The overall pollution load corresponding to each washing is estimated on the basis of the hydraulic load for the respective step and the corresponding COD concentration.

Potential for Constructed Wetlands Aimed at Sustainable ...

Table 3 The indicative quantities of the pollution loads (basis: 1,000 kg batch of fabric) generated during washing steps performed at the end of each stage in wet processing of textile dyeing and printing; estimated in this study. The reference data on emissions corresponding to the various washing steps are sourced from [33]

Sr. No	Wash effluents in textile dyeing process	Estimated volume[a] (L)	Typical wash effluent characteristics[b]			Estimated step-wise pollution load[a, b]
			pH^b	COD^b (mg/L)	BOD^b (mg/L)	COD (Kg/batch)
1	After bleaching	10,000	8–9	50–100	10–20	0.5–1
2	After acid rinsing	10,000	6.5–7.6	120–250	25–50	1.2–2.5
3	After dyeing (hot wash)	10,000	7.5–8.5	300–500	100–200	3–5
4	After dyeing (acid & soap wash)	10,000	7.5–8.64	50–100	25–50	0.5–1
5	After dyeing (final wash)	10,000	7–7.8	25–50	–	0.25–0.5
6	Printing washing	10,000	8–9	250–450	115–150	2.5–4.5
7	Blanket washing of rotary printer	10,000	7–8	100–150	25–50	1–1.5
	Total	70,000	7.1–8.1	128–229	43–74	8.95–16

[a]Estimated for processing 1,000 kg fabric based on 1:10 MLR as reported in Raja et al. [90]
[b]Values reported in [33]

As reported in Table 3, the cumulative wastewater resulting from several washings during wet processing of 1,000 kg cloth adds up to 70,000 L. Thus, if all the wash effluents were combined the effective pollutant load would have been between 8.95 kg and 16 kg in 70,000 L of wastewater. When the wash effluents are considered individually, we can see the effect of segregation of wastewaters. The wash effluent after bleaching has a volume of 10,000 L and COD concentration of 50–100 mg/L. Thus, by multiplying volume and concentration, the pollution load from wash effluent after bleaching is estimated which works out to be 0.5–1 kg in 10,000 L. Similarly, the wash effluent after acid rinsing has a pollution load of 1.2–2.5 kg in 10,000 L. The wash effluent generated after dyeing in a hot wash and soap and acid wash have pollution loads of 3–5 kg in 10,000 L and 0.5–1 kg in 10,000 L, respectively. The final wash effluent has a pollution load of 0.25–0.5 kg in 10,000 L. The wash effluent after printing has a pollution load of 2.5–4.5 kg in 10,000 L. Lastly, the wash effluent blanket washing of the rotary printer has a pollution load of 1–1.5 kg in 10,000 L.

7.2 Potential for Using Constructed Wetlands for Treating Segregated Streams

It is well known that wastewaters containing high concentrations of mixtures of dyes and associated pollutants are challenging to treat. It is argued by the authors of this Chapter that based on the analysis presented in Sect. 7.1 (Table 3), it is evident that the segregation of dilute rinsing and washing wastewaters from the other concentrated wastewater streams in textile wet processing offers a possibility of treating them separately using suitable technologies.

Accordingly, it is proposed in this research that these rinsing and washing wastewaters (mixed dilute stream) can be suitably treated with the help of engineered CWs. Alternatively, the mixed concentrated stream may be first subjected to the appropriate advanced physico-chemical process to fragment, decolourize, and detoxify the xenobiotic dye molecules in secondary treatment. Further, the resulting partially treated wastewater may be subjected to engineered CW for rendering incremental secondary treatment.

The effort is made here to establish the benefit of implementing the "preventive environmental management" (PEM) as endorsed by [6] to enhance water reuse and recycle on one hand and also achieve waste minimization and cleaner production on the other. The crux of the so-called PEM strategy is in integrating water conservation and reuse with the production process as well as in implementing eco-friendly nature-based solutions through incorporating engineered CWs for rendering biological treatment. Thus, this research illustrates the significance of incorporating natural treatment technology by replacing the conventional biological wastewater treatment like activated sludge process or anaerobic contact reactor—which have rather large carbon footprint on account of the requirement of chemicals and electrical power.

As discussed earlier in Sect. 3, it is well known that high concentration mixed dye wastewaters are difficult to treat using conventional biological treatment methods. After segregation, the dilute wastewaters will have much lower concentrations of dyes and the associated pollutants and it is expected that they will be fairly in the range where CW systems can handle the load.

Upon spending the stipulated residence time in the CW, the treated water may be reused or disposed of to the legally designated receiving body. The following strategies explore the potential of the use of engineered CWs for dye wastewater treatment in the dyestuff as well as textile processing industry:

CWs for Dilute Wastewaters in Dyestuff Industry: The segregated dilute wastewater streams in the dyestuff production process are mainly the reactor cleaning wastewater and the filter washing wastewater as depicted in Fig. 2. The schematic flow diagram of a conceptual strategy for secondary treatment of the segregated dilute wastewater streams generated in the dyestuff manufacturing process using engineered Constructed Wetlands and advanced physico-chemical treatment for the segregated concentrated streams (the first stage of secondary treatment) is presented in Fig. 8. This strategy manages to separate the dilute wash waters containing low concentrations of dyes and associated pollutant molecules and subject them directly

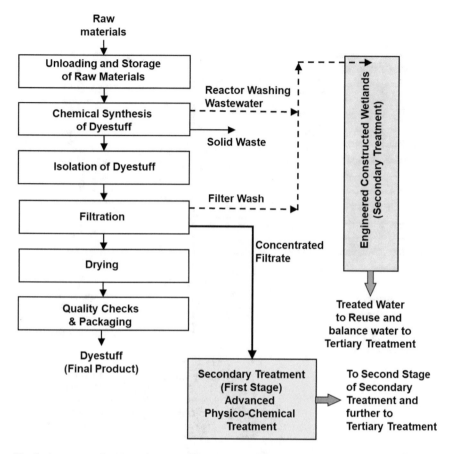

Fig. 8 A conceptual strategy for secondary treatment of the segregated dilute wastewater streams generated in the dyestuff manufacturing process using engineered Constructed Wetlands and advanced physico-chemical treatment for the segregated concentrated streams (the first stage of secondary treatment)

to treatment in CW beds. After the secondary treatment, this treated water can be used for reactor washing, filter washing, floor flushing, or for several other purposes in the industry.

CWs for Dilute Wastewaters in Textile Processing Industry: The segregated dilute wastewater streams in textile processing are mostly the large amounts of washing wastewaters generated after every processing stage as depicted in Fig. 3. The schematic flow diagram of a conceptual strategy for secondary treatment of the segregated diluted wastewater streams generated in the textile processing using engineered Constructed Wetlands and advanced physico-chemical treatment for the segregated concentrated streams (the first stage of secondary treatment) is presented in Fig. 9. This strategy manages to separate the diluted wash waters containing low concentrations of dyes and associated pollutant molecules and subjecting them

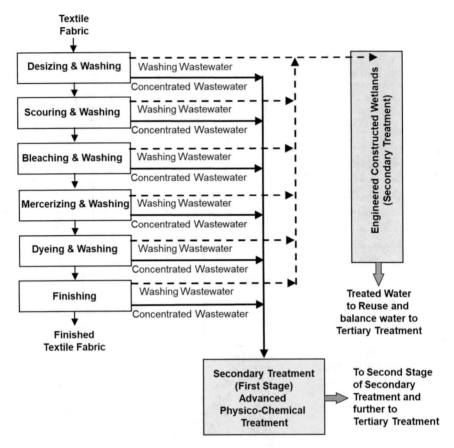

Fig. 9 A conceptual strategy for secondary treatment of the segregated dilute wastewater streams generated in the textile processing using engineered Constructed Wetlands and advanced physico-chemical treatment for the segregated concentrated streams (the first stage of secondary treatment)

directly to treatment in CW beds. After the secondary treatment, this treated water can be used for reactor washing, filter washing, floor flushing or for several other purposes in the industry.

7.3 Advanced Treatment Technologies for Concentrated Fractions of Dye Wastewaters

As described in Sect. 7.2, the proposed strategies for the treatment of segregated dilute wastewater streams in dyestuff industry and in textile processing industry were depicted in Figs. 8 and 9, respectively. One of the high points of these proposed strategies is in use of engineered CWs for rendering secondary treatment to these segregated dilute wastewaters. However, the use of NTS may not be possible for treatment of the other mixed concentrated streams from dyestuff and textile processing industry—owing to the xenobiotic nature of dye molecules and the reported difficulty of biodegradation of the mixture of dyes.

Therefore, a hybrid approach is recommended in this research; wherein the first stage of secondary treatment could comprise of a suitable advanced physico-chemical treatment process and the resulting effluent can possibly be subjected to Coagulation and Flocculation. If required, additionally, filtration (or sedimentation) could be provided to remove the settleable and suspended solids. A limited objective of this sub-section is to present a brief discussion on some of the commonly used advanced treatment technologies for treatment of dye-bearing wastewaters:

Membrane Filtration: Membrane Filtration is a novel and modern technology, recently gaining traction in the dye-related industries, which is capable of separating molecules in water based on their molecular size. It is well known that dye molecules have large molecular sizes making them easy to separate using physical separation such as in membrane filtration. The dye wastewaters are usually highly coloured and have large amounts of low molecular weight inorganic salts as part of their dye baths in textile processing or reaction liquor in dyestuff manufacturing.

The membranes are capable of retaining most of the dye molecules while allowing the other low molecular weight compounds to pass through. Thus, the reject stream from membrane filtration applied to dye wastewaters becomes the more concentrated stream which will be subjected to further treatment and the permeate stream becomes the low concentration water which has the potential to be reused after some treatment [26, 104]. In addition, the low concentration streams contain most of the inorganic salt and thus, membrane separation facilitates in recovery of salt in the form of the permeate stream. The permeate stream will be significantly less coloured than the reject stream as the membrane will sufficiently prevent the escape of dye molecule into the permeate. Usually, Reverse Osmosis (RO) and Nanofiltration (NF) membrane filtration systems are employed for separation of dye wastewater into concentrated and dilute steams. In recent times, the integration of membrane filtration systems into the dyestuff production processes has eliminated the use of precipitation and filtration of dyes. Thus, membrane filtration systems are advanced technologies that can be used in the treatment stage as well as be integrated directly into the dyestuff production processes [58].

Fenton's Treatment: Fenton's treatment is an oxidation process widely used in decolourization of dye-containing wastewater. The Fenton Treatment makes use of the combination of Fe^{2+} and H_2O_2 (Fenton's Reagent) for achieving decolourization along with COD reduction through oxidation of the toxic organic pollutants. This method has extensively been applied to wastewater treatment of dye wastewaters in the dye and dye intermediate sector [69].

Consequently, Fenton's treatment can also be used for treating textile wastewater. Since dye compounds are fairly resistant to mineralization through biodegradation, a chemical oxidant like Fenton's Reagent works well in decolourizing dyes. The performance of the Fenton Treatment, typically, is governed by the interplay of a variety of parameters including the inlet concentration of dyes, pH, concentration of FeSO$_4$, and concentration of H$_2$O$_2$. Although the chemicals for Fenton's reagent are relatively cheap and easily available, there is usually a slim operating range. The complexity of the system is known to be extremely high when one is addressing a mixture of dyes. Further, another serious issue to be addressed is the lack of calibration of addition of quantity of H$_2$O$_2$. To say the least, the Fenton Treatment is known to be a complex web of inorganic and organic reactions. The net result of Fenton's treatment would be the destruction of substitution on the organic molecules (referred to as the "chromophores") resulting in decolourization of the wastewater in addition to COD reduction. The Fenton-treated wastewater, however, will have a low pH, high TDS, and some residual COD. There have also been studies which integrate Fenton's treatment with advanced photochemical processes like UV [48, 71, 72].

Other Advanced Oxidation Processes: Advanced Oxidation Processes (AOPs) are chemical processes that produce the oxidizing entity in-situ with the addition of certain chemical agents and oxidation promoters. These oxidizing entities are extremely unstable and do not exist in the free state for more than a couple of seconds. They are capable of attacking the colour-imparting bonds in the dye molecule and breaking them, leading to decolourization. AOPs usually comprise some combination of oxidizing generating species. The most common setup for AOPs is some combination of Ozonation, Ultraviolet irradiation, and H$_2$O$_2$. The highest efficiency decolourization is obtained when all three are used together [18].

AOPs generate hydroxyl radicals which oxidize the organic matter and lead to their decolourization and degradation. Although AOPs are capable of decolourizing the dye-bearing wastewaters in a short span of time, they alone are insufficient in adequately treating the wastewater to the regulatory limits. Apart from this, the associated electrical costs are high and the efficiency of decolourization and degradation for a mixture of dyes is low. The "optimum conditions" for various dye compounds might be different according to their respective kinetics of degradation. Thus, AOPs are extremely capable and a powerful means to treat dye wastewater, but they usually cannot be used as a standalone treatment strategy [11, 88]. This in fact, strengthens the case for source-separated wastewater treatment, whose significance was made clear in the earlier section.

Coagulation and Flocculation: Coagulation and Flocculation are two distinct steps performed in succession in order to remove suspended material in wastewater treatment. Coagulation is a chemical process that involves changing surface charges of suspended material and in effect lead to destabilization of the solution. Flocculation is a physical operation involving the addition of a flocculant and slow stirring in order to agglomerate the suspended material into clusters. These clusters are eventually settled in a separate tank and the suspended material is removed.

Coagulation and Flocculation treatment for dye wastewaters is preferred in the case of insoluble dyes as the removal efficiency is relatively lesser for soluble dyes. Coagulation and Flocculation are capable in reduction of dye concentration and thereby bring about decolourization. Thus, as a result of the removal of dyes, the "difficult to biodegrade" organic content in the wastewaters is reduced making it more conducive to biological treatment downstream. Multiple studies have examined combinations of different coagulants and flocculants and their results have shown successful dye removal and decolourization, and in some cases, even complete decolourization has been observed [1, 35, 100]. These studies have concluded that although a particular combination of coagulant and flocculant may work very well in decolourization of one dye, it is unlikely that the same combination may work for another dye. As a result, Coagulation and Flocculation become challenging for a mixture of dyes and depends largely on the selection of the appropriate combination of coagulant and flocculant [68].

Finally, at the end of the first stage, the filtered effluent may be subjected to engineered CWs for rendering incremental treatment to the segregated concentrated stream (second stage of secondary treatment).

7.4 A Conceptual Framework for Preventive Environmental Management

By and large, the conventional approach for treating dye-bearing wastewaters consists of combining the wastewaters from all steps and providing treatment to the entire mixed wastewater in the centralized wastewater treatment plant. In contrast, in this research, the so-called "Preventive Environmental Management" (PEM) approach is recommended—which incorporates the greener strategies such as "segregation" of concentrated wastewaters from the dilute washing streams as well as the use of engineered CWs (a natural treatment system).

Accordingly, Fig. 10 depicts the integrated effluent segregation, treatment, and reuse strategy for dye-bearing wastewaters from the textile processing industry; inspired by the so-called PEM approach. The high point of this approach is in adopting the greener strategies such as segregation and rendering separate treatment to the dilute washing streams by carefully giving the suitable pre-treatments to each stream and then subjecting them to an engineered Constructed Wetland (eco-friendly secondary biological treatment).

The resulting treated water could be of reuse quality and thereby minimize the water footprint appreciably. In the events of persistence of pollutant molecules after the secondary treatment, the treated water may further be subjected to tertiary treatment in another engineered CW (refer to Fig. 10). The present state-of-art of designing CWs offers possibilities for improving treatment efficiencies through carefully selected vegetation, media, flow regimes, and creation of the combination of aerobic, facultative, and anaerobic zones in wetland beds.

It is envisaged in the strategy proposed in Fig. 10 that the concentrated wastewaters from the textile processing industry would be treated by adopting a hybrid approach. As explained earlier (refer to Sect. 7.3), it is recommended in this research that a suitable advanced physico-chemical treatment process (the first stage) and engineered CWs (the second stage) shall be employed to render secondary treatment.

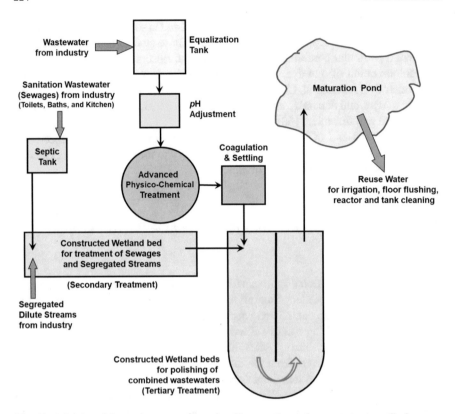

Fig. 10 Adoption of the greener strategies such as "segregation and separate treatment" of concentrated wastewaters from the dilute washing streams in the textile processing industry and by incorporating engineered Constructed Wetlands (a natural treatment system) to enhance recycle and reuse of treated water and minimize the water footprint, carbon footprint, and consumption of chemicals and electrical energy

Subsequently, as shown in Fig. 10, the secondary treated wastewaters from treatment of dilute as well as concentrated streams shall be combined in a separate engineered CW for rendering tertiary treatment. The engineered CW employed for tertiary treatment shall be suitably designed to achieve further removal of dyes as well as associated pollutants. Finally, the treated water from the final CW can be stored in a Maturation Pond—which "polishes" the treated water, acts as an additional buffer layer, and also gives opportunity for storage for the routine and occasional reuse of treated water. This treated water can be reused for reactor washing, filter washing, floor flushing or at multiple other places and the remaining unused water could be discharged to the legally designated receiving body.

Incorporating a robust and eco-friendly NTS such as engineered Constructed Wetlands for dye wastewater treatment in the textile processing industries, thus, provides several benefits such as enhancement of recycle and reuse of treated water and thereby minimizing the water footprint, carbon footprint, and consumption of chemicals and electrical energy. It need not be re-emphasized that an analogous approach may be employed for treatment of dye-bearing wastewaters in the dyestuff industries.

8   Summary and Conclusions

India has a robust textile industrial sector—which is the ready market for dyestuff production base. Thus, the dyestuff and textile processing sectors are part of a well-knit industrial ecosystem having close market relations and an appreciable share in India's international trade. However, it is rarely appreciated that their respective environmental concerns, too, are comparable. It is therefore envisaged in this research that these industries can indeed be mutually inspired by the solutions developed for treatment and reuse of dye-bearing wastewaters.

(1)   It is well known that decolourization and degradation of mixture of dyes is rather difficult using the conventional biological treatment processes. The carbon footprint, requirements of chemicals, and electrical power are typically high for advanced physico-chemical treatment processes leading to high capital as well as operating costs. To break this impasse, a logical alternative could be to segregate wastewater streams in production plants and render appropriate treatment to them separately and make efforts to minimize the costs of environmental protection.

(2)   The engineered constructed wetlands are typically classified on the basis of planted vegetation, system architecture and flow paths. A wetland bed may have a free water surface flow on top of the media or a subsurface flow or a combination of the two flow patterns. The movement of water inside the media can be horizontal or vertical. Alternatively, the vegetation in the engineered CW can be floating, submerged under water, or rooted in the bed having a shallow unsaturated layer on top of the bed and the roots are in the saturated zone underneath the unsaturated layer.

(3)   The industrial applications of CWs have become possible due to certain innate virtues of the "ecosystems" present in wetlands including sedimentation, adsorption of pollutants at media-water interface as well as interfaces of biofilms and plant roots, microbial degradation, and plant uptake.

(4)   In this research, yet another innovative approach is proposed; which is inspired by the principles of so-called "Preventive Environmental Management". The crux of this approach hinges on a four-pronged strategy, namely:

   (a)   at-source segregation of dilute and concentrated wastewater streams during textile wet processing,

   (b)   rendering separate targeted pre-treatment to dilute wastewaters and further subject the streams to suitably designed engineered Constructed Wetlands and achieve recycling and reuse of water,

(c)   rendering a specialized secondary treatment to the segregated concentrated wastewaters by combining advanced physico-chemical treatment with incremental biological treatment in engineered Constructed Wetlands, and

(d)   it may be possible to enhance reuse and recycling of water by further providing tertiary treatment using Natural Treatment System such as engineered Constructed Wetlands.

(5)   It is recommended in this research that adoption of this integrated greener strategy will enhance recycle and reuse of treated water on one hand and minimize the water footprint, carbon footprint, and consumption of chemicals and electrical energy on the other. Clearly, the judicious combination of physico-chemical and biological wastewater treatment for the two segregated wastewater streams would minimize the treatment costs significantly.

(6)   In addition, incorporation of the NTS (may also be referred to as Nature-based Solution) makes the treatment strategy eco-centric and innovative. In all, the liability of treatment of wastewater generated by textile wet processing is converted into an opportunity by the production of "reuse-quality water" through the adoption of the so-called PEM approach.

References

1. Allegre C, Maisseu M, Charbit F, Moulin P (2004) Coagulation—flocculation—decantation of dye house effluents : concentrated effluents. J Hazard Mater 116(1–2):57–64. https://doi.org/10.1016/j.jhazmat.2004.07.005
2. Allègre C, Moulin P, Maisseu M, Charbit F (2006) Treatment and reuse of reactive dyeing effluents. J Membr Sci 269(1–2):15–34. https://doi.org/10.1016/j.memsci.2005.06.014
3. Álvarez JA, Ávila C, Otter P, Kilian R, Istenič D, Rolletschek M, Molle P, Khalil N, Ameršek I, Mishra VK, Jorgensen C, Garfi A, Carvalho P, Brix H, Arias CA (2017) Constructed wetlands and solar-driven disinfection technologies for sustainable wastewater treatment and reclamation in rural India: SWINGS project. Water Sci Technol 76(6):1474–1489. https://doi.org/10.2166/wst.2017.329
4. Arceivala SJ, Asolekar SR (2012) Environmental studies: a practitioner's approach. McGraw Hill Education (India) Pvt. Ltd., New Delhi
5. Arceivala SJ, Asolekar SR (2018) Wastewater treatment for pollution control (3rd edn). McGraw Hill Education (India) Pvt. Ltd., New Delhi (Originally published in 2006)
6. Asolekar SR, Gopichandran R (2005) Preventive environmental management—an indian perspective. Foundation Books Pvt. Ltd., New Delhi (the Indian associate of Cambridge University Press, UK)
7. Asolekar SR, Kalbar PP, Chaturvedi MKM, Maillacheruvu K (2014) Rejuvenation of rivers and lakes in India: balancing societal priorities with technological possibilities. In: Ahuja A (ed) Comprehensive water quality and purification. Elsevier, USA, Waltham, pp 181–229. https://doi.org/10.1016/B978-0-12-382182-9.00075-X
8. Aydın Temel F, Avcı E, Ardalı Y (2018) Full scale horizontal subsurface flow constructed wetlands to treat domestic wastewater by *Juncus acutus* and *Cortaderia selloana*. Int J Phytorem 20(3):264–273. https://doi.org/10.1080/15226514.2017.1374336

9. Balachandran S, Rudramoorthy R (2008) Efficient water utilisation in textile wet processing. Journal of the Institution of Engineers (India), Part TX: Textile Engineering Division, 89(August), pp 26–29

10. Benny C, Chakraborty S (2020) Continuous removals of phenol, organics, thiocyanate and nitrogen in horizontal subsurface flow constructed wetland. J Water Process Eng 33:101099. https://doi.org/10.1016/j.jwpe.2019.101099

11. Bessegato GG, De Souza JC, Cardoso JC, Zanoni MVB (2018) Assessment of several advanced oxidation processes applied in the treatment of environmental concern constituents from a real hair dye wastewater. J Environ Chem Eng 6(2):2794–2802. https://doi.org/10.1016/j.jece.2018.04.041

12. Brimecombe MJ, Leij FAAM, Lynch JM (2001) Rhizodeposition and microbial populations. In: Pinton R, Varanini Z, Nannipieri P (eds) The rhizosphere: biochemistry and organic substances at the soil-plant interface. Marcel Dekker, New York, pp 74–98

13. Bulc TG, Ojstršek A (2008) The use of constructed wetland for dye-rich textile wastewater treatment. J Hazard Mater 155(1–2):76–82. https://doi.org/10.1016/j.jhazmat.2007.11.068

14. Central Pollution Control Board (2011) Final report on inventorization of 17 category/GPI/red category industries. https://cpcb.nic.in/cepi-technical-reports/

15. Central Pollution Control Board (2014) Standards for emission or discharge of environmental pollutants from various industries. https://cpcb.nic.in/effluent-emission/

16. Central Pollution Control Board (2016) Standards for emission or discharge of environmental pollutants from various industries. https://cpcb.nic.in/effluent-emission/

17. Chaturvedi MKM, Asolekar SR (2009) Wastewater treatment using natural systems: the Indian experience. In: Nair J, Furedy C (eds) Technologies and management for sustainable biosystems. Nova Science Publishers, ISBN: 978–1–60876–104–3

18. Chung J, Kim JO (2011) Application of advanced oxidation processes to remove refractory compounds from dye wastewater. Desalin Water Treat 25(1–3):233–240. https://doi.org/10.5004/dwt.2011.1935

19. Correia VM, Stephenson T, Judd SJ, Correia VM, Stephenson T, Judd SJ (1998) Characterisation of textile wastewaters: a review. 15(10):917–929. https://doi.org/10.1080/09593339409385500

20. Cui L, Li W, Zhang Y, Wei J, Lei Y, Zhang M, Pan X, Zhao X, Li K, Ma W (2016) Nitrogen removal in a horizontal subsurface flow constructed wetland estimated using the first-order kinetic model. Water 8(11):514. https://doi.org/10.3390/w8110514

21. Cumnan S, Yimrattanabovorn J (2012) The use of constructed wetland for azo dye textile wastewater. Int J Civil Eng Build Mater 2(4):150–158. http://green.sut.ac.th/wp-content/uploads/2013/10/the-use-of-constructed-wetland-for-azo-dye.pdf

22. Datta A, Singh HO, Raja SK, Dixit S (2021) Constructed wetland for improved wastewater management and increased water use efficiency in resource scarce SAT villages: a case study from Kothapally village, in India. Int J Phytoremediation 1–10. https://doi.org/10.1080/15226514.2021.1876627

23. Davies TH, Cottingham PD (1994) The use of constructed wetlands for treating industrial effluent (textile dyes). Water Sci Technol 29(4):227–232. https://doi.org/10.2166/wst.1994.0197

24. Dogdu G, Yalcuk A (2016) Evaluation of the treatment performance of lab-scaled vertical flow constructed wetlands in removal of organic compounds, color and nutrients in azo dye-containing wastewater. Int J Phytorem 18(2):171–183. https://doi.org/10.1080/15226514.2015.1073672

25. Dogdu G, Yalcuk A (2016) Indigo dyeing wastewater treatment by eco-friendly constructed wetlands using different bedding media. Desalin Water Treat 57(32):15007–15019. https://doi.org/10.1080/19443994.2015.1070290

26. Donkadokula NY, Kola AK, Naz I, Saroj D (2020) A review on advanced physico-chemical and biological textile dye wastewater treatment techniques. Rev Environ Sci Biotechnol 19(3):543–560. https://doi.org/10.1007/s11157-020-09543-z

27. Dubey SK, Yadav R, Chaturvedi RK, Yadav RK, Sharma VK, Minhas PS (2010) Contamination of ground water as a consequence of land disposal of dye waste mixed sewage effluents: a case study of Panipat District of Haryana, India. Bull Environ Contam Toxicol 85(3):295–300. https://doi.org/10.1007/s00128-010-0073-2

28. Fang Z, Song HL, Cang N, Li XN (2015) Electricity production from Azo dye wastewater using a microbial fuel cell coupled constructed wetland operating under different operating conditions. Biosens Bioelectron 68:135–141. https://doi.org/10.1016/j.bios.2014.12.047

29. Ferreira RA, Duarte JG, Vergine P, Antunes CD, Freire F, Martins-Dias S (2014) Phragmites sp. physiological changes in a constructed wetland treating an effluent contaminated with a diazo dye (DR81). Environ Sci Pollut Res 21(16):9626–9643. https://doi.org/10.1007/s11 356-014-2988-3

30. Fibbi D, Doumett S, Colzi I, Coppini E, Pucci S, Gonnelli C, Lepri L, Del Bubba M (2011) Total and hexavalent chromium removal in a subsurface horizontal flow (h-SSF) constructed wetland operating as post-treatment of textile wastewater for water reuse. Water Sci Technol 64(4):826–831. https://doi.org/10.2166/wst.2011.548

31. Gajewska M, Skrzypiec K, Jóźwiakowski K, Mucha Z, Wójcik W, Karczmarczyk A, Bugajski P (2020) Kinetics of pollutants removal in vertical and horizontal flow constructed wetlands in temperate climate. Sci Total Environ 718:137371. https://doi.org/10.1016/j.scitotenv.2020. 137371

32. García-Valero A, Martínez-Martínez S, Faz N, Terrero MA, Muñoz MN, Gómez-López MD, Acosta JA (2020) Treatment of wastewater from the tannery industry in a constructed wetland planted with phragmites Australis. Agronomy 10(2):176. https://doi.org/10.3390/agronomy1 0020176

33. Ghaly AE, Ananthashankar R, Alhattab M, Ramakrishnan V (2013) Production, characterization and treatment of textile effluents: a critical review. J Chem Eng Process Technol 05(01):1–18. https://doi.org/10.4172/2157-7048.1000182

34. Ghosh A, Dastidar MG, Sreekrishnan TR (2017) Bioremediation of chromium complex dyes and treatment of sludge generated during the process. Int Biodeterior Biodegradation 119:448–460. https://doi.org/10.1016/j.ibiod.2016.08.013

35. Golob V, Vinder A, Simonič M (2005) Efficiency of the coagulation/flocculation method for the treatment of dyebath effluents. Dyes Pigm 67(2):93–97. https://doi.org/10.1016/j.dyepig. 2004.11.003

36. Gomes AC, Silva L, Albuquerque A, Simões R, Stefanakis AI (2018) Investigation of lab-scale horizontal subsurface flow constructed wetlands treating industrial cork boiling wastewater. Chemosphere 207:430–439. https://doi.org/10.1016/j.chemosphere.2018.05.123

37. Gregory P (2004) Dyes and dye intermediates. In: Kirk-othmer concise encyclopedia of chemical technology (5th edn, vol 9, pp 230–300). https://doi.org/10.1002/0471238961

38. Güyer GT, Nadeem K, Dizge N (2016) Recycling of pad-batch washing textile wastewater through advanced oxidation processes and its reusability assessment for Turkish textile industry. J Clean Prod 139:488–494. https://doi.org/10.1016/j.jclepro.2016.08.009

39. Haddis A, van der Bruggen B, Smets I (2020) Constructed wetlands as nature based solutions in removing organic pollutants from wastewater under irregular flow conditions in a tropical climate. Ecohydrol Hydrobiol 20(1):38–47. https://doi.org/10.1016/j.ecohyd.2019.03.001

40. Haritash AK, Sharma A, Bahel K (2015) The potential of *Canna lily* for wastewater treatment under Indian conditions. Int J Phytorem 17(10):999–1004. https://doi.org/10.1080/15226514. 2014.1003790

41. Haydar S, Anis M, Afaq M (2020) Performance evaluation of hybrid constructed wetlands for the treatment of municipal wastewater in developing countries. Chin J Chem Eng 28(6):1717–1724. https://doi.org/10.1016/j.cjche.2020.02.017

42. Hayward J, Jamieson R (2015) Derivation of treatment rate constants for an arctic tundra wetland receiving primary treated municipal wastewater. Ecol Eng 82:165–174. https://doi. org/10.1016/j.ecoleng.2015.04.086

43. Hijosa-Valsero M, Matamoros V, Martín-Villacorta J, Bécares E, Bayona JM (2010) Assessment of full-scale natural systems for the removal of PPCPs from wastewater in small communities. Water Res 44(5):1429–1439. https://doi.org/10.1016/j.watres.2009.10.032

44. Hussain Z, Arslan M, Malik MH, Mohsin M, Iqbal S, Afzal M (2018) Treatment of the textile industry effluent in a pilot-scale vertical flow constructed wetland system augmented with bacterial endophytes. Sci Total Environ 645:966–973. https://doi.org/10.1016/j.scitot env.2018.07.163

45. Hussein A, Scholz M (2017) Dye wastewater treatment by vertical-flow constructed wetlands. Ecol Eng 101:28–38. https://doi.org/10.1016/j.ecoleng.2017.01.016

46. Hussein A, Scholz M (2018) Treatment of artificial wastewater containing two azo textile dyes by vertical-flow constructed wetlands. Environ Sci Pollut Res 25(7):6870–6889. https://doi.org/10.1007/s11356-017-0992-0

47. Imfeld G, Braeckevelt M, Kuschk P, Richnow HH (2009) Monitoring and assessing processes of organic chemicals removal in constructed wetlands. Chemosphere 74(3):349–362. https://doi.org/10.1016/j.chemosphere.2008.09.062

48. Islam S, Shaikh IA, Firdous N, Ali A, Sadef Y (2019) A new approach for the removal of unfixed dyes from reactive dyed cotton by Fenton oxidation. J Water Reuse Desalination 9(2):133–141. https://doi.org/10.2166/wrd.2019.011

49. IUCN (2016) Nature-based solutions to address global societal challenges. In: Cohen-Shacham E, Walters G, Janzen C, Maginnis S (eds) IUCN international union for conservation of nature. https://doi.org/10.2305/IUCN.CH.2016.13.en

50. Jamieson R, Gordon R, Wheeler N, Smith E, Stratton G, Madani A (2007) Determination of first order rate constants for wetlands treating livestock wastewater in cold climates. J Environ Eng Sci 6(1):65–72. https://doi.org/10.1139/s06-028

51. Jamwal P, Raj AV, Raveendran L, Shirin S, Connelly S, Yeluripati J, Richards S, Rao L, Helliwell R, Tamburini M (2021) Evaluating the performance of horizontal sub-surface flow constructed wetlands: a case study from southern India. Ecol Eng 162:106170. https://doi.org/10.1016/j.ecoleng.2021.106170

52. Kadlec RH, Wallace S (2009) Treatment wetlands. CRC Press

53. Kaseva ME, Mbuligwe SE (2010) Potential of constructed wetland systems for treating tannery industrial wastewater. Water Sci Technol 61(4):1043–1052. https://doi.org/10.2166/wst.2010.474

54. Khan NA, el Morabet R, Khan RA, Ahmed S, Dhingra A, Alsubih M, Khan AR (2020) Horizontal sub surface flow constructed Wetlands coupled with tubesettler for hospital wastewater treatment. J Environ Manage 267:110627. https://doi.org/10.1016/j.jenvman.2020.110627

55. Khan S, Nawab J, Waqas M (2020) Constructed wetlands: a clean-green technology for degradation and detoxification of industrial wastewaters. Bioremediation Indus Waste Environ Saf. https://doi.org/10.1007/978-981-13-3426-9_6

56. Khandare RV, Govindwar SP (2015) Phytoremediation of textile dyes and effluents: current scenario and future prospects. Biotechnol Adv 33(8):1697–1714. https://doi.org/10.1016/j.biotechadv.2015.09.003

57. Khandare RV, Kabra AN, Kadam AA, Govindwar SP (2013) International biodeterioration & biodegradation treatment of dye containing wastewaters by a developed lab scale phytoreactor and enhancement of its efficacy by bacterial augmentation. Int Biodeterior Biodegradation 78:89–97. https://doi.org/10.1016/j.ibiod.2013.01.003

58. Kim TH, Park C, Kim S (2005) Water recycling from desalination and purification process of reactive dye manufacturing industry by combined membrane filtration. J Clean Prod 13(8):779–786. https://doi.org/10.1016/j.jclepro.2004.02.044

59. Knight RL, Payne VW, Borer RE, Clarke RA, Pries JH (2000) Constructed wetlands for livestock wastewater management. Ecol Eng 15(1–2):41–55. https://doi.org/10.1016/s0925-8574(99)00034-8

60. Konnerup D, Koottatep T, Brix H (2009) Treatment of domestic wastewater in tropical, subsurface flow constructed wetlands planted with Canna and Heliconia. Ecol Eng 35(2):248–257. https://doi.org/10.1016/j.ecoleng.2008.04.018

61. Kumar D, Asolekar SR (2016) Experiences with laboratory and pilot scale constructed wetlands for treatment of sewages and effluents. In: Wintgens T, Nattorp A, Elango L, Asolekar SR (eds) Natural water treatment systems for safe and sustainable water supply in the indian context: saph pani. IWA Publishing, Gb, London, pp 143–160

62. Kumar D, Sharma SK, Asolekar SR (2016) Constructed wetlands and other engineered natural treatment systems: India status report. In: Wintgens T, Nattorp A, Elango L, Asolekar SR (eds) Natural water treatment systems for safe and sustainable water supply in the indian context: saph pani. IWA Publishing, Gb, London, pp 127–142

63. Kumar D, Sharma SK, Asolekar SR (2016) Significance of incorporating constructed wetlands to enhance reuse of treated wastewater in India. In: Wintgens T, Nattorp A, Elango L, Asolekar SR (eds) Natural water treatment systems for safe and sustainable water supply in the indian context: saph pani. IWA Publishing, Gb, London, pp 161–176

64. Kumar D, Asolekar SR, Sharma SK (2015) Post-treatment and reuse of secondary effluents using natural ltreatment systems: the Indian practices. Environ Monitoring Assessment 187(10). https://doi.org/10.1007/s10661-015-4792-z

65. Lekshmi B, Sharma S, Sutar RS, Parikh YJ, Ranade DR, Asolekar SR (2020) Circular economy approach to women empowerment through reusing treated rural wastewater using constructed wetlands. In: Ghosh SK (ed) Waste management as economic industry towards circular economy. Springer, Singapore, pp 1–10. https://doi.org/10.1007/978-981-15-1620-7_1

66. Lekshmi B, Sutar RS, Ranade DR, Parikh YJ, Asolekar SR (2020) Enhancement of water reuse by treating wastewater in constructed wetlands: minimization of nutrients and fecal coliform. In: Reddy KR, Agnihotri AK, Aksoy YY, Dubey BK, Bansal A (eds) Sustainable environmental geotechnics. Springer, Cham, Switzerland, pp 213–223. https://doi.org/10.1007/978-3-030-51350-4_23

67. Li J, Zhu K, Li R, Fan X, Lin H, Zhang H (2020) The removal of azo dye from aqueous solution by oxidation with peroxydisulfate in the presence of granular activated carbon: performance, mechanism and reusability. Chemosphere 259:127400. https://doi.org/10.1016/j.chemosphere.2020.127400

68. Liang CZ, Sun SP, Li FY, Ong YK, Chung TS (2014) Treatment of highly concentrated wastewater containing multiple synthetic dyes by a combined process of coagulation/flocculation and nanofiltration. J Membr Sci 469:306–315. https://doi.org/10.1016/j.memsci.2014.06.057

69. Lin SH, Lo CC (1997) Fenton process for treatment of desizing wastewater. Water Res 31(8):2050–2056. https://doi.org/10.1016/S0043-1354(97)00024-9

70. Lin YF, Jing SR, Lee DY, Wang TW (2002) Nutrient removal from aquaculture wastewater using a constructed wetlands system. Aquaculture 209(1–4):169–184. https://doi.org/10.1016/s0044-8486(01)00801-8

71. Liu X, Qiu M, Huang C (2011) Degradation of the reactive black 5 by Fenton and Fenton-like system. Procedia Engineering 15:4835–4840. https://doi.org/10.1016/j.proeng.2011.08.902

72. Lucas MS, Peres JA (2007) Degradation of reactive black 5 by fenton/UV-C and ferrioxalate/H_2O_2/solar light processes. Dyes Pigm 74(3):622–629. https://doi.org/10.1016/j.dyepig.2006.04.005

73. Madhav S, Ahamad A, Singh P, Mishra PK (2018) A review of textile industry: wet processing, environmental impacts, and effluent treatment methods. Environ Qual Manage 27:31–41. https://doi.org/10.1002/tqem.21538

74. Maine MA, Sanchez GC, Hadad HR, Caffaratti SE, Pedro MDC, Mufarrege MM, Di Luca GA (2019) Hybrid constructed wetlands for the treatment of wastewater from a fertilizer manufacturing plant: microcosms and field scale experiments. Sci Total Environ 650:297–302. https://doi.org/10.1016/j.scitotenv.2018.09.044

75. Masters K (1988) Spray drying of dyestuffs and pigments. J Soc Dyers Colour 104:79–85. https://doi.org/10.1111/j.1478-4408.1988.tb01148.x

76. Mirbolooki H, Amirnezhad R, Pendashteh AR (2017) Treatment of high saline textile wastewater by activated sludge microorganisms. J Appl Res Technol 15(2):167–172. https://doi.org/10.1016/j.jart.2017.01.012

77. Mohanty S, Dafale N, Rao NN (2006) Microbial decolorization of reactive black-5 in a two-stage anaerobic-aerobic reactor using acclimatized activated textile sludge. Biodegradation 17(5):403–413. https://doi.org/10.1007/s10532-005-9011-0

78. Nawab B, Esser KB, Jenssen PD, Nyborg ILP, Baig SA (2018) Technical viability of constructed wetland for treatment of dye wastewater in Gadoon industrial estate, Khyber Pakhtunkhwa, Pakistan. Wetlands 38(6):1097–1105. https://doi.org/10.1007/s13157-016-0824-x

79. Nilratnisakorn S, Thiravetyan P, Nakbanpote W (2009) A constructed wetland model for synthetic reactive dye wastewater treatment by narrow-leaved cattails (*Typha angustifolia* Linn.). Water Sci Technol 60(6):1565–1574. https://doi.org/10.2166/wst.2009.500

80. Noonpui S, Thiravetyan P (2011) Treatment of reactive azo dye from textile wastewater by burhead (*Echinodorus cordifolius* L.) in constructed wetland: effect of molecular size. J Environ Sci Health Part A Toxic/Hazardous Substances Environ Eng 46(7):709–714. https://doi.org/10.1080/10934529.2011.571577

81. Ojstršek A, Fakin D, Vrhovšek D (2007) Residual dyebath purification using a system of constructed wetland. Dyes Pigm 74(3):503–507. https://doi.org/10.1016/j.dyepig.2006.10.007

82. Ong SA, Uchiyama K, Inadama D, Ishida Y, Yamagiwa K (2010) Treatment of azo dye Acid Orange 7 containing wastewater using up-flow constructed wetland with and without supplementary aeration. Biores Technol 101(23):9049–9057. https://doi.org/10.1016/j.biortech.2010.07.034

83. Ong SA, Uchiyama K, Inadama D, Yamagiwa K (2009) Simultaneous removal of color, organic compounds and nutrients in azo dye-containing wastewater using up-flow constructed wetla. J Hazardous Mater 165(1–3):696–703. https://doi.org/10.1016/j.jhazmat.2008.10.071

84. Oon YL, Ong SA, Ho LN, Wong YS, Dahalan FA, Oon YS, Teoh TP, Lehl HK, Thung WE (2020) Constructed wetland–microbial fuel cell for azo dyes degradation and energy recovery: influence of molecular structure, kinetics, mechanisms and degradation pathways. Sci Total Environ 720:137370. https://doi.org/10.1016/j.scitotenv.2020.137370

85. Ozturk E, Cinperi NC, Kitis M (2020) Green textile production: a chemical minimization and substitution study in a woolen fabric production. Environ Sci Pollut Res 27(36):45358–45373. https://doi.org/10.1007/s11356-020-10433-8

86. Ozturk E, Koseoglu H, Karaboyaci M, Yigit NO, Yetis U, Kitis M (2016) Sustainable textile production: cleaner production assessment/eco-efficiency analysis study in a textile mill. J Clean Prod 138:248–263. https://doi.org/10.1016/j.jclepro.2016.02.071

87. Pagga U, Brown D (1986) The degradation of dyestuffs: Part II behaviour of dyestuffs in aerobic biodegradation tests. Chemosphere 15(4):479–491. https://doi.org/10.1016/0045-6535(86)90542-4

88. Papić S, Koprivanac N, Božić AL, Vujević D, Dragičević SK, Kušić H, Peternel I (2006) Advanced oxidation processes in azo dye wastewater treatment. Water Environ Res 78(6):572–579. https://doi.org/10.2175/106143006x101665

89. PCI Magazine (2015) World demand for dyes & organic pigments to reach $19.5 Billion | 2015–05–03 | PCI Magazine. https://www.pcimag.com/articles/100559-world-demand-for-dyes-organic-pigments-to-reach-195-billion. Accessed on 12 June, 2021

90. Raja ASM, Arputharaj A, Saxena S, Patil PG (2019) Water requirement and sustainability of textile processing industries. In: Muthu SS (ed) Water in textiles and fashion. Woodhead Publishing, pp 155–173. https://doi.org/10.1016/b978-0-08-102633-5.00009-9

91. Ramprasad C, Smith CS, Memon FA, Philip L (2017) Removal of chemical and microbial contaminants from greywater using a novel constructed wetland: GROW. Ecol Eng 106:55–65. https://doi.org/10.1016/j.ecoleng.2017.05.022

92. Rampuria A, Gupta AB, Brighu U (2020) Nitrogen transformation processes and mass balance in deep constructed wetlands treating sewage, exploring the anammox contribution. Biores Technol 314:123737. https://doi.org/10.1016/j.biortech.2020.123737

93. Rani N, Maheshwari RC, Kumar V, Vijay VK (2011) Purification of pulp and paper mill effluent through Typha and Canna using constructed wetlands technology. J Water Reuse Desalination 1(4):237–242. https://doi.org/10.2166/wrd.2011.045

94. Raphael OD, Ojo SIA, Ogedengbe K, Eghobamien C, Morakinyo AO (2019) Comparison of the performance of horizontal and vertical flow constructed wetland planted with *Rhynchospora corymbosa*. Int J Phytorem 21(2):152–159. https://doi.org/10.1080/15226514.2018.1488809

95. Riva V, Mapelli F, Syranidou E, Crotti E, Choukrallah R, Kalogerakis N, Borin S (2019) Root bacteria recruited by phragmites australis in constructed wetlands have the potential to enhance azo-dye phytodepuration. Microorganisms 7(10):384. https://doi.org/10.3390/microorganisms7100384

96. Rosi OL, Casarci M, Mattioli D, De Florio L (2007) Best available technique for water reuse in textile SMEs (BATTLE LIFE Project). Desalination 206(1–3):614–619. https://doi.org/10.1016/j.desal.2006.06.010

97. Rousseau DP, Vanrolleghem PA, de Pauw N (2004) Model-based design of horizontal subsurface flow constructed treatment wetlands: a review. Water Res 38(6):1484–1493. https://doi.org/10.1016/j.watres.2003.12.013

98. Rugaika AM, Kajunguri D, van Deun R, van der Bruggen B, Njau KN (2018) Mass transfer approach and the designing of horizontal subsurface flow constructed wetland systems treating waste stabilisation pond effluent. Water Sci Technol 78(12):2639–2646. https://doi.org/10.2166/wst.2019.031

99. Saeed T, Sun G (2013) A lab-scale study of constructed wetlands with sugarcane bagasse and sand media for the treatment of textile wastewater. Biores Technol 128:438–447. https://doi.org/10.1016/j.biortech.2012.10.052

100. Sarasa J, Roche MP, Ormad MP, Gimeno E, Puig A, Ovelleiro JL (1998) Treatment of a wastewater resulting from dyes manufacturing with ozone and chemical coagulation. Water Res 32(9):2721–2727. https://doi.org/10.1016/S0043-1354(98)00030-X

101. Schierano MC, Panigatti MC, Maine MA, Griffa CA, Boglione R (2020) Horizontal subsurface flow constructed wetland for tertiary treatment of dairy wastewater: removal efficiencies and plant uptake. J Environ Manage 272:111094. https://doi.org/10.1016/j.jenvman.2020.111094

102. Shehzadi M, Afzal M, Khan MU, Islam E, Mobin A, Anwar S, Khan QM (2014) Enhanced degradation of textile effluent in constructed wetland system using Typha domingensis and textile effluent-degrading endophytic bacteria. Water Res 58:152–159. https://doi.org/10.1016/j.watres.2014.03.064

103. Shukla R, Gupta D, Singh G, Mishra VK (2021). Performance of horizontal flow constructed wetland for secondary treatment of domestic wastewater in a remote tribal area of Central India. Sustain Environ Res 31(1). https://doi.org/10.1186/s42834-021-00087-7

104. Singh RP, Singh PK, Gupta R, Singh RL (2019) Treatment and recycling of wastewater from textile industry. In: Singh R, Singh R (eds) Advances in biological treatment of industrial waste water and their recycling for a sustainable future. Applied environmental science and engineering for a sustainable future. Springer, Singapore. https://doi.org/10.1007/978-981-13-1468-1_8

105. Srivastava P, Abbassi R, Garaniya V, Lewis T, Yadav AK (2020) Performance of pilot-scale horizontal subsurface flow constructed wetland coupled with a microbial fuel cell for treating wastewater. J Water Process Eng 33:100994. https://doi.org/10.1016/j.jwpe.2019.100994

106. Statista (2019) India—value of chemicals and related product exports by type 2019. https://www.statista.com/statistics/652786/export-value-of-chemicals-and-related-products-by-type-india/. Accessed on 21 June 2021

107. Statista (2020) India—dyes and pigments production volume. https://www.statista.com/statistics/726947/india-dyes-and-pigments-production-volume/. Accessed on 21 June 2021

108. Stein O, Hook P, Biederman J, Allen W, Borden D (2003) Does batch operation enhance oxidation in subsurface constructed wetlands? Water Sci Technol 48(5):149–156. https://doi.org/10.2166/wst.2003.0306

109. Stone K, Poach M, Hunt P, Reddy G (2004) Marsh-pond-marsh constructed wetland design analysis for swine lagoon wastewater treatment. Ecol Eng 23(2):127–133. https://doi.org/10.1016/j.ecoleng.2004.07.008

110. Sutar RS, Lekshmi B, Kamble KA, Asolekar SR (2018) Rate constants for the removal of pollutants in wetlands: a mini-review. Desalin Water Treat 122:50–56. https://doi.org/10.5004/dwt.2018.22638

111. Sutar RS, Lekshmi B, Ranade DR, Parikh YJ, Asolekar SR (2021) Towards enhancement of water sovereignty by implementing the 'constructed wetland for reuse' technology in gated community. In: Reddy KR, Agnihotri AK, Yukselen-Aksoy Y, Dubey BK, Bansal A (eds) Sustainable environment and infrastructure, vol 90. Springer, Cham, Switzerland, pp 157–163. https://doi.org/10.1007/978-3-030-51354-2_15

112. Szpyrkowicz L, Zilio-Grandi F, Canepa P (1996) Performance of a full-scale treatment plant for textile dyeing wastewaters. Toxicol Environ Chem 56(1–4):23–34. https://doi.org/10.1080/02772249609358347

113. Tahir U, Yasmin A, Khan UH (2016) Phytoremediation: potential flora for synthetic dyestuff metabolism. J King Saud Univ Sci 28(2):119–130. https://doi.org/10.1016/j.jksus.2015.05.009

114. Verlicchi P, Galletti A, Petrovic M, Barceló D, Al Aukidy M, Zambello E (2013) Removal of selected pharmaceuticals from domestic wastewater in an activated sludge system followed by a horizontal subsurface flow bed — Analysis of their respective contributions. Sci Total Environ 454–455, 411–425. https://doi.org/10.1016/j.scitotenv.2013.03.044

115. Verma Y (2008) Acute toxicity assessment of textile dyes and textile and dye industrial effluents using *Daphnia magna* bioassay. Toxicol Ind Health 24(7):491–500. https://doi.org/10.1177/0748233708095769

116. Vymazal J (2009) The use constructed wetlands with horizontal sub-surface flow for various types of wastewater. Ecol Eng 35(1):1–17. https://doi.org/10.1016/j.ecoleng.2008.08.016

117. Vymazal J (2011) Constructed wetlands for wastewater treatment: five decades of experience. Environ Sci Technol 45(1):61–69. https://doi.org/10.1021/es101403q

118. Vymazal J (2014) Constructed wetlands for treatment of industrial wastewaters: a review. Ecol Eng 73:724–751. https://doi.org/10.1016/j.ecoleng.2014.09.034

119. Wang CC, Li JR, Lv XL, Zhang YQ, Guo G (2014) Photocatalytic organic pollutants degradation in metal-organic frameworks. Energy Environ Sci 7(9):2831–2867. https://doi.org/10.1039/c4ee01299b

120. Wintgens T, Nättorp A, Lakshmanan E, Aselokar SR (eds) (2016) SaphPani Handbook entitled. "Natural water treatment systems for safe and sustainable water supply in the Indian context". IWA Publishing, London, UK

121. World Bank Group (1998) Pollution prevention and abatement handbook

122. Wu S, Wallace S, Brix H, Kuschk P, Kirui WK, Masi F, Dong R (2015) Treatment of industrial effluents in constructed wetlands: challenges, operational strategies and overall performance. Environ Pollut 201:107–120. https://doi.org/10.1016/j.envpol.2015.03.006

123. Xu F, Ouyang DL, Rene ER, Ng HY, Guo LL, Zhu YJ, Zhou LL, Yuan Q, Miao MS, Wang Q, Kong Q (2019) Electricity production enhancement in a constructed wetland-microbial fuel cell system for treating saline wastewater. Biores Technol 288:121462. https://doi.org/10.1016/j.biortech.2019.121462

124. Yaseen DA, Scholz M (2018) Treatment of synthetic textile wastewater containing dye mixtures with microcosms. Environ Sci Pollut Res 25(2):1980–1997. https://doi.org/10.1007/s11356-017-0633-7

125. Yukseler H, Uzal N, Sahinkaya E, Kitis M, Dilek FB, Yetis U (2017) Analysis of the best available techniques for wastewaters from a denim manufacturing textile mill. J Environ Manage 203:1118–1125. https://doi.org/10.1016/j.jenvman.2017.03.041

126. Zhang Z, Rengel Z, Meney K (2009) Kinetics of ammonium, nitrate and phosphorus uptake by Canna indica and Schoenoplectus validus. Aquat Bot 91(2):71–74. https://doi.org/10.1016/j.aquabot.2009.02.002

Printed in the United States
by Baker & Taylor Publisher Services